23562

# A History of Engineering and Science in the Bell System

# A History of Engineering and Science in the Bell System

## Electronics Technology
## (1925-1975)

Prepared by Members of the Technical Staff, AT&T Bell Laboratories.

F. M. Smits, Editor.

AT&T Bell Laboratories

Inquiries may be made to the following address:
AT&T Technologies
Commercial Sales Clerk
Select Code 500-472
PO Box 19901
Indianapolis, Indiana 46219
1-800-432-6600

First Printing, 1985

International Standard Book Number: 0-932764-07-X

Library of Congress Catalog Card Number: 84-73157

Printed in the United States of America

# Contents

## 3. Electron Tubes .............................. 133

## 4. Optical Devices ............................... 197

## 5. Magnetic Memories ........................... 223

## 6. Piezoelectric Devices .......................... 255

## 7. Relays and Switches .......................... 285

## 8. Capacitors and Resistors ...................... 311

---

\* Trademark of AT&T.

# Acknowledgments

This volume on electronics technology is the sixth book in the series on the *History of Engineering and Science in the Bell System* to be published by AT&T Bell Laboratories. The manuscripts for each chapter were prepared by one or more principal authors, who in turn received significant help from many of their colleagues. All principal authors are recognized authorities on their topics.

Chapter 1 on the history of the transistor was prepared by J. A. Hornbeck. He drew on detailed background insights provided by A. E. Anderson. S. O. Ekstrand provided guidance relating to developments in Pennsylvania, W. Shockley reviewed the section on the early history of the transistor, and D. S. Peck critically examined sections on reliability.

The history of integrated circuits was prepared by J. M. Goldey, with the assistance of J. H. Forster and B. T. Murphy. In eliminating overlap, some material initially prepared for Chapter 1 by J. A. Hornbeck was moved to Chapter 2.

Chapter 3 on electron tubes is the work of V. Rutter, L. Von Ohlsen, and W. Van Haste; S. O. Ekstrand provided major editorial guidance.

L. A. D'Asaro and L. K. Anderson are the principal authors of Chapter 4, on optical devices. J. E. Geusic made major contributions, with additional material provided by M. DiDomenico, A. A. Bergh, B. D. DeLoach, Jr., and H. A. Watson.

The history of magnetic memories (Chapter 5) was prepared by A. H. Bobeck and L. W. Stammerjohn. R. A. Sykes and T. R. Meeker provided Chapter 6, on piezoelectric devices; E. J. Alexander assisted greatly in editing that chapter.

Chapter 7, covering relays and switches, was authored by S. J. Elliott; editorial comments were provided by J. M. Morabito. Chapter 8, on capacitors and resistors, is the work of C. T. Goddard. Chapter 9, the story of thin-film circuits, was drafted by R. W. Berry and D. Gerstenberg, with additional input from H. Basseches.

The Bell Laboratories work on devices and components draws heavily on related research work. The material covered in the present volume therefore complements the history of *Physical Sciences* and *Communication Sciences*, volumes 4 and 5 of this series. S. Millman, the editor of these volumes, critically reviewed the present volume and provided me with invaluable editorial guidance. The reader should also recognize that in many cases, additional information on the work of individual scientists, often including a photograph, can be found in the companion volumes. I attempted in all these instances to include a reference to the other volumes.

As I found out soon after starting my editorial task, it takes a lot of effort to convert drafts of manuscripts into a book. This task was greatly aided by the outstanding assistance of a number of people:

D. Chappell not only entered the text into the UNIX* text processing system, but took it through what must have seemed to her to be interminable revisions. S. Lipton, R. L. Stumm, and C. Biczak did exacting, important work on long lists of references. V. W. Hashizume, D. M. Solomon, and C. Biczak skillfully prepared all illustrations, while C. Yerger, M. Sanok, and A. Schillinger did the artwork. E. Warner set the type for the illustrations. D. Mann and others at Tapsco, Incorporated did a fine job of typesetting the text and preparing mechanicals. R. J. Walsh, on short notice, provided additional photographs where needed. E. T. Foreman of Design in Mind designed the dust jacket. S. Annunziata managed to bring the disparate parts of the manuscript to final form through artful scheduling. Last, but not least, M. Heimerdinger patiently handled mountains of editorial detail with skill and insight.

I would also like to thank G. E. Schindler, Jr., for helpful discussions and assistance with many writing and editing aspects of this book; N. J. Miller and D. McGrew for helpful suggestions that resulted in significant improvements; and W. Feik for his expert and comprehensive job of indexing.

F. M. Smits
Editor

---

* Trademark of AT&T Bell Laboratories.

# Introduction

# Electronics Technology

During the period from 1925 to 1975, the Bell Laboratories effort devoted to electronics technology was shaped by its being an integral part of the overall Bell Laboratories effort, the ultimate purpose of which was the design and development of communications systems. Scientists and engineers in electrical and electronics device groups developed a clear sense of mission and an understanding of opportunities and unfulfilled system needs through close interaction with groups in basic research on the one hand and groups in systems development on the other. As a consequence, new devices and components again and again made possible new systems. To give just a few examples, transistors led the way to modern computers, quartz crystal filters and frequency control helped push transmission systems to ever-higher capacities, and work on lasers led to a new family of optical transmission systems.

Dominating this volume of the *History of Engineering and Science in the Bell System* is the story of the transistor and its remarkable progeny, silicon integrated circuits (SICs). By 1960, we had about a decade of experience with the transistor, and it had found a wide variety of applications in switching, transmission, station apparatus, and power systems. But the explosive growth of the SIC had only begun. From 1960 to 1975, the number of interconnected components on a silicon chip doubled every year—a factor of one thousand per decade—with corresponding spectacular improvements in cost, size, performance, power consumption, and reliability. Such prolonged exponential growth was unprecedented, and led to the birth of entire new industries. Since the mid-1970s, growth in the number of interconnected components per chip has slowed to perhaps a factor of one hundred per decade, still enough to keep the industry in turmoil.

Many of the technological innovations making transistors and integrated circuits possible have come from Bell Laboratories. A few of the major ones are zone refining, diffusion, oxide masking, epitaxial deposition, oxide protection, photolithography, thermocompression bonding, ion implantation, metal-oxide-semiconductor technology, and the self-aligned silicon gate structure. The story of these and other developments, as told in this book, is unique in many ways, not the least in the vigorous effort put into educating rival companies in the art and science of the new field of semi-conductors during the formative years of the transistor. One wonders if we will ever again see such an interindustry cooperative effort voluntarily undertaken in the private sector.

For the period covered by this history, the capabilities and limitations of communications systems stemmed from the capabilities and limitations of active devices. Transistors, SICs, and diode lasers have especially dom-inated the last 25 years. One must not forget, however, that a very so-phisticated and powerful technology based on the electron tube existed before 1947, when the transistor was invented at Bell Laboratories. Electron tubes were used with great versatility for a large number of applications: as transmitters and amplifiers, rectifiers, visual indicators, and for displays. Those old enough to remember how often we had to change electron tubes in home radio receivers will be impressed to learn that the first transatlantic cable laid in 1956 was removed from service after 22 years, and not one of its 306 tubes had ever malfunctioned. In addition to the undersea application, electron tubes were used underground, in aircraft, and in space, as well as in more normal environments. In view of the inherent problems associated with hot cathodes, vacuum envelopes, and sizable power requirements, the achievements were remarkable. Indeed, electron tubes in the form of cathode ray tubes, klystrons, and traveling wave tubes are still very much a part of the contemporary scene.

Nothing can compete with the exponential growth in capabilities of SICs, but in recent years, optical devices have come closest. Most of their growth has occurred in the decade following 1975 and hence is beyond the period covered by this history. But the pioneering developments de-scribed here laid the foundation for explosive progress, for which the diode laser was a key element. In conjunction with increasingly sensitive optical detectors and ever-lower losses in the intervening fiber, the product of capacity times distance (megabits per second times kilometers) has already been doubling every year for many years, and it looks as if this progress will continue. The true potential has so far hardly been tapped, especially since work on monolithic integrated optoelectronics is still in its infancy. If past experience with SIC technology is an indication, one would expect integrated photonics to usher in a new era of capability and cost effectiveness in the application of light waves to the communications needs of society.

Another active device—the electromagnetic relay—has played a large role in the technology of communications for many years. After early direct-dial systems based on the step-by-step switch, memory, logic, and control functions were increasingly incorporated in relay form into many generations of panel and crossbar telephone switching systems. Today, most logic functions are realized in silicon, but reed contacts are still used in large numbers for the talking-path circuits in electronic switching systems.

Furthermore, relays continue to play an important role in the per-line circuits used to interface telephone lines with central offices.

Throughout the period covered by this book, active devices have been the more glamorous, since they tended to pace system performance. Such devices, however, must be complemented with passive components of high performance and quality. Thus, considerable effort was devoted to passive components, including magnetic devices, quartz crystals and other piezoelectric devices, resistors and capacitors, and tantalum film circuits.

While transistors had a major impact on the development of digital computers starting in the first half of the 1950s, it was only around 1970 that integrated circuits could perform the memory function economically. In the interim, computer memories relied primarily on magnetic devices. Bell Laboratories contributed its share to a variety of magnetic memories that made the early application of stored program control to telephone switching possible. Another contribution to magnetic technology is the magnetic bubble, which supports large serial memories, all solid state and nonvolatile.

Quartz resonators were first used in filters for frequency-division multiplexed telephone systems. During World War II, the requisite technical and manufacturing information was shared with some fifty other companies, and led to the creation of the quartz crystal industry. Subsequently, considerable progress was made towards functional devices—primarily the monolithic crystal filter. Instead of complex quartz structures, however, we tend today to transfer the complexity to silicon circuits, leaving quartz once again to its traditional function as a simple oscillator that provides a precision frequency reference.

Work on capacitors and resistors had its origin in the early days of telephony. These traditional components benefited greatly from the introduction of new material systems and technologies leading to significantly improved stability and to much reduced size.

Starting in the early 1960s, tantalum film circuit technology grew out of the effort on capacitors and resistors to became a natural complement to SIC technology. Tantalum films provide resistors and capacitors of a precision beyond the capability of silicon. The ceramic substrate can be batch fabricated with the precision film components, as well as the interconnection paths, to which SICs are to be applied, leading to a high degree of integration at the system level. It was widely predicted that tantalum film circuits would rapidly decline as a result of the increasing role of digital electronics, but this prediction has turned out to be an overstatement.

This broad range of efforts was made possible by Bell Laboratories having been an integral part of the large, financially strong telecommunications entity known as the Bell System. AT&T's top officers encouraged a long-range outlook and were able to provide for continuity of funding. These were critical ingredients to the sustained success.

The service orientation of AT&T led to an emphasis more on life-cycle costs than on first costs—an emphasis that in turn led to very high device reliability. There was a great reluctance to commit any device to service until it was fully understood. The search for this knowledge, however, had to be pursued with an eye to schedules, since the belief that the not-understood is not fit to be deployed can lead to missed opportunities in the marketplace.

This environment—pushing for the ultimate in performance while simultaneously assuring very high reliability—explains the deeply ingrained credo that sustained success depends on searching for, and gaining, fundamental understanding. This stubborn quest was the root of many innovations, such as the transistor and other semiconductor-based devices that depended on a detailed understanding of device physics and its relation to the structure of crystals and the energy states of their electrons.

AT&T is now in a new age of competitive pressures and opportunities in which an important element of our future performance will be how well we can capitalize on the strengths that made Bell Laboratories great. Perhaps history does not help very much with our day-to-day problems, but we should be wiser for knowing something about our predecessors and what they did, since continuing and reinforcing the best features of our heritage should help us in new marketplaces.

K. D. Bowers
January 28, 1985

# Chapter 1

# The Transistor

*The search for solid-state amplification led to the invention of the transistor. It was immediately recognized that major efforts would be needed to understand transistor phenomena and to bring a developed semiconductor technology to the marketplace. There followed a period of intense research and development, during which many problems of device design and fabrication, impurity control, reliability, cost, and manufacturability were solved. An electronics revolution resulted, ushering in the era of transistor radios and economic digital computers, along with telecommunications systems that had greatly improved performance and that were lower in cost. The revolution caused by the transistor also laid the foundation for the next stage of electronics technology—that of silicon integrated circuits, which promised to make available to a mass market infinitely more complex memory and logic functions that could be organized with the aid of software into powerful communications systems.*

## I. INVENTION OF THE TRANSISTOR

### 1.1 Research Leading to the Invention

As World War II was drawing to an end, the research management of Bell Laboratories, led by then Vice President M. J. Kelly (later president of Bell Laboratories), was formulating plans for organizing its postwar basic research activities. Solid-state physics, physical electronics, and microwave high-frequency physics were especially to be emphasized. Within the solid-state domain, the decision was made to commit major research talent to semiconductors. The purpose of this research activity, according to an internal document authorizing the funding of the work, was to obtain "new knowledge that can be used in the development of completely new and improved components . . . of communication systems." Kelly was convinced that advances in the communications art, leading to new, better,

---

Principal author: J. A. Hornbeck

and cheaper services, were closely tied to advances in the understanding of materials (e.g., conductors, semiconductors, dielectrics, and insulators) and of components (e.g., vacuum, piezoelectric, and magnetic) from which communications systems are assembled.

In succeeding months, an able group of scientists and technicians was assembled by Kelly, assisted by J. B. Fisk, both from within Bell Laboratories and by recruitment from outside, under the coleadership of W. Shockley and S. O. Morgan, in the Physical Research area.

An important decision of the semiconductor group in January 1946 was to focus its attention on the two simplest semiconductor materials, crystals of silicon and germanium, and to ignore then technologically important materials such as selenium, with its imperfect atomic arrangements, and compounds such as copper oxide. The basis for this decision was two-fold. First, at Bell Laboratories prior to 1945, J. H. Scaff, H. C. Theuerer, and E. E. Schumacher had experience with these materials in the preparation and discovery of p-n junctions in silicon;[1] and during the war, Scaff and R. S. Ohl developed crystal rectifiers for radar application.[2] Second, there had been extensive wartime research and development in crystal rectifiers at the Radiation Laboratory of the Massachusetts Institute of Technology (MIT) and at other laboratories in the United States and England. The reason for the decision was the desire to improve opportunities, as research progressed, to link theory and experiment. Exploitation of this linkage required not only materials more amenable to physical calculation, like the elementary semiconductors, but also the existence of in-house laboratory expertise in the preparation, formulation, purification, and control of these materials. This expertise was already present, or could be readily developed, in the chemical and metallurgical laboratories.

The two decisions mentioned above—the choice of semiconductors as a primary field of research and the choice of silicon and germanium as materials within the field on which to concentrate—were almost necessary prerequisites to the early discovery of solid-state amplification. A third factor in the discovery process was a challenging proposal advanced by Shockley that guided the direction the research took, the choices of experiments, and the line of thinking, and no doubt thereby significantly shortened the time required for the invention to take place.[3] Shockley's substantive challenge was a straightforward quantitative calculation based on a simplified physical model of a semiconductor. He showed that an external electric field applied at the surface of a thin germanium slab (by making the slab one plate of a capacitor, for example) should create a space charge layer within the volume of the semiconductor, thus altering (modulating electronically) the conductance of the slab. [Fig. 1-1] Further, if the germanium slab in the capacitor configuration was connected in series with a load and an external battery, Shockley concluded that an appreciable power gain from the electric-field-modulated conductivity should result, provided there were no losses in the dielectric through which

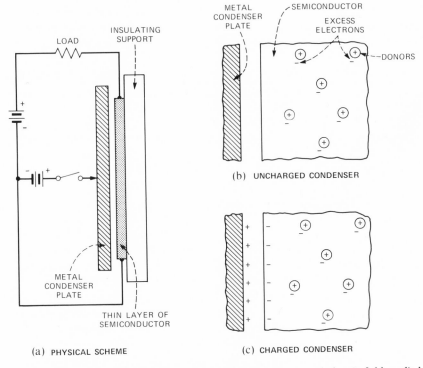

Fig. 1-1.   Schematic of W. Shockley's field-effect idea. An external electric field applied at the surface of a thin germanium slab creates a space charge layer within the volume of the semiconductor, modulating the slab conductance electronically. [Shockley, Electrons and Holes in Semiconductors (1950): 30.]

the external field is applied.* To Shockley and others in the group, these calculations represented an **existence theorem** that field-effect amplification was possible in semiconductors. By stating the goal of solid-state amplification in such understandable and concrete terms, the group leadership provided a focus that both quickened the pace of semiconductor research and enhanced its productivity. Within a remarkably short time—two years—the transistor discovery took place.

The field-effect experiment at Bell Laboratories was tried in many ways and the results were disappointing. Negligible changes in conductance were produced, so it became obvious that there was a major discrepancy between theory and reality for the field effect.

To explain this discrepancy, J. Bardeen proposed that not all the field-

---

* The basic concept of a field-effect transistor and the importance of an active solid-state circuit element undoubtedly occurred to many workers in the early days of solid-state physics. In particular, a 1935 patent[4] shows a structure very similar to Shockley's concept. Other patent applications date back to the 1920s,[5] but apparently all attempts to realize these concepts were futile.

induced charge was free to participate in changing the slab conductance.[6] Rather, he theorized, a portion was immobile, tightly bound in energy states localized at the semiconductor surface. A low density of surface states, many times less than one per surface atom, would be sufficient to shield the interior of the semiconductor from even a large field applied to the surface. Even with improvements in technique, later experimental results by Shockley and G. L. Pearson suggested that 90 percent or more of the induced charge went into surface states.[7] These results, however, were significant in that they did show a finite field effect of the proper sign and that the conductivity change was proportional to the applied field.

Bardeen's theory prompted the group to initiate experiments on the surface properties of semiconductors to verify the predictions of the theory. One of these predictions indicated that a double charge layer existed at the free surface of a semiconductor, a property of the semiconductor independent of any contact. The theory was generally confirmed, and it explained an assortment of apparently unrelated experimental facts about semiconductor surfaces, e.g., the insensitivity of rectification characteristics between germanium and metal contacts to the difference in contact potential between metals. As Brattain explained it, "Further experiments along these lines (designed to change the surface potential) led to the use of an electrolyte to bias the surface, and it was during the course of this work by J. Bardeen and W. H. Brattain[8] that the point-contact transistor was born."[9]

More specifically, in late 1947, at the suggestion of R. B. Gibney, Brattain applied an electrolyte, actually a strange-looking chemical material (glycol borate) commonly called Gu, to the surface of a single crystal of germanium to which some wires and meters were attached. When he applied a voltage to the Gu—to cause a strong electric field at the surface—he found that the current flow between a metal contact and the germanium was affected. Working with Bardeen, he then replaced the Gu, first with an evaporated gold spot and a nearby reverse-biased point contact, next by two gold line contacts made by cementing a single ribbon of gold foil over the sharp edge of an insulating plastic wedge and cutting the foil with a razor blade along the sharp edge. In each case, forward biasing the one contact increased the reverse current in the other contact. On December 16, 1947, he spaced the two contacts about 4 mils apart and observed a small power gain— the phenomenon now known as transistor action or the transistor effect. Figure 1-2 shows the original transistor structure. Figure 1-3 is a reproduction of a page from Brattain's lab notebook recording the power gain and the birth of the transistor.[10]

Arrangements were made to demonstrate the new effect to Bell Laboratories officials. During the following week, H. R. Moore and Brattain assembled the semiconductor device and other components to form an audio amplifier. On the afternoon of December 23, this arrangement am-

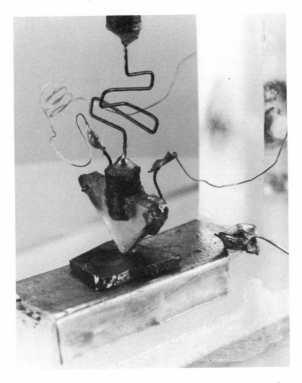

Fig. 1-2.   The original point-contact transistor structure, invented in 1947.

plifying speech—a power gain of 18 times—was demonstrated by Brattain and Moore to R. Bown, director of Research, H. Fletcher, director of Physical Research, Shockley, Bardeen, Pearson, and Gibney.[11] [Fig. 1-4] Intrigued, Fletcher asked, "Will it oscillate?" Oscillation was an unambiguous proof of the existence of power gain. It did! The arrangement was operated as an oscillator the next day, Christmas Eve of 1947, and management was convinced that something really significant had been discovered.

The initial interpretation was based on Bardeen's theory of surface states as holes* flowing from the gold point into a p-type (inversion) layer on

---

* A hole is a theoretical abstraction that emerged from the band theory of solids, beginning with the work of A. H. Wilson in 1931. It represents the collective action produced by the removal of an electron from the filled valence-band structure of a crystal, a deficit in negative charge thus behaving as a positively charged hole. Experimentally, it should behave like a free single positive charge with its own effective mass (different from that of a free electron). In silicon or germanium, weak concentrations of special chemical impurities cause electrical conductivity by producing holes, while different impurities produce electrons. For the former case, the current carriers are positive and the conductivity is called p-type, in contrast to the negative n-type for the electron case. Understanding of these facts had developed during wartime research on crystal detectors.

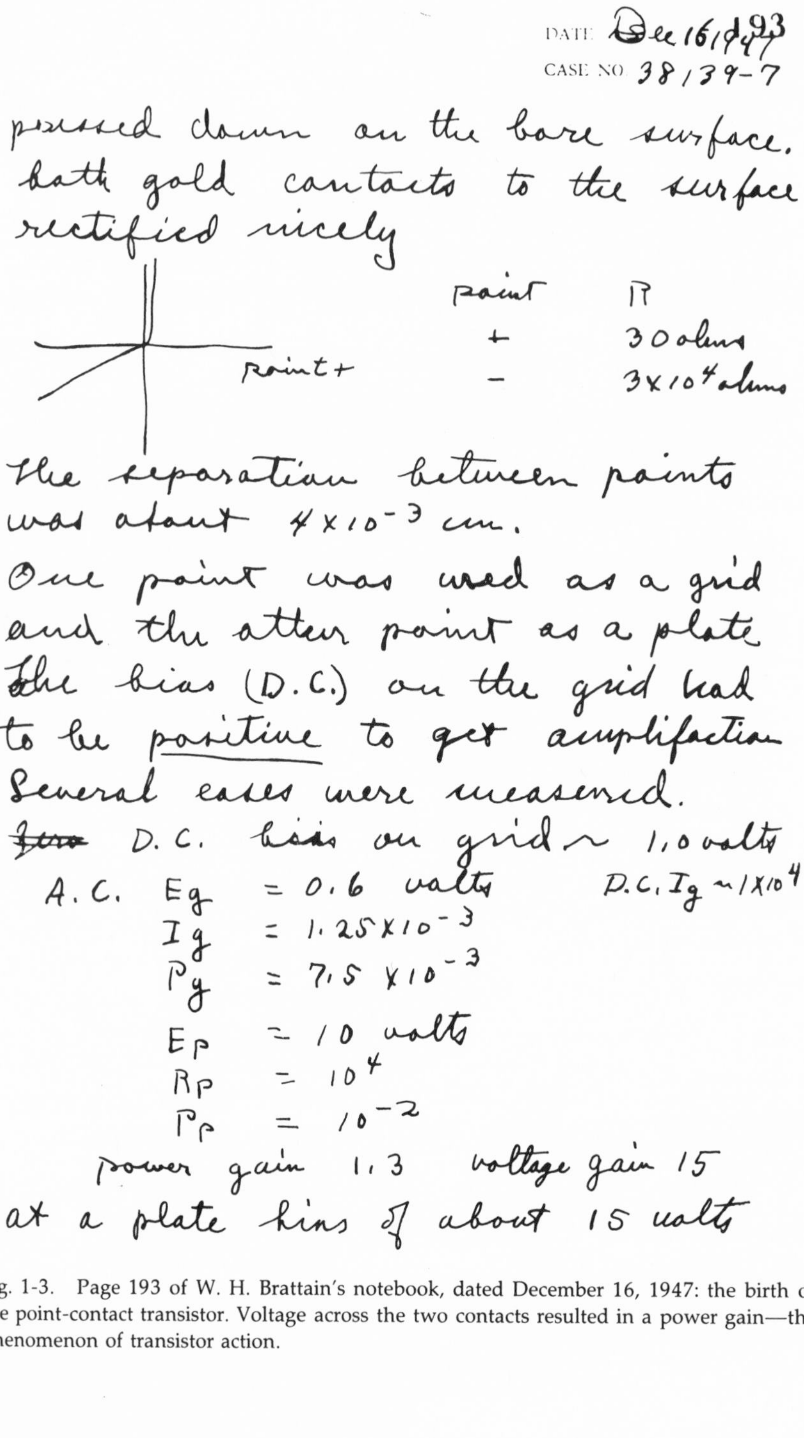

DATE Dec 16, 1947
CASE NO. 38/39-7

pressed down on the bare surface.
both gold contacts to the surface
rectified nicely

|            | R            |
|------------|--------------|
| point  +   | 30 ohms      |
| point +  − | $3 \times 10^4$ ohms |

the separation between points
was about $4 \times 10^{-3}$ cm.
One point was used as a grid
and the other point as a plate.
The bias (D.C.) on the grid had
to be _positive_ to get amplification
Several cases were measured.
D. C. bias on grid ~ 1,0 volts

A.C.  $E_g = 0.6$ volts        $D.C. I_g \sim 1 \times 10^4$
      $I_g = 1.25 \times 10^{-3}$
      $P_y = 7.5 \times 10^{-3}$

      $E_p = 10$ volts
      $R_p = 10^4$
      $P_p = 10^{-2}$

      power gain 1.3     voltage gain 15
at a plate bias of about 15 volts

Fig. 1-3.   Page 193 of W. H. Brattain's notebook, dated December 16, 1947: the birth of
the point-contact transistor. Voltage across the two contacts resulted in a power gain—the
phenomenon of transistor action.

DATE Dec 19 1947

CASE No. 3P139-7

Two points on surface of this
unit less than $\frac{1}{64}$" apart

1 wire     100 + $10^6$ ohms
2 point    100 + $2 \times 10^6$ ohms

very little resistivity

Dec 24 1947

using the ge surface (see
top of page 197 N.B. 18194 and
the gold contacts according to
B.A. 240026 the following circuit
was set up

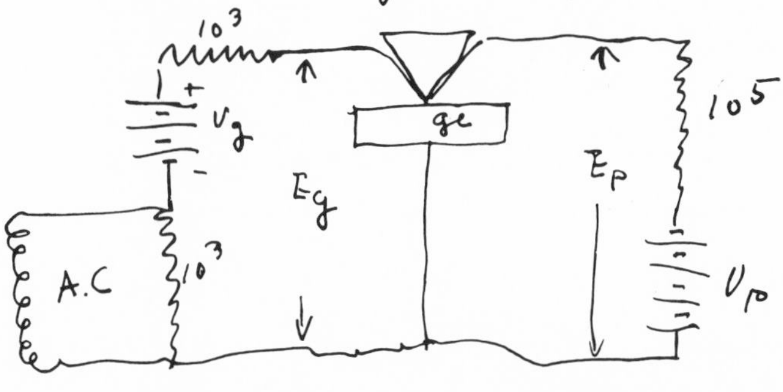

with $V_g \sim 3$ volts    $V_p = 90$ volts
$I_g \sim 4 \times 10^{-4}$ amps    ~~$I_p \sim 4 \times 10^{-4}$ amps~~
$I_p \sim 4.5 \times 10^{-4}$ amps

the above being D.C. values

Fig. 1-4 (a).   Page 6 of W. H. Brattain's notebook, dated December 24, 1947: a description of the December 23 in-house demonstration amplifying speech with a power gain of 18 times.

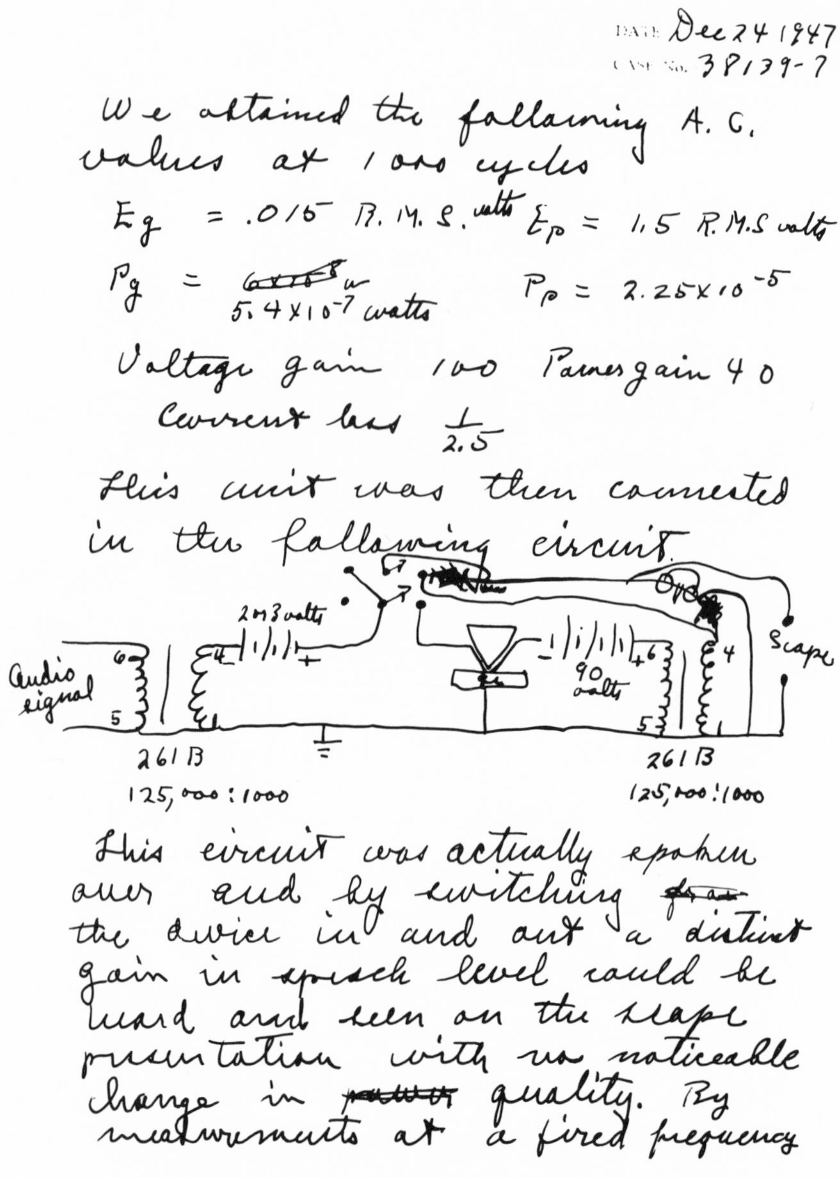
DATE Dec 24 1947
CASE No. 38139-7

We obtained the following A. C. values at 1000 cycles

Eg = .015 R. M. S. volts Ep = 1.5 R.M.S volts

Pg = ~~6×10⁻⁸~~ w 5.4×10⁻⁷ watts Pp = 2.25×10⁻⁵

Voltage gain 100 Power gain 40

Current loss 1/2.5

This unit was then connected in the following circuit.


Audio signal
6
5
4
1
2 or 3 volts
+
90 volts
+6
−
5
4
1
Scope
261 B
125,000 : 1000
261 B
125,000 : 1000


This circuit was actually spoken over and by switching ~~the~~ the device in and out a distinct gain in speech level could be heard and seen on the scope presentation with no noticeable change in ~~pattern~~ quality. By measurements at a fixed frequency

Fig. 1-4 (b). Page 7 of Brattain's notebook.

the surface of the n-type germanium, and along the surface layer to the second point contact, thus increasing the reverse current of the contact.

Observation of the transistor effect led almost simultaneously to the questions: Is the initial interpretation correct? How does one explain it in

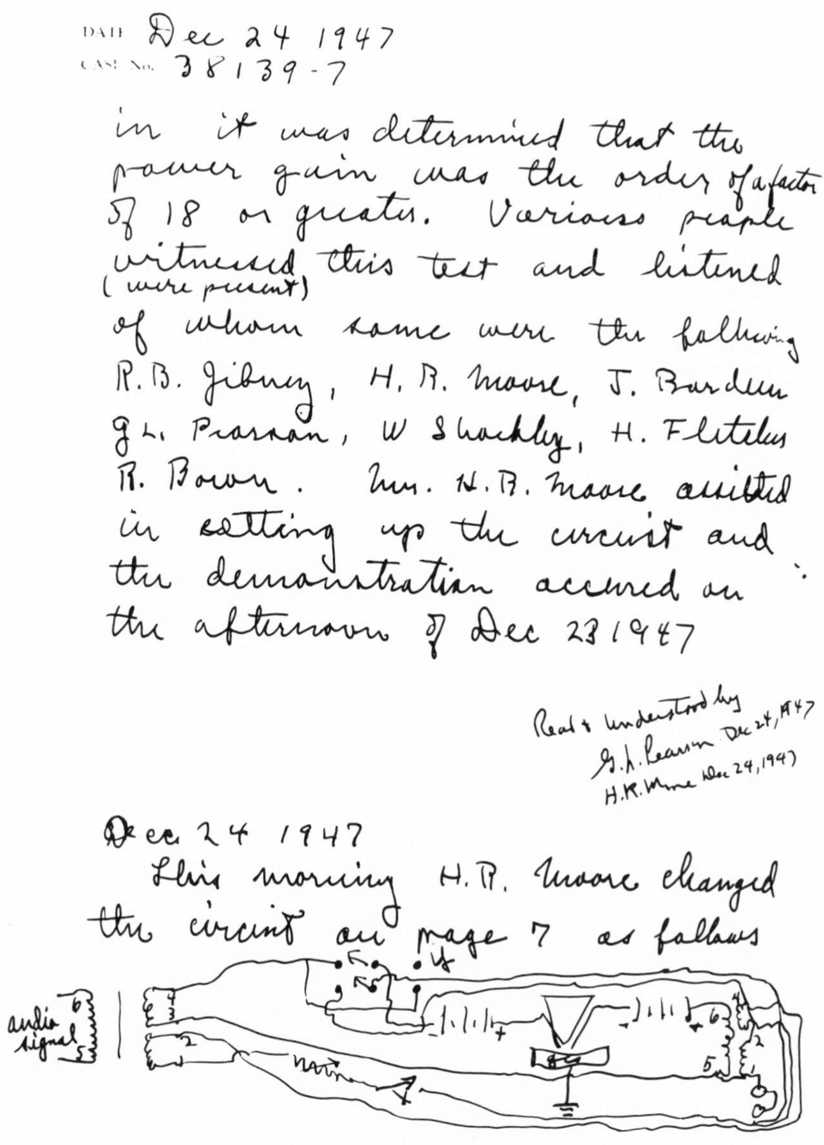

DATE Dec 24 1947
CASE No. 38139-7

*in* it was determined that the
power gain was the order of a factor
of 18 or greater. Various people
witnessed this test and listened
( were present)
of whom some were the following
R. B. Gibney, H. R. Moore, J. Bardeen
G. L. Pearson, W Shockley, H. Fletcher
R. Bown. Mrs. H. R. Moore assisted
in setting up the circuit and
the demonstration occured on
the afternoon of Dec 23 1947

Read + understood by
G. L. Pearson Dec 24, 1947
H. R. Moore Dec 24, 1947

Dec 24 1947
This morning H. R. Moore changed
the circuit on page 7 as follows

audio
signal

norm

Fig. 1-4 (c).    Page 8 of Brattain's notebook.

detail? The responses to these questions by individuals in the research
group and others at Bell Laboratories led in the following months and
years to a new understanding of semiconductor phenomena, thereby creating the field of research and development we call today transistor electronics.

The creative basis of the new understanding was the formulation by Shockley of his p-n junction theory, conceptually elegant in its recognition of the central role played by minority carrier injection at p-n junctions and in body transport phenomena. Using his new concepts, which were developed while he was trying to design experiments to determine the

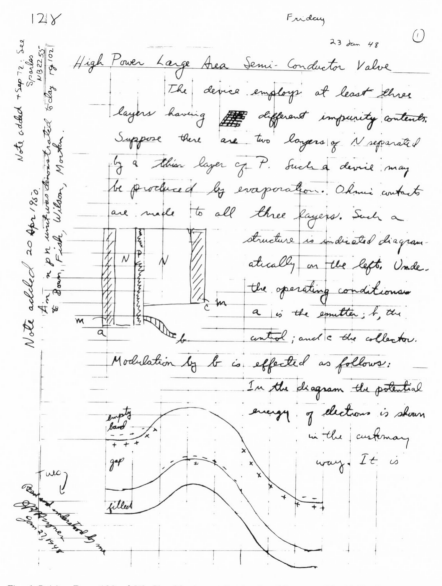

Fig. 1-5 (a).   Page 128 of W. Shockley's notebook, dated January 23, 1948: the original description of the junction transistor, then called the semiconductor valve.

mechanisms of the transistor effect, Shockley immediately invented the p-n junction transistor in which minority carriers (holes), injected by the forward current of a p-n junction, flow through the body of n-type germanium and are collected at an n-p junction.[12,13] Figure 1-5 is a reproduction of two pages of Shockley's lab notebook entry dated January

129
(2)

to be observed that there is a potential barrier over which electrons must climb in order to go from a to b. This barrier is produced by the acceptor impurities in the P layer. The P layer is so thin or so slightly excess in P impurities that it ~~is no~~ does not produce a very high potential barrier. If now a positive potential is applied at b, whose contact is such that holes flow easily into the P layer, these holes will flow into and throughout the P layer thus lowering its potential for electrons. This will increase the flow of electrons over the barrier exponentially. Since the region to the right of the P layer is being operated in the reverse direction, ~~there is so to~~ all practically all of the electrons crossing the barrier reach it ~~so that~~ so that the output is essentially high impedance. This will lead to voltage and power gains.

Read and understood by me [signature] Jan 27, 1948

Fig. 1-5 (b). Page 129 of Shockley's notebook.

23, 1948, in which the junction transistor (nee the semiconductor valve) is described. Confirmation and general acceptance of the new understanding took more than a year.

The Bardeen-Brattain experiments of 1947 and early 1948 continued to suggest to them that, for the point-contact transistor, the hole flow from emitter to collector took place in a surface layer.[9,14] Surface states played an essential role in their theory of the hole-emission process.

In early February 1948, J. N. Shive obtained evidence for point-contact transistor action taking place through the body of the semiconductor.[15] Shive's "double-surface" point-contact transistor was prepared with two phosphor-bronze cat whiskers, emitter and collector, connected to *opposite* sides of a thin slab (0.01 cm) of germanium. This configuration and the respectable power gain he realized are recorded in Fig. 1-6, a reproduction of his laboratory notebook dated February 13, 1948. (Note that toward the bottom of the page he refers to Brattain's "surface states" effect.) The long surface path between emitter and collector of the double-surface transistor ruled out a surface conduction layer as a reasonable purveyor of the transistor action; Shive concluded that a body (volume) effect was involved. Shockley immediately proposed an explanation of the experiment using the concepts of his junction transistor invention made 26 days earlier: penetration of excess holes from the metal emitter point into the body of the n-type germanium (hole injection); and diffusion of a thin stream of excess minority carriers, embedded in an ocean of majority carriers (electrons), across the slab into the electric field at the collector point, where they were "collected" as increased collector current.[12]

In the six months following the discovery of the transistor, not enough experiments had been performed to explain definitively the true nature of point-contact transistor action—whether it was essentially a volume effect, or whether, as proposed later by Bardeen and Brattain, when two point contacts are placed close together on a properly conditioned plane surface of germanium, holes may flow either in a surface layer or through the interior of the germanium.[16]

The technological importance of the transistor discoveries, however, had become much clearer. The performance of the point-contact transistor had improved remarkably. Specifically, W. G. Pfann had modified the Western Electric 1N26 shielded point-contact (silicon) diode to include two spring-loaded cat whisker point contacts, making a three-electrode configuration with good electrical amplifying properties. This configuration became known as the Type A transistor. The promise for the future was bright both for it and for other devices, yet to be constructed, based on Shockley's formulation of p-n junction theory and two-carrier transport phenomena. The three inventors, subsequently winners of the 1956 Nobel prize in physics, are shown together in Fig. 1-7.

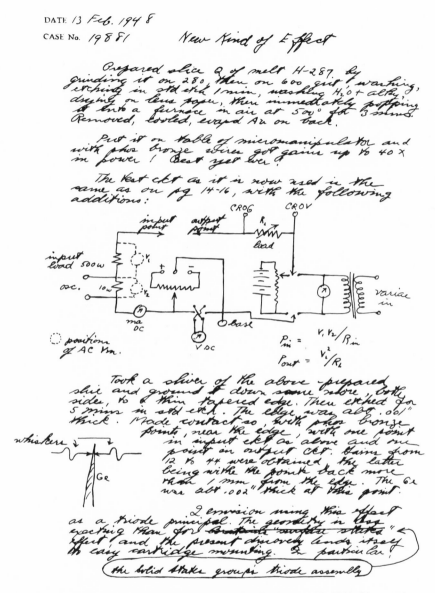

DATE *13 Feb. 1948*

CASE No. *19881*

*New Kind of Effect*

*Prepared slice Q of melt H-287 by grinding it on 280, then on 600 grit, washing, etching in std etch 1 min, washing $H_2O$ + alc, drying on lens paper, then immediately popping it into a furnace in air at 500° for 3 mins. Removed, cooled, evapd Au on back.*

*Put it on table of micromanipulator and with phos bronze wire got gains up to 40× in power! Best yet ever!*

*The test ckt as it is now used is the same as on pg 14-16, with the following additions:*

*P_in = V_1 V_2 / R_in*

*P_out = V_1^2 / R_L*

*Took a sliver of the above-prepared slice and ground it down some more, both sides, to a thin tapered edge. Then etched for 5 mins in std etch. The edge was abt .001" thick. Made contacts so, with phos bronze points, near the edge, with one point in input ckt as above and one point on output ckt. Gains from 12 to 44 were obtained, the latter being with the points back more than 1 mm from the edge. The Ge was abt .002" thick at this point.*

*I envision using this effect as a triode principal. The geometry is less exacting than for _____ "surface states" effect, and the present discovery lends itself to easy cartridge mounting. In particular the solid state groups triode assembly*

Fig. 1-6. Page 30 from J. N. Shive's notebook, dated February 13, 1948: his experiment suggested a volume effect rather than surface conduction.

## 1.2 Public Announcement

From the earliest days, weighty nontechnical questions concerning disclosure faced Bell Laboratories management. Who at Bell Laboratories

Fig. 1-7.    The three inventors of the transistor (left to right): W. Shockley, W. H. Brattain, and J. Bardeen.

should learn of the discovery? When to make a public announcement? What to do about publication and adequate patent protection? These questions could be answered sensibly, it was decided, only after more was known about the new device and its implications. Public announcement should be deferred, it was decided, until the scientists felt they understood the device better and could give other scientists enough information to enable them to reproduce the work. Also, patent applications had to be filed to protect the interests of the shareowners.

As it turned out, these activities required some six months, and public disclosure took place by means of several carefully orchestrated actions in June 1948. The patent application for the point-contact transistor was filed on June 17 by Bardeen and Brattain. The transistor was disclosed to a meeting of the technical staff in the research area in the Bell Laboratories auditorium at Murray Hill, New Jersey on June 22. The following day, a disclosure was made to the military, the National Military Establishment (later, the Department of Defense) having concurred with a Bell Laboratories recommendation that the transistor be unclassified. On June 25, the first scientific publications, in the form of three letters to the editor were sent to the *Physical Review*, where they were published in the July 15, 1948 issue. These letters were entitled, respectively, "The Transistor, a Semiconductor Triode" by Bardeen and Brattain;[8] "Nature of the Forward Current in Germanium Point Contacts" by Brattain and Bardeen;[14] and "Modulation of Conductance of Thin Films of Semiconductors by Surface Charges" by Shockley and Pearson.[7] The two Bardeen-Brattain letters espouse the surface-effect interpretation of transistor action. On June 26, Shockley filed his junction transistor patent application, and finally, on June 30, the transistor was demonstrated to the press in the Bell Laboratories auditorium at the West Street location in New York City, at which time Bown exhibited Type A transistors with respectable performance. Power output was 50 milliwatts (mW), frequency response went to 10 megahertz (MHz), and power gain was 100 times. Voice frequency and TV video signal amplification were demonstrated, as was a superheterodyne radio containing only semiconductor elements. Significant progress had been made in six months. Figure 1-8 is a photograph taken of the press conference.

The conference, while well attended, generated little attention in the public press. *The New York Times* included it with other items in a column

Fig. 1-8.   The June 30, 1948 press conference demonstrating the transistor to the public. R. Bown, director of research, addresses the audience in the auditorium of the Bell Laboratories facility on West Street in New York City.

devoted to events in radio. The *New York Herald Tribune* carried an article. The reporter, while noting engineers were predicting a minor revolution in the electronics industry, went on to say: "However, aside from the fact that a transistor radio works instantly without waiting to warm up, company experts agreed that the spectacular aspects of the device are more technical than popular."[17]

From the start, as the *Herald Tribune* implied, the transistor was faced with intense, initially overwhelming, competition from the centerpiece of electronics, the electron tube. The thermionic vacuum tube, or electron tube, beginning with L. de Forest's audion, had become the base of a vigorous industry, creating the first electronics era. This era was marked by a greatly expanded communications capability: transcontinental telephony in 1914, and television transmitted between Washington, D. C. and New York City in 1928. It was also marked by the advent of consumer electronics, which offered a radio to every home. Further, more recent high-frequency developments, notably klystrons, traveling wave tubes, and the close-spaced triode, had made possible radio relay systems, home television, and various radar applications, such as navigational aids. (A more detailed discussion of Bell Laboratories contributions to tube development appears in Chapter 3.) When the transistor was discovered, in the public's opinion, electronics technology was mature, sophisticated, well understood, and complete.

The trade press, on the other hand, had more vision. For example, the September 1948 issue of *Electronics* magazine made the transistor its cover story. The accompanying article stated: "Because of its unique properties, the Transistor is destined to have far-reaching effects on the technology of electronics and will undoubtedly replace conventional electron tubes in a wide range of applications."[18] And indeed, the transistor very soon found special applications, such as in hearing aids, and in time, the transistor was to become the dominant factor in electronics technology.

Initially, what did the transistor have in its favor? The transistor offered much lower power consumption and, in addition, miniaturization. Since no inherent wear-out mechanism, such as that associated with thermionic emission from the hot cathode of a vacuum tube, was perceived in the transistor, it was seen to offer the promise of indefinite service life. High reliability was also hoped for because it was solid state—no cathode destruction due to a leaky vacuum, for example. But more specifically, to displace the vacuum tube, the transistor had to be able to compete on the bases of cost and/or performance, and its performance initially was severely restricted. As it turned out, for the transistor to compete broadly, mastery had to be achieved of both the physics of transistor action and the subtle, complex art of processing the semiconductors, initially germanium and then silicon. As the technology of transistor electronics matured through breakthroughs in these two areas, lower cost, higher performance, and better reliability would follow. The result would be the second electronics era, one based on the technology associated with the discrete transistor.

### 1.3 Establishing Fundamental Understanding

Within a year after the transistor announcement, great strides in understanding had been taken through a series of seminal experiments designed to elucidate the behavior of holes and electrons in germanium. First there was the question of how the point-contact transistor really worked—a volume and/or a surface effect? By the time of the transistor announcement, Shockley and J. R. Haynes had already initiated a classic investigation to settle this point and establish the reality of minority carrier injection.

In the prototype setup of Haynes and Shockley, a longitudinal electric field was established by a battery in a thin rod, or filament, of high-purity, single-crystal n-type germanium.[19] [Fig. 1-9] Point contacts were made to

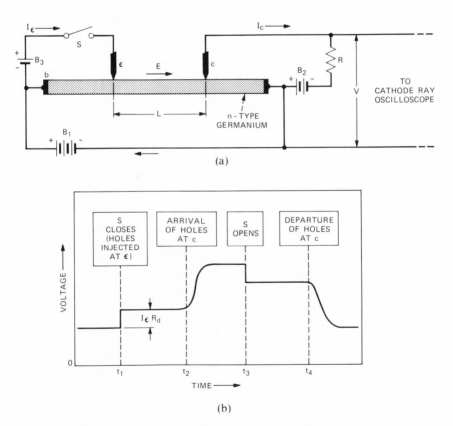

(a)

(b)

Fig. 1-9.   Schematic of the Haynes-Shockley experiment. Placing emitter and collector points on opposite sides of the filament showed that the holes move in the bulk material—a volume effect.

the rod, separated by a variable distance. One contact was biased as the emitter and the other as the collector, as in a Type A transistor. In the experiment (and as the experiment established), a pulse of holes was injected into the filament by forward biasing the emitter point for a short period of time. The holes drifted in the electric field down the rod toward the collector. Their initial, and subsequent, arrival at the collector was detected by an increase in the reverse current of the collector circuit. Placing emitter and collector points on opposite sides of the filament showed that the holes moved in the bulk material—a volume effect. Measurement of the transit time from emitter to collector and the distance between the points yielded the drift velocity of the holes. The ratio of the drift velocity to the applied longitudinal electric field gave the hole (drift) mobility. By observing the variation of signal amplitudes with time, and as the experimental parameters and configuration were changed, Haynes, Shockley, and Pearson gained quantitative information concerning hole injection, hole mobility, hole diffusion, hole lifetime, surface recombination velocity, and the nature of conductivity modulation in semiconductors.[20] [Fig. 1-10]

In his classic book, Shockley emphasizes the important role of excess minority carriers (here, holes) in the conductivity process and in transistor

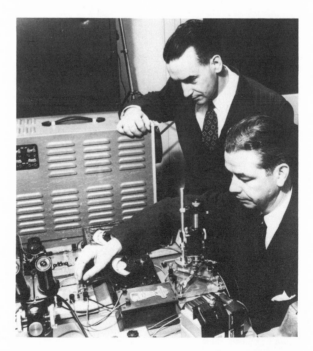

Fig. 1-10.   G. L. Pearson (left) and J. R. Haynes (right) performing their study on minority carrier injection in filamentary transistors.

ability from chemists and metallurgists of large single crystals of germanium—crystals of high chemical purity and of a high degree of crystal perfection (relative absence of grain boundaries and lattice dislocations). By modifying the Czochralski process, a single-crystal growth technique first used by J. Czochralski in 1917,[23] G. K. Teal and J. B. Little succeeded in October 1948 in growing a single crystal of germanium by slowly withdrawing a seed crystal from a melt of very pure germanium.[24] Volume recombination of excess holes with electrons, as differentiated from recombination at the filament surfaces, in single-crystal material was characterized by a lifetime of more than 100 microseconds ($\mu$s), 20 to 100 times that characteristic of polycrystalline material. This was a major step forward in terms of minority carrier lifetime and other crystal properties.

## II. INITIAL TRANSISTOR DEVELOPMENT

### 2.1 Development of the Point-Contact Transistor

Simultaneous with the public disclosure of the transistor in June 1948, a new transistor department in the device development area was set up under J. A. Morton [Fig. 1-11], reporting to J. R. Wilson, director of Electron Apparatus Development. The new device group worked in close partnership with colleagues in chemical and metallurgical research, as well as with Shockley's device-oriented group in physical research. The chemists and metallurgists contributed vital materials and/or processing know-how and device creativity. Also at that time, a Bell Laboratories group in Allentown, Pennsylvania, located within the Western Electric Allentown Works, started work on transistors in anticipation of production.

The early point-contact transistors worked, but not too well. No two performed exactly alike, and they were easily destroyed by inexperienced experimenters. For the Type A transistor, a special method was devised of electrically forming the emitter and collector point contacts made to the n-type germanium chip.[25] A phosphor-bronze wire was used for the collector contact spring. Phosphorus from the spring wire was associated later with the phenomenon of current multiplication at the collector contact (intrinsic $\alpha$ very much greater than unity), presumably through a "hook" collector action.[3]

By the summer of 1949, a fair amount of progress had taken place. The Type A transistor was then made in two different versions, the M1729 for small-signal linear amplification and the M1698 for large-signal (ON-OFF) switching applications. Small-signal circuitry was given a firm basis via four-pole analysis[26] analogous to the prior electron tube analysis of F. B. Llewellyn and L. C. Peterson.[27] Telephone system applications that seemed attractive included a tone ringer for the telephone handset, a tone generator for toll signaling, and a photodetector-amplifier for a card translator used in direct distance dialing. The photodetector was a light-detecting pho-

action by the summary statements: "In a semiconductor containing stantially only one type of carrier, it is impossible to increase the carrier concentration by injecting carriers of the same type; however, increases can be produced by injecting the opposite type since the s<sub>]</sub> charge of the latter can be neutralized by an increased concentratio: the type normally present. *Thus we conclude that* the existence of processes of electronic conduction in semiconductors, corresponding spectively to positive and negative mobile charges, is a major feature several forms of transistor action."[21] Here, Shockley taught the two-car concept of semiconductor transport phenomena and rectification.

The hole injection experiments measured the fraction of the emit current carried by holes, and it was found that this fraction could approa unity—essentially all the emitter current is holes. Quantitative studies the presence of holes near a collector point determined properties, su< as the intrinsic $\alpha$ of a point—that is, the ratio of change in collector curre: per unit change in hole current actually arriving at the collector. Ampl fication in a Type A transistor was attributed to the modulation of th collector current by the holes in the emitter current. Thus the essentia features of the point-contact transistor and the nature of transistor actio were explained in simple physical terms.

The filamentary experiments were extended by H. Suhl and Shockle by placing the rod, in which the holes drifted in the longitudinal electric field, in a transverse magnetic field.[22] The experiments gave direct evidence that the excess minority carriers (holes) were subjected to a sidewise thrust (the Suhl effect) in the same direction as were the majority carriers (electrons). In essence the experiment was an extension of the Hall effect (single charge carrier) to the bipolar case (both positive and negative charge carriers present simultaneously).

Thus, based on the Shockley theory, experiments with filaments—those of Haynes on drift mobility, of Pearson and Haynes on conductivity modulation, and of Suhl on magnetic concentration of holes and electrons— provided the linkage between theory and experiment that gave "operational reality" (Shockley's words) to holes and electrons as positive and negative carriers of current in semiconductors. By their study of transient phenomena, Haynes, Pearson, and Suhl showed that an excess hole actually does drift in an electric field with a drift velocity, that it undergoes the random thermal motion of diffusion, that it is deflected sidewise by a magnetic field, and that it does behave as if it had a positive charge equal in magnitude to an electron's negative one. A similar statement could be made about excess electrons in p-type semiconducting materials.

The experiments also provided clarification of the functioning of the point-contact transistor. They put the action of both emitter and collector points on a quantitative basis (with the principal exception of explaining a collector intrinsic $\alpha$ greater than unity).

These experiments could not have been carried out without the avail-

Fig. 1-11.   J. A. Morton, who headed the first transistor development group in 1948 and later became vice president of Electronics Technology.

totransistor, essentially a point-contact transistor lacking an emitter.[28] The emitter function was replaced by light (photons) falling on the germanium wafer, being absorbed, and thereby creating excess electron-hole pairs, one pair for each absorbed photon.[29] Some excess holes drifted to the collector region, creating a current at the collector enhanced by the collector's current multiplication factor.

Military applications and designs were also investigated. In June 1949, a joint services (Army, Navy, and Air Force) contract was initiated to enable Bell Laboratories to undertake specific developments related to transistors for the military and to keep the military informed of newly discovered information. Prominent workers in this activity were the early transistor developers A. E. Anderson, Shive, and R. M. Ryder, shown in Fig. 1-12. The first task of this study covered work, to quote from the contract, "aimed at the creation of plug-in packages suitable for standardization; combining features of ruggedness, long life, small bulk and weight, and low power requirements. The immediate objective is the development of transistor package units suitable for application in the AN/TSQ-1 Data Receiver-Transmitter Set and other circuits in which considerable use is made of 'flip-flops,' 'and' circuits, 'not-and' circuits, and the like."[30]

Fig. 1-12.  Early developers (left to right) A. E. Anderson, J. N. Shive, and R. M. Ryder, who worked on military applications of plug-in packages combining ruggedness, long life, small size, and low power requirements.

This work required the cooperation of transistor device developers with circuit and system designers, and a task force of such people was organized for the purpose. By the end of 1950, the group was able to demonstrate all needed functions for the system.[31] Circuit packages no bigger than a contemporary miniature vacuum tube were encapsulated in plastic. Some packages were interchangeable and therefore could be separately coded, i.e., identified as procurable parts. This work demonstrated the feasibility of building an all-transistorized digital data transmission system suitable for use in the field. This led in 1951 to the start of the development under the direction of J. H. Felker of TRADIC (TRAnsistor DIgital Computer), which became the first successful airborne digital computer. (For a comprehensive discussion of TRADIC, see a companion volume in this series subtitled *National Service in War and Peace (1925-1975)*, Chapter 13, section 2.2.)

### 2.2 Crystal Growth and Demonstration of the Junction Transistor

Formulation of the concepts of the p-n junction transistor by Shockley created a new challenge to the research and development community of chemists, physicists, and metallurgists at Bell Laboratories. How could the new device structures be made? There were essential things that had to be known before sophisticated devices could be fabricated. Importantly, these things involved making large single crystals of germanium, and later silicon, of extraordinarily high purity (i.e., sufficiently free from the effects of imperfections and unwanted impurities); and making p-n junctions in the crystals in a controlled way, junctions situated in a device such that they were relatively unaffected by undesirable surface effects. P-n junctions in silicon had been discovered and named by Ohl and Scaff before World War II. (The work by Ohl and Scaff is covered from a materials research point of view in another volume in this series subtitled *Physical Sciences (1925-1980)*, pp. 417-420.) Experiments by Theuerer had shown the junction to be the result of the segregation of unknown impurities during the freezing of a molten silicon charge in a crucible of fused silicon.[1] The concentrations of the impurities present were too low to be detected by the techniques of chemical analysis. Theuerer discovered that adding small amounts of boron increased the p-type conductivity of the ingots. He concluded that phosphorus produced n-type conductivity, having suspected earlier that phosphorus was present because he smelled a trace of phosphine. He concluded further that phosphorus opposed boron and that the p-n junction occurred during solidification because phosphorus segregated differently from boron. Thus from Theuerer's ingenious contributions came the beautifully simple picture that elements of column III of the periodic table produce p-type conductivity in the column IV semiconductors silicon and germanium; that elements from column V produce n-type conductivity; and that these column III and V impurities can compensate each other in the semiconductor crystal in the sense that the resultant conductivity is proportional to the difference in their atomic concentrations.

Both silicon and germanium have four valence electrons and four bonds to nearest neighbor atoms in their (diamond-type) crystal structure. Substituting a phosphorus atom (five valence electrons) for a silicon atom leaves one electron additional to the four tied up in nearest neighbor bonds, and it is free to conduct. With its negative charge, it gives n-type conductivity, and the phosphorus atom is a *donor* of a free electron to the system. Analogously, substituting a boron atom (three valence electrons) for a silicon atom leaves the deficit of one electron, i.e., one hole that is free to conduct with its associated positive charge. Hence we have p-type conductivity, and boron is an *acceptor* of an electron from the collective system.

The success in 1948 of Teal and Little[24] in growing single crystals of germanium led Morton, in keeping with his transistor development responsibility, to encourage and support increased crystal-growing activity. P-n junctions in principle could be made in the crystals if methods were found to control accurately the donor-acceptor impurity imbalance in the melt during crystal growth. This control had to be achieved in the presence of very small quantities of other unknown and unwanted impurities, which always were present. The growth of germanium single crystals containing p-n junctions[32] was followed by the fabrication of the first grown-junction n-p-n transistor in April 1950.[33] [Fig. 1-13] It was made by a double-doping technique in which pellets of gallium and antimony alloys of germanium were added in quick succession to the melt of a growing n-type crystal.[34] This created a thin p layer between two n-type regions, the first lightly doped and the second heavily doped. The crystal was then sawed into many small n-p-n rods to which metal contacts were attached to form three-electrode transistor structures. (This work is covered in greater depth and with additional illustrations in *Physical Sciences (1925-1980)*, Chapter 19, section 1.2.)

The devices were extensively characterized by R. L. Wallace and W. J. Pietenpol.[35] [Fig. 1-14] A satisfying feature of the new germanium grown-

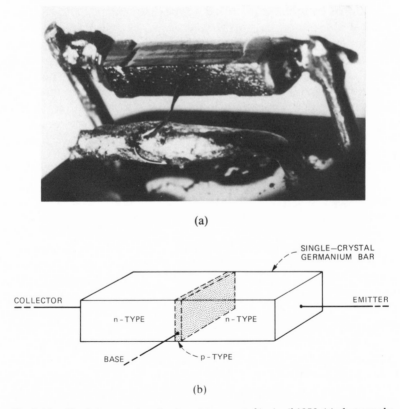

(a)

(b)

Fig. 1-13.    The first grown-junction transistor, created in April 1950: (a) photograph; (b) schematic.

Fig. 1-14. R. L. Wallace (seated) and W. J. Pietenpol (standing), who characterized the first grown-junction transistors.

junction devices, both the p-n junctions and the n-p-n transistors, was that their electrical characteristics agreed quantitatively with the earlier theory, thus confirming it. From this linkage, transistor electronics had indeed acquired a solid foundation.

A marked departure from theory was observed, however, in the reverse characteristic of the grown p-n junction. As the reverse-biased voltage was increased, the reverse current initially approached a constant saturation value independent of voltage as predicted by p-n junction theory. However, as the reverse voltage was increased, a point was reached where the current suddenly increased very sharply, almost without limit, in turn limiting the voltage to that value at which current breakdown set in. Obviously some new effect, which Shockley found "reminiscent of field emission," was occurring at the high reverse electric field. The first experimental investigation[36] of this phenomenon concluded that it was the quantum mechanical Zener effect, i.e., internal field emission tunneling by majority carriers through the barrier at the junction, according to C. Zener's earlier theory of electrical breakdown in solid dielectrics.[37] Further investigation showed this explanation of the effect to be incorrect, concluding that in fact at high electric fields and high reverse voltage, electron-hole multiplication by impact ionization was taking place: electrons drifting in the electric field at high fields gain enough energy to generate electron-hole pairs.[38] The newly formed charge carriers can also acquire energy from the field, in turn creating more pairs. Above a certain threshold field, an avalanche of free charge is created and avalanche breakdown occurs. (The true Zener effect was later found in very narrow silicon junctions at low applied voltage [too low for impact ionization] but still very high electric field strength.)[39] Avalanche breakdown has been widely applied in circuitry in the form of the Zener diode, so called first by Shockley, for voltage limiting and regulation.

The new junction transistors behaved differently from, and in many respects better than, the Type A transistor. Because they had a simple reverse-biased p-n junction as a collector, no current multiplication was associated with collector action. From a circuit viewpoint, this meant that the negative resistance properties of the point-contact transistor were not present, and the junction transistors were unconditionally stable. This made them easier to use in many applications. As reported by Shockley, Sparks, and Teal, "They have operated with gains of 50 dB and noise figures of about 10 to 15 dB at 1000 cycles per second . . . improvement of several orders of magnitude over point-contact transistors . . . [with] full gain at voltages higher than 0.1 volt . . . [and] may be operated at 48 to 49 percent of the theoretical (Class A) maximum efficiency . . . an oscillator . . . operates on 0.6 microwatt input."[40]

Evidence had accumulated over a long period of time for the existence of effects in germanium other than those caused by the donor and acceptor atoms introduced into the crystal-growing apparatus to make p- or n-type material. The body lifetime of excess minority carriers seemed to be relatively insensitive to the presence of the donor or acceptor impurities. Heat treatment of germanium samples, however, could decrease this lifetime markedly. This effect was attributed to the presence of some unknown impurity, called "deathnium," which acted as a hole-electron recombination center. Heat treatment could also change the sample resistivity, tending to convert n-type to p-type, an effect attributed to another unknown impurity, "thermium." Bombardment by nuclear particles was found to produce p-type centers, which apparently also acted as recombination centers. Grain boundaries, besides adding electrical resistance, also exhibited some of the characteristics of p-type regions. The surface recombination process, as distinct from body recombination, appeared to depend on the chemistry of the surface as well as the electrical field at or just inside the surface.

In 1951, Pfann invented a technique for purifying germanium far more accurately than any existing methods.[41,42] It is called zone refining, as described in greater detail with illustrations in *Physical Sciences (1925-1980)*, pp. 419, 597-600. This invention removed electrically active (i.e., measurable) impurities from germanium to a concentration of less than one atom in $10^{10}$ atoms of germanium. It established a new base from which the search for very small concentrations of impurities and other imperfections could proceed. The problems associated with unwanted impurities and their sources became one mostly concerned with recontamination from vessels and chemicals used in subsequent processing—on its own, a set of nontrivial problems. Deathnium and thermium in germanium were later shown to be due to copper at low trace levels, derived largely through recontamination from furnaces, jigs, reagents, "pure" water, etc.[43] Potassium cyanide cleaning to remove trace copper impurities had previously been anticipated.[44]

Pfann's zone-refining technique consisted of melting a short section, or zone, of germanium material placed in a long, open boat of very pure graphite. Melting was achieved by a radio frequency (RF) induction heating coil surrounding the boat. The molten zone was made to traverse the length of the germanium material by moving the boat longitudinally along the axis of the coil. The impurities tended to segregate selectively in the molten zone, and thus were removed from the solid phase as the zone passed through. Multiple zones, and thus multiple passes, could be made by using an array of heating coils.

Pfann extended the zone-refining technique to zone leveling, a process by which a desired impurity was introduced uniformly into a single crystal. Starting with a zone-refined germanium charge in the graphite (or graphite-coated) boat, he added a seed crystal at the starting end together with a small pellet of, for example, antimony-germanium alloy. The initial molten zone was arranged to melt the tip end of the seed crystal along with the pellet. The zone was moved along the charge, leaving behind a single crystal, attached to the seed, in which a constant amount of antimony was incorporated into the lattice per unit length. Measurements of Pfann and K. M. Olsen showed that the uniformity in antimony impurity concentration improved by a factor of ten with this technique over that associated with normal freezing from the melt.[45]

### 2.3 Information Disclosure and Teaching

Historically, the Bell Laboratories policy on information disclosure had been to publish information on advances in science and technology at the "earliest time," the major limitation in the determination of the earliest time being the time prior to publication required to prepare and file patent applications for design and process information thought to be of a patentable nature. With the advent of the transistor and its wealth of new technology, this policy was not only followed but in practice was consciously turned into a procedure of active teaching. Leading in this direction was the publication in 1950 of Shockley's treatise, *Electrons and Holes in Semi-conductors with Applications to Transistor Electronics.*[3] It comprised a thoughtful presentation not only of the fundamentals but also of almost all that was known about the subject. As Shockley said in the book's preface, "This material (Part I) is intended to be accessible to electrical engineers or undergraduate physicists with no knowledge of quantum theory or wave mechanics. It should serve as a basis for understanding the operation of transistor devices and for elementary design considerations. Part III, at the other extreme, is intended to show how fundamental quantum theory leads to the abstractions of holes and electrons. . . . Problems follow many of the chapters." By this time, also, more than 3000 point-contact transistors, small "packages of a scientific phenomenon," had been made at Bell Laboratories and distributed for study to other organizations throughout the world.

The advances from 1948 to 1951 in transistor understanding and in knowledge of transistor processing technology placed Bell Laboratories and Western Electric in a unique position of know-how vis-à-vis the rest of the world, despite the many publications. To communicate the results of the transistor circuit task force and other recent information, and in fulfillment of Task 3 of the Joint Services Contract, in September 1951, Bell Laboratories sponsored at Murray Hill a five-day "Symposium on Properties and Applications of Transistors." The symposium was attended by 121 members of the military services and their contractors, 41 university people, and 139 industry representatives. The papers delivered at the symposium were assembled into a book, and 5500 copies were delivered to the military for their distribution.[46] Highlights of the meeting were talks by Shockley and others describing the first experimental grown-junction transistors and the experiments confirming the nature of transistor action in terms of the flow of holes and electrons in the body of the transistor.

A second cornerstone of the Bell Laboratories information disclosure policy, one in addition to that of early and full publication, was the granting of nonexclusive patent licenses to all comers at a reasonable royalty. In April 1952, a second symposium was held to disclose transistor technology in an orderly and efficient way to transistor licensees of Western Electric, in practice, the effective owner of the Bell Laboratories patents. Admission to the symposium cost each licensee a fee of $25,000, a mechanism designed to limit attendance to those companies seriously interested in the transistor business. The fee would be a credit to any future royalties to the Bell System. Twenty-six domestic and fourteen foreign licensees, many of these newly signed, attended the symposium, which convened at Murray Hill for six days and at the Western Electric-Bell Laboratories location in Allentown for two final days, where the point-contact transistor had been introduced into Western Electric manufacture in October 1951.

The symposium seems unique in the history of information transmitted between potentially competitive members of the free world's industrial society. Bell Laboratories and Western Electric undertook to teach in eight days the art and science of transistor technology, theory, and practice as it was known to the host institutions. This included: point-contact technology through device fabrication, characterization, and manufacture; the advances in materials, e.g., germanium oxide reduction, zone melting and purification, crystal growing, and testing; the current status of the infant grown-junction technology (controls for the n-p-n crystal-growing machine), surface treatments, and gold bonding; Zener diode characteristics; and noise and reliability testing. The introduction to the program agenda carried these words: "It is hoped that the material to be presented will be sufficient to enable qualified engineers to set up equipment, procedures, and methods for the manufacture of these products."[47] The proceedings of the symposium were published in September 1952.

Shortly after the transistor technology symposium, in June 1952, Bell Laboratories researchers held a six-day "summer school" for teachers. Participants came from more than 30 universities. The object was to encourage the introduction and teaching of transistor physics as part of the scholastic curriculum. The course at Murray Hill included both classroom and laboratory work. [Fig. 1-15] One result of this activity was a paper published in the *Physical Review*, written by and signed "The Transistor Teachers Summer School."[48] The paper was extraordinary in that it was coauthored by 60 authors from 33 academic institutions. The laboratory experiment, set up and monitored by Haynes, which was performed by

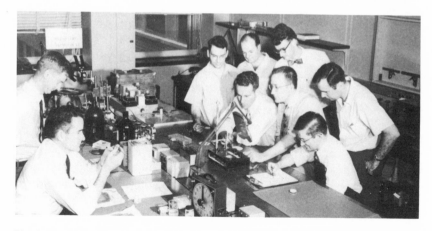

Fig. 1-15. A laboratory session from the 1952 "summer school," held to disseminate transistor technology to professors from 30 universities so that transistor physics could become a part of the university curriculum.

the "students" a number of different times, confirmed the Einstein relationship between mobility and the diffusion constant for excess holes.

The symposium succeeded as an information-transfer mechanism, creating almost overnight a broadly scattered base of know-how from which
the transistor industry could grow and flourish, both in the U.S. and
abroad. It also set a standard for the freer interchange of information in
the semiconductor arena, a standard that prevailed for many years. The
teachers summer school supplemented the symposium by planting seeds
of the new transistor electronics widely in the university environment,
thus encouraging the growth of new science, new technology, and new
graduates educated in the art. It seems reasonable to connect the initial
nearly exponential growth of the semiconductor business with the Bell
Laboratories practices of open publication, granting of nonexclusive licenses,
and of active teaching. These practices made it possible for others to make
their own contributions sooner. As an example, the simple, inexpensive
technique for making junction transistors by alloying metal pellets to germanium was discovered by J. E. Saby and others at the General Electric
Research Laboratories before mid-1951.[49]

### III. EARLY TRANSISTOR MANUFACTURE

The first point-contact transistor to go into manufacture in 1951 at
Allentown, the Type A, consisted of a small wafer of germanium soldered
to a metallic base, with two point contacts on the opposite surface spaced
about 0.001 inch apart. Its manufacture had been delayed in part by
problems involved in reliability and reproducibility. To improve the gain
and stability, the practice was to form (i.e., electrically pulse) the phosphor-
bronze collector point. Forming, a good example of the manufacturing
technique known as "black art" in electron device manufacture, probably
introduced impurities and imperfections into the immediate vicinity of the
collector point, resulting in current amplification, as has been noted, as
well as voltage amplification. The first designs were not sealed to exclude
the atmosphere.

With the Type A transistor, Western Electric was faced with handling
a completely new technology and its accompanying set of yet-to-be-un-
derstood manufacturing problems. Transistors were placed in a nonair-
conditioned area (actually the gate house of the main building) to see
whether their electrical characteristics would change under ambient temperature and humidity conditions. They did. Problems were also encountered with the range of performance.

It was to solve and minimize problems similar to these in the manufacture
of electron tubes that electron device manufacture had been moved in
1947 from Hudson Street in New York City to a new Western Electric
facility in Allentown, which was specially designed to permit controlled

manufacturing conditions. Kelly had also established there, with enthusiastic Western Electric concurrence, a branch of Bell Laboratories, which consisted of a group of resident Bell Laboratories device engineers, under the direction of V. L. Ronci, with their own equipped laboratories. [Fig. 1-16] The group was intended to work closely with Western Electric manufacturing engineers. The branch lab concept was an attempt by management to create an even better interface between development and manufacture, one the responsibility of Bell Laboratories, the other of Western Electric. At this interface many basic questions arise, among them: Is the Bell Laboratories device design inherently manufacturable? How does it need to be changed to make it more so? Will the manufacturing changes affect its circuit compatibility and function? In anticipation of transistor manufacture, as mentioned before, the Allentown Laboratories staff had been augmented by semiconductor device engineers. By 1954, the effort had grown to a level where it was moved from Ronci's responsibility into a separate department headed by Pietenpol. That department became part of Morton's organization, who by then was director of transistor development with three additional departments at Murray Hill. Together with their Bell Laboratories colleagues, the Western Electric engineers went to work on the common problem of producing an initial transistor product that would meet Bell System requirements.

Fig. 1-16.  Left to right: V. L. Ronci, W. K. Wiggins, J. R. Wilson, M. J. Kelly, and F. E. Hansen at the Western Electric facility in Allentown, Pennsylvania in 1947. Ronci directed a group of Bell Laboratories device engineers to work with Western Electric in order to promote a smooth transition from development to manufacture.

To free the transistor from ambient variations in humidity, vacuum-tight encapsulation was found to be necessary for long-term, reliable operation. The first five years with this enclosure in the card translator application gave a system reliability history of 1 million socket-hours per transistor failure. This failure rate compared favorably with electron tube experience.

The Type A transistor, together with the photodiode, was manufactured for some ten years at Allentown for the card translator and other applications. Over a somewhat shorter span, it was also manufactured for military applications at the Western Electric Laureldale, Pennsylvania shop for the AN/TSQ data set. Each of these systems performed satisfactorily in the field. The card translator was finally replaced by a new system after 20 years. The data set was made a North Atlantic Treaty Organization (NATO) standard and remained in service for many years. The satisfied customers made it difficult for Western Electric to discontinue manufacture of point-contact technology until long after it had been made obsolete by junction technology.

## IV. THE EMERGENCE OF GERMANIUM JUNCTION TRANSISTORS

### 4.1 Junction Transistors: Concepts, Uses, Structures, and Technology

The new grown-junction transistor soon came to be appreciated as a more understandable, more reproducible electron device than its earlier cousin, the Type A. Its wider range of parametric properties extended the number of potential applications. Several codes were made, with power dissipation up to 50 mW and with frequency response to several megahertz. The grown-junction transistor would also operate as a microwatt device. The first large-scale use in the Bell System was in the telephone handset. A single transistor amplified received voice signals, in one case to benefit people with impaired hearing, and in another to improve service at noisy locations. It had never been practical to provide amplification in telephone sets, primarily because of the difficulty of supplying power required by vacuum tubes over telephone lines from the central office.

Outside the Bell System, following the transistor licensee symposium, many industrial organizations entered into commercial manufacture of grown-junction devices. There were 35 domestic and foreign licensees in 1953. The first significant commercial product was the portable, miniature transistor radio, still a ubiquitous sign of our technological society. In this application, the lower power demand of the transistors improved battery life by a factor of ten over the vacuum-tube version. The same factor-of-ten improvement in battery life was obtained with hearing aids. The first transistorized hearing aid was built at Bell Laboratories to demonstrate this feature. Before long, commercial manufacture of transistors caused the complete replacement of the vacuum tube in this application. Commercial applications of the grown-junction transistor did not grow more rapidly, however, primarily because of cost, but also because of limited performance parameters.

Soon accompanying the junction transistor idea were other device concepts employing p-n junctions. In 1952, J. J. Ebers carried out the first analytic treatment of what he called the four-terminal p-n-p-n transistor.[50,51] He showed that it could be represented by an interconnection of an n-p-n transistor and a p-n-p transistor, a concept that he attributed to Shockley.[52] These devices have a current gain of greater than unity because the third junction acts as a "hook" collector. The studies on these devices helped in understanding the behavior of the point-contact transistor and were the basis for subsequent work on p-n-p-n switching devices.

Shockley also arranged p-n junctions to invent a volume-type field-effect transistor, later generally known as the junction field-effect transistor (JFET), in which the electric field was applied within the interior of the semiconductor crystal.[53,54] This was in contrast to the earlier field-effect concept in which the field was applied at the surface. In a JFET, the conductive channel of the semiconductor, through which majority current flowed, was made to be one side of a p-n junction. When a reverse bias was applied to the p-n junction (called the gate), free carriers were swept away from the junction, leaving a space charge layer devoid of free carriers (a depletion layer) in the region of the junction. As the reverse bias was increased, the depletion layer spread, thus reducing the dimensions of the conductive channel (i.e., modulating its conductance). The channel current flowed through ohmic electrodes (contacts) attached to either end of the channel, which Shockley called *source* (of free current carriers) and *drain* (of free current carriers). The carriers involved are the majority carriers of the channel conduction type—holes if p-type, electrons if n-type. Hence, Shockley called field-effect transistors *unipolar*, involving (primarily) one sign of charge carrier, to be distinguished from bipolar transistors, junction transistors in which excess minority carriers play an important role together with majority carriers. Unipolar transistors do not have the minority carrier storage or transit-time effects associated with bipolar transistors. In Shockley's junction design, the transistor was expected to be relatively free of variable surface effects.

The first JFET was realized in germanium by G. C. Dacey and I. M. Ross in 1952,[55,56] as described in a companion volume in this series subtitled *Physical Sciences (1925-1980)*, pp. 76-78. Their experimental results on the device shown in Fig. 1-17 confirmed Shockley's original analysis. They reported a device with transconductance "as high as 0.3 mA/volt whose characteristics were very stable . . . (and) flat frequency response up to about 3 Mc/sec."[57] The transconductance achieved, however, was low, hardly competitive with electron tubes; and the power dissipation was high compared to bipolar devices. These unattractive characteristics fol-

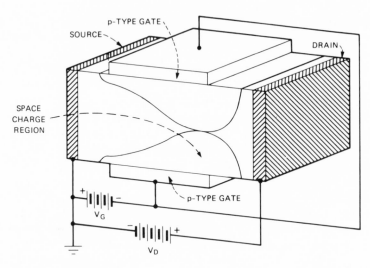

Fig. 1-17. Schematic of the G. C. Dacey - I. M. Ross junction field-effect transistor (JFET).

lowed directly from the (too large) linear dimensions and spacings associated with the junction technology of that time. Also at that time, Pearson realized a silicon JFET[58] by cutting a slot into the p-type material of a p-n junction, leaving just a thin p-type channel near the junction, as shown in Fig. 1-18. The transconductance of his device was even lower than the one reported by Dacey and Ross. Thus the JFET was a constructive and

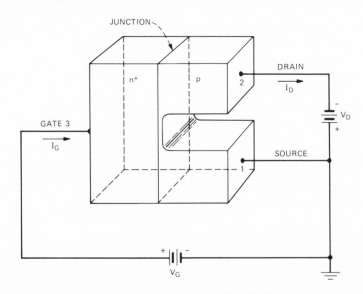

Fig. 1-18. G. L. Pearson's silicon JFET.

interesting piece of device research, but it was not worth pursuing for development with then existing fabrication techniques. Dacey and Ross, however, foresaw remarkable performance characteristics, "1000 Mc/sec cutoff . . . transconductance of 70 mA/volt . . . and low power operations," provided techniques could be found to create a JFET with significantly smaller dimensions.[59]

About the time of the first JFET, S. Darlington proposed that a pair of transistors could be combined in a manner to give improved characteristics.[60] Such a combination became known as a Darlington pair. Moreover, Darlington further proposed that the two transistors could be merged in a common chip, presaging the integrated circuit (IC).

In spite of the superior properties of the grown-junction device, its limitations precluded use in many possible applications. Perhaps foremost of these was frequency response. With a frequency cutoff of approximately 1 MHz, it was not capable of amplifying very high frequencies or very broad bands. Efforts to improve this deficiency led to the conception of the junction tetrode transistor.[61] By adding a fourth electrode (to decrease the active area of the emitter junction) and by using germanium bars of very small cross section (to reduce collector capacitance), it was possible to raise the operating frequency of the grown-junction structure to the 100-MHz range. This was the first successful high-frequency transistor. About 2000 of these devices were fabricated and applied in experimental transmission circuits.

J. M. Early [Fig. 1-19] brought additional insight to the high-frequency design problem. In an analysis of the effects of space charge layer widening on frequency response in junction transistors, Early pointed the way towards operation at thousands of megahertz rather than the one hundred or so already demonstrated.[62] To achieve these results, Early conceived the intrinsic barrier transistor, a p-n-i-p (or n-p-i-n) triode structure in which an intrinsic (i) region was introduced into the single crystal between the base region and the collector, effectively decoupling the two. (An intrinsic semiconductor is essentially devoid of either donor or acceptor impurities, i.e., it is neither n type nor p type. Its natural resistivity depends on the width of the gap between the valence and conduction bands.) The abstract of Early's paper presenting theoretical and experimental results read, in part, "This structure will permit simultaneous achievement of high alpha cut-off frequency, low ohmic base resistance, low collector capacitance, and high collector breakdown voltage. . . . Oscillations as high as 3000 MHz may be possible."[63] Crude fabrication techniques still permitted Early to demonstrate experimentally a structure with frequency performance of about 95 MHz. Perhaps more importantly, his intrinsic barrier design theory correlated well with the properties of the experimental transistor, thus validating the potential of high-frequency performance with better construction techniques.

Fig. 1-19.   J. M. Early and J. A. Wenger testing an early model of the
p-n-i-p triode—the intrinsic barrier transistor—developed to increase
high-frequency capability.

## 4.2 Alloy Junction Technology

A major contribution to transistor technology originating outside of Bell
Laboratories came from the General Electric Research Laboratories in 1951:
the previously mentioned alloy junction transistor of Saby.[49] It was fabricated by placing two small pellets of the metal indium on opposite sides
of a thin, n-type germanium wafer and heating the parts to about 450
degrees C. At this temperature, molten indium dissolves germanium, forming an alloy. On cooling, part of the dissolved germanium, together with
some indium, recrystalizes onto the unmelted germanium crystal forming
indium-rich p-type regions on either side of the n-type material, i.e., a p-
n-p structure. The remaining indium solidifies (together with the remaining
dissolved germanium) into soft, metallic buttons in intimate contact with
the newly formed and heavily doped p-regions to which external connections are easily made. Contacting the base region presents no problem,
as the base is physically a large piece of germanium, in contrast to the
small piece of germanium comprising the base region wedged between
emitter and collector of the grown-junction structure. [Fig. 1-20]

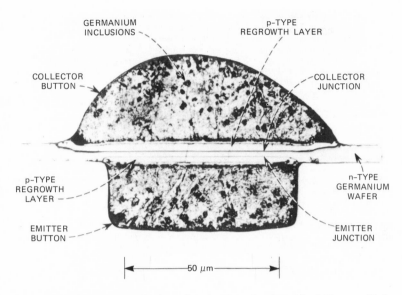

GERMANIUM
INCLUSIONS

p-TYPE
REGROWTH LAYER

COLLECTOR
BUTTON

COLLECTOR
JUNCTION

p-TYPE
REGROWTH
LAYER

n-TYPE
GERMANIUM
WAFER

EMITTER
BUTTON

EMITTER
JUNCTION

50 μm

Fig. 1-20.   Photomicrograph of an alloy transistor developed by J. Ebers and J. J. Kleimack. Alloy technology allowed low-cost manufacture and sped the entry of the transistor in the marketplace.

The recrystallized p-regions of the emitter and collector, narrow and heavily doped (low resistivity), together with the indium contact, gave series resistance much lower than the grown-junction technique and therefore superior performance, particularly in ON-OFF switching applications where the voltage drop in the ON condition had to be held to a minimum. Low series resistance also lent itself naturally to designs for power applications.

The common alloy process for making n-p-n transistors consisted of using lead doped with arsenic as the alloying metal (in place of indium) and starting with a p-type germanium wafer. The alloy techniques were accompanied by an unexpected but welcome by-product. It turned out that, during the liquid phase of the alloying process, the metals indium and lead would getter undetected minute amounts of copper[43] (deathnium and thermium in germanium) from the germanium surface and prevent it from diffusing into the crystal during the heating cycles. Earlier heating experiments with germanium crystals at Bell Laboratories had been accompanied by unexplained, gross changes in resistivity (due to thermium) and by a loss in minority carrier lifetime (deathnium). These uncontrolled results prevented for years the successful application of high-temperature diffusion as a process for making p-n junctions in semiconductor wafers with the precise dimensional control demanded for high-frequency performance.

Alloy germanium transistors accelerated the growth of the transistor industry and the appearance of transistorized products in the marketplace. Alloy transistors were potentially inexpensive to manufacture. They were efficient users of the entire "pulled" single-crystal material, not just the slice containing the junctions, as in the grown-junction process. A large-area, low-cost contact could easily be made to the base region. Although frequency response was limited, a range of characteristics could be provided readily. Relatively high-power audio output devices were designed for the portable radio and low-power-drain devices for hearing aids. Other devices were provided for both linear and digital uses. Alloy transistor manufacture, much by fledgling companies, spread in the U.S. and to many foreign countries, forming a part of the fast early growth of the transistor industry.

Even as the alloy germanium technology was beginning to provide the commercial base for the solid-state electronics era in 1952, technological events of potentially greater future importance to the industry were occurring. Teal and E. Buehler reported to the May 1952 meeting of the American Physical Society in Washington, D.C. their success in growing large single crystals of silicon and p-n junctions.[64] With this material, Pearson and B. Sawyer made the prototype alloy silicon diode that had remarkable properties compared to germanium (primarily because of its larger bandgap).[65] [Fig. 1-21] With a reverse current less than 1 nanoampere (nA), 1000 times less than that of germanium, a high and stable reverse voltage, and a good forward characteristic, the new diode demonstrated at once some of the very superior characteristics of silicon as a semiconductor material. It could also be operated at 300 degrees C. The diode was made by alloying an aluminum wire into one face of an n-type silicon chip, thus forming a p-n junction, and similarly a gold wire, possibly antimony doped, into the other face to make an ohmic contact. Based on applications of this diode, a separate branch of the growing semiconductor industry came into existence.

### 4.3 Germanium Alloy Transistors in the Bell System

The early germanium alloy junction development days were both fruitful and frustrating. Particularly fruitful was the formulation of the analytic design theory for bipolar devices and circuits. The basic model to enable mathematical device description and the prediction of circuit performance was contributed by Ebers and J. L. Moll.[66] Moll extended this work to the transient response of switching transistors.[67] Transistor design theory completed the modeling activity.[62,68,69]

Efforts to obtain good junctions and to keep them good were frustrating. In the alloy process, considerable mechanical damage surrounded the metal button, caused by cracks that occurred while cooling. Chemical etching removed debris and turned electrically "soft" junctions into "hard" junc-

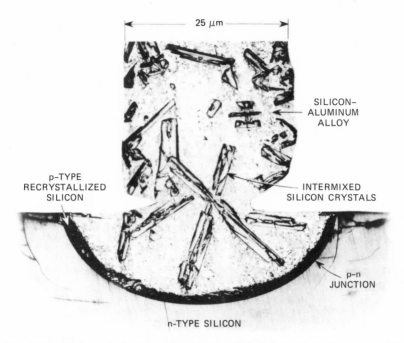

Fig. 1-21. First practical silicon alloy diode, fabricated by G. L. Pearson and B. Sawyer. The new diode, 1000 times smaller than its germanium predecessor, demonstrated the superiority of silicon as a semiconductor material.

tions, which were clean and smooth in appearance. The quality and uniformity of the junctions depended on the process art and the skill of the processor. An array of empirical etches soon developed, and junction etching became and remained a critical process step.

As with grown-junction transistors, the stability of the alloy junctions was tied to residual ionic impurities on the surface and to the presence of moisture there. Again, the "solution" to the stability problem was to encapsulate the device in vacuum-tight enclosures (with a metal-to-glass hermetic seal), often sealing in a moisture getter after employing repeated drying cycles, i.e., flushing with a dry gas. Hermetic encapsulation was deemed necessary to achieve the high reliability required for telephone and military applications. It added substantially to the unit cost, however, and was a grave disappointment for that reason; inelegantly, it also made transistors more like vacuum tubes, which also required hermetic seals. Even so, the low-temperature drying cycles by gas flushing did not consistently remove all the moisture, and reliability suffered in some units.

High-frequency response was constrained by limitations of geometry and structure, principally base width and the area of the metal alloy dots.[69]

In practice, control of chip thickness, dot size, and the alloy temperature cycle limited frequency response to a few megahertz. This was not nearly high enough for most broadband transmission and high-speed digital applications required in the Bell System.

Nonetheless, a first major application in the Bell System of the alloy germanium transistor was in E-type repeaters.[70] The transistorized negative impedance repeater replaced a vacuum tube version because it offered savings in space, power, and maintenance. It was an ideal first application: only four transistors per assembly and a minimum of operating problems in the event of difficulty; unit transistor cost was not a primary consideration, and, assuming success, a large production would be scheduled. Success did come. Reliability was excellent, and millions of transistors were manufactured over the years by Western Electric in Allentown, beginning in 1955, for use in a variety of applications.

Applications in the Bell System in the early 1960s included D1 and D2 channel banks, A5 channel modems, series E inband signaling, MF and JR receivers, 43 B1 carrier, 404C and 404D tone generators, 800 private branch exchange (PBX), 103 E6 data set, 1A data station, and N2 carrier. The applications were nearly all linear. The alloy junction transistor was not employed extensively in logic or digital circuits, although it was often used in control circuits. It was used, however, in logic circuits in the Morris, Illinois field trial of the pre-prototype electronic switching system, where it performed well. By 1968, use grew to some 400 applications.

## V. THE APPEARANCE OF SILICON

### 5.1 Diffusion and Silicon

With the successful efforts of Teal and Buehler in growing large single crystals of silicon, a second single-element transistor material became available, a potential rival to germanium, and it begged for research and development effort to determine and explore its possibilities. The new material, however, displayed low (excess) minority carrier lifetime, resistivity far from intrinsic, and other evidences of impurity contamination, some of which came from the fused silicon crucible in which the crystal was grown. For example, while measuring the (deathnium) lifetime of specimen crystals by a new photoconductive technique, Haynes and J. A. Hornbeck went on to identify the existence of two sets of temporary (nonrecombining) traps for electrons in p-type silicon.*[71,72] The traps manifested themselves

---

\* In these traps, a trapped electron sits for a time and then is ejected back into the conduction band as an excess minority carrier. There, the excess carrier faces the chance of being retrapped (immobilized in one of the two kinds of temporary traps), or of recombining with a positive hole and thereby disappearing from the conduction process. Similar trapping was found in n-type silicon, but in germanium, such trapping was observed only at low temperature (below −80 degrees C).

as an apparently anomalously long minority carrier lifetime. This characteristic was not observed in germanium at room temperature. While zone refining succeeded almost from the beginning in purifying germanium, molten silicon was so reactive that no satisfactory material for a crucible was to be found, and the silicon ingots were loaded with oxygen as a principal contaminant. More generally, silicon was an attractive new material, but a technology for its use had to be developed.

Solid-state diffusion as a technique for making p-n junctions with fine dimensional control was an attractive technology for some time. In 1947, Scaff and Theuerer had proposed, in particular, diffusion from the vapor phase.[73] This form of diffusion involved heating a single crystal wafer of the semiconductor in the presence of the vapor of the desired impurity to introduce enough of the impurity (donor or acceptor) into a surface layer of the wafer to modify (i.e., dope) the conductivity of the wafer a predetermined amount. No melting took place in this process. One keystone of this technology to be mastered was the art of diffusing the desired impurity without the simultaneous diffusion of contaminants, which were usually present in very small quantities as well.

In 1952, C. S. Fuller provided a sound basis for diffusion technology with his studies of the diffusion of donors and acceptors in germanium.[74] He extended the diffusion experiments to single-crystal silicon, when it became available.[75] The first significant breakthrough for diffusion came in the fabrication of large-area silicon diodes. Pearson and Fuller applied these results successfully to make large-area p-n junction diodes with remarkable electrical properties, which they saw as useful for the large-quantity Bell System applications of power rectification and lightning protection.[76] The technology was perhaps more spectacularly applied as a p-n junction photocell for converting sunlight directly into electrical power.[77] As reported by D. M. Chapin at the annual meeting of the American Institute of Mining and Metallurgical Engineers in February 1954, its conversion efficiency was more than 6 percent. This number, perhaps 15 times larger than the best previous solar energy converter, was high enough to suggest the feasibility of isolated solar-powered transmission equipment and relay stations. (Work on solar cells is also discussed in *Physical Sciences (1925-1980)*, Chapter 11, section 8.1.) The combination of silicon as a material and diffusion as a process offered diode designers a vast range of desirable electrical properties in devices that were potentially of low cost— low cost because diffusion naturally lends itself to batch processing, which is often more economical than one-at-a-time processing. The latter generally requires a high degree of mechanization to be cheap.

The successes with silicon and diffusion, highlighted by the Bell Solar Battery, were announced through a press conference held at Murray Hill in the spring of 1954. In contrast to the small splash made by the 1948 announcement of the transistor, this one was covered widely and well reported. Stories were written by science editors of the large New York dailies, by the national wire services, and also by trade magazine specialists. Demonstrations of silicon diodes in several applications accompanied an oral presentation by Fisk, then vice-president, Research and later president of Bell Laboratories. The Bell Solar Battery was introduced by a demonstration in which a portable FM radio transmitter, powered by sunlight through a cloudy, overcast sky, transmitted to a receiver in the press room inside a Bell Laboratories building. All could hear the reception and, at the same time, observe the demonstrator and transmitter designer, D. E. Thomas, walking around outside with the transistor radio transmitter and its solar power supply. [Fig. 1-22] The latter was a small array of individual quarter-sized p-on-n silicon diodes, which connected together formed the solar battery. A similar array was used to power a miniature water pump which, with energy from the light of a bright, incandescent lamp, pumped a stream of water from a lower reservoir into a sandy surface pool. Each p-n junction diode (or cell) gave photovoltaic output of about 0.5 volt (V),

Fig. 1-22.   A 1954 demonstration at Murray Hill, New Jersey in which sunlight falling on a solar cell powered a small radio transmitter. The voice of D. E. Thomas (foreground) was received in the press room inside the building.

so that the voltage supplied by the solar battery was about one-half the number of series-connected cells in the array.

A field demonstration of a Bell Solar Battery in a remote location was begun about 18 months later. In October 1955, a 20-V, 1-watt (W) solar conversion power supply was installed on a telephone pole near Americus, Georgia in connection with a field trial of a transistorized rural carrier transmission system. The demonstration of the battery turned out to be an engineering success while predicting an economic failure. The combination of then experimental nickel-cadmium batteries charged by the solar array performed satisfactorily as a power supply, but was judged to be more costly than competing alternatives.

### 5.2 Diffused-Base Transistors

A diffused-base method of fabricating transistors with high-frequency response was discussed at a departmental conference Shockley held in December 1953. Successful fabrication of the structure in germanium was achieved in July 1954 by C. A. Lee.[78] [Fig. 1-23] He diffused a 1.5-micrometer ($\mu$m) layer of arsenic into p-type germanium, forming a base layer; an

Fig. 1-23.   M. Tanenbaum (left) and C. A. Lee (right), who fabricated the first diffused-base transistors in silicon and germanium, respectively.

emitter layer about 0.5 µm thick was then made by alloying lightly a thin layer of evaporated aluminum into the base region. [Figs. 1-24 and 1-25] Lee employed pure germanium material that was essentially devoid of unwanted critical impurities, i.e., those which during the diffusion heat treatment would also diffuse and interfere with the formation of the desired n-p junction. He found it essential, also, to use painstaking methods to clean trace impurities both from the surface of the germanium and from the diffusion furnace.[43] The transistor was disclosed at the June 1955 Solid State Device Research Conference held at the University of Pennsylvania. Lee reported nearly unity current gain ($\alpha_0 = 0.98$) and a cutoff frequency of 500 MHz while predicting even higher values. The related patent was broad, describing virtually every diffused transistor structure, including double diffusion, field effect, and intrinsic layers (i.e., n-p-i-n), in a variety of materials—germanium, silicon, and Groups III-V compounds.[79]

Successful fabrication of the first diffused silicon transistor took somewhat longer. Ultrapure material was not available, and silicon diffusion required higher temperatures than germanium. As a result, the control of silicon resistivity at diffusion temperatures was poor, and minority carrier lifetime was predictably small. During high-temperature diffusion experiments, drastic pitting and roughening of the surface occurred, for reasons unknown at the time, which prevented the formation of smooth, planar junctions near the surface. In silicon solar cells and other large-area diodes, the junction was formed deeper into the material to avoid the effects of surface roughness. These devices tolerated the higher-resistive components associated with deeper diffusion.

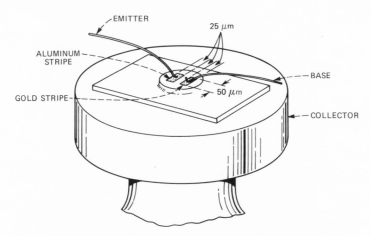

Fig. 1-24.   Details of the germanium transistor. Two metal stripes are evaporated on the raised mesa area; wires from the headers are bonded to these stripes to form the transistor's base and emitter connections.

Fig. 1-25.   Internal structure of a diffused-base germanium transistor.

In the face of these difficulties, in March 1955, M. Tanenbaum [Fig. 1-23] and Thomas were able to fabricate a diffused silicon transistor with superior high-frequency properties (for silicon).[80] They accomplished this fabrication using double-diffused silicon wafers prepared by Fuller and J. A. Ditzenberger. By simultaneously diffusing both a donor element, antimony, and an acceptor element, aluminum, into a wafer of n-type silicon, the smaller atom, aluminum, moved more rapidly, ahead of the antimony atoms to form a p-type base region where it overcompensated the original n-type silicon. By starting with a greater surface concentration of antimony, an emitter n-layer formed behind the advancing aluminum. The result of the double diffusion, then, was an n-p-n structure with base layer width of about 4 $\mu$m. The transistor had a high-frequency cut-off of 120 MHz and an $\alpha_0$ of 0.97, which might be correlated with about a 3-percent loss by recombination of minority carriers (electrons) in traversing the base region. This would correspond to a base lifetime, damaged by the heat treatments, of the order of 0.1 $\mu$s. These results were disclosed, together with Lee's results on diffused germanium, at the same June 1955 Solid State Device Research Conference. From this work it was clear that, in the application of diffusion technology, the less reactive germanium was more tractable and controllable than silicon for transistor development.[81] Overall, the details of the diffusion process were found to be much more complicated than originally perceived.

Because the current gain in silicon transistors increases with current, p-n-p-n silicon devices have bistable characteristics.[82] At low currents, a p-n-p-n diode typically exhibits the high impedance of a reverse-biased junction. Once breakdown is reached, the current increases, and with it, the internal current gain. Once the current gain exceeds unity, the device switches into a low-impedance state characteristic of a forward-biased junction.

At Bell Laboratories, such diodes were extensively studied for potential solid-state switches because the ON-OFF impedance ratio is adequate for telephone switching applications. However, because of the inability of the diode to handle the high currents and voltages required for the telephone ringer, solid-state switching proved uneconomical in the 1950s.

If connections are made to at least one internal region of a p-n-p-n diode, a three-terminal p-n-p-n switch results. Today, this device is called a thyristor (derived from "thyratron" and "transistor"; a thyratron is a gas-filled, grid-controlled tube, with two stable states, functionally similar to a p-n-p-n device). Thyristors have found widespread application in power switching and control.

P-n-p-n diodes were later manufactured by a semiconductor company started by Shockley in Palo Alto, California in 1955, at which time they became known as Shockley diodes. (This company spawned what became Silicon Valley—the area between Palo Alto and San Jose, California— which became a major center of the electronics industry.)

The successful demonstration of diffused transistor structures by members of the research area of Bell Laboratories prompted a reaffirmation of semiconductor development goals. Morton was convinced that the most promising paths toward high-performance, low-cost devices for Bell System applications involved exploitation of diffusion technology and, ultimately, silicon as the preferred material. This interest was translated into committing development effort to diffused silicon diodes, diffused germanium transistors, and diffused silicon transistors, as pieces of technology critical to their development became available through the effort of both research and development groups.* Low cost would follow from batch processing, a natural use of diffusion; and the electrical properties of silicon junctions were vastly superior to germanium, particularly as the temperature of operation rose. Germanium alloy transistor designs and alloy silicon diodes, both healthy products, were to be supported to meet customer demand,

---

* The remarkable, perhaps unique, electrical properties of a clean thin film of silicon dioxide on a clean silicon (single-crystal) surface, i.e., the clean system Si-SiO$_2$, of course, were then unknown and unanticipated. These properties, when discovered, insured that silicon was the right choice.

i.e., development effort would be confined to supporting devices already committed to manufacture. No further effort, however, was to be devoted to grown-junction technology, silicon or germanium, alloy germanium diodes, or possible variants of germanium alloy transistors; a small effort on alloy silicon transistors was canceled.

Work was to continue on diodes designed for microwave applications. Stimulated by Shockley's suggestion of using the transit time delay associated with diffusion of free carriers to construct a transit time (negative resistance) oscillator,[83] W. T. Read, Jr., in 1954 proposed that microwave oscillation and amplification could be achieved by a structure employing both avalanche multiplication and transit time delay in a low-field drift region to yield negative resistance.[84] The technology and ingenuity of the time were unable to make these proposals work. Progress was striking, however, in understanding and utilizing the nonlinear voltage-dependent capacitance of a p-n diode to perform circuit functions. In studying frequency conversion effects, A. Uhlir, Jr. [Fig. 1-26] realized that the variable reactance of the diode normally led to power gain in the upward frequency shift induced by mixing a signal with a local beating oscillator.[85] Using a gold alloy diode, experimentally he obtained an up-conversion gain of 5.7 dB between 75 MHz and 6175 MHz output. M. E. Hines and H. E. Elder

Fig. 1-26.  A. Uhlir, Jr. (standing) and P. I. Sandsmark (seated), who carried on early work with microwave diode modulators.

demonstrated parametric amplification at microwave frequencies.[86] They placed a varactor (variable reactance) diode in a waveguide circuit such that it could be pumped at 12 gigahertz (GHz); they observed negative resistance gain near 6 GHz, as predicted by the fundamental theoretical work on parametric amplification codified by J. M. Manley and H. E. Rowe.[87]

In keeping with the Morton philosophy of diffusion and silicon, late in 1955, the diffused silicon diode was taken to the Allentown Laboratory for final development and design for manufacture. It was intended to serve the Pennsylvania location as a vehicle for learning diffused silicon technology. The project was organized into a closely knit effort involving Western Electric and Bell Laboratories. Prior Murray Hill work, led by K. D. Smith, was published by M. B. Prince, who initially supported the new activity from New Jersey.[88]

Diffusion and silicon were the theme of a second symposium held in January 1956 for licensees of Western Electric to bring them up to date on recent technical breakthroughs at Bell Laboratories. As a group, the licensees on this occasion were considerably more sophisticated and knowledgeable than they had been four years earlier. Most were conducting profitable semiconductor enterprises, and many had acquired more manufacturing experience than Western Electric. The commercial transistor business rested upon the manufacture and sale of germanium alloy transistors. The silicon alloy diode also served as a production base for both new and established manufacturing operations. About 1 million transistors valued at $4.8 million were manufactured in 1954, and production value grew to about $14 million the following year. By 1956, silicon alloy diodes were manufactured by perhaps 15 companies in the United States and half that number abroad. There was a ready, expanding market for both these products, the technology was relatively simple, and the price was right.

From the point of view of the licensees, in the short term, diffusion did not appear to be timely for the needs of the commercial market except as an adaptive add-on technique for improved frequency response, where this was required. Thus, the symposium aroused little enthusiasm among the licensees to undertake and master the major change in technology represented by diffusion.

The record shows, however, that while the first diffused-base transistor was manufactured by Western Electric, the first to be marketed generally— a high-frequency device—was introduced by Texas Instruments in the spring of 1957, barely a year after the symposium. The first diffused-silicon transistor marketed widely was offered in 1959 by Fairchild Semiconductor, a subsidiary of Fairchild Camera and Instrument Corporation. Western Electric in Allentown had introduced earlier a double-diffused silicon transistor into pilot line manufacture. It was a switching-type transistor and not generally available.

For several reasons, applications of the new, more sophisticated diffused transistors surfaced more rapidly in military development than in Bell System telephone development. From the viewpoint of the Bell System, it was going to be difficult to achieve significant economic gains by transistorizing existing products in the telephone plant, many of which had benefited from years of cost-reduction activities by Western Electric. Also, the new communications systems going into large-scale manufacture, microwave radio as an example, were being made possible by advances in vacuum tube technology, notably the traveling wave tube and the close-spaced triode, both of which could operate at frequencies and power outputs unattainable with transistors. The U.S. military, on the other hand, needed new systems to perform functions, often digital, not then being performed. This need led to different cost considerations and to a different technical confrontation between tubes and transistors.

A major segment of military systems development at Bell Laboratories after World War II was in the Nike ground-to-air radar-guided missile systems, under contract to the U.S. Army in cooperation with Douglas Aircraft Company, the missile contractor. Nike Ajax and Nike Hercules, both antiaircraft systems, were designed, manufactured, and deployed during the late 1940s and 1950s. (See *National Service in War and Peace (1925-1975)*, for a discussion of Bell System involvement with the armed forces.) Concern over the potential Soviet ballistic missile threat led the Army to ask Western Electric and Bell Laboratories to develop a radar missile system designed to intercept incoming ballistic missiles armed with nuclear warheads. The new antiballistic missile (ABM) was called Nike Zeus. Bell Laboratories military designers at the Whippany, New Jersey location worked with transistor developers to arrive at transistor circuit functions, many digital in nature, that optimized the design specification of the transistor to the circuit. Functions that these circuits performed were to convert analog quantities accurately, like radar measurement of range or the pointing of an antenna, to digital numbers that a computer could then manipulate. Thus the military system requirements, passed on to the diffused-base transistor developers, were for transistor oscillators capable of operation in the hundreds of megahertz range, radar intermediate-frequency (IF) amplifiers, and video amplifiers—all of very high reliability.

By means of much thoughtful planning, the system design needs were evolved in parallel with the diffused-base transistor development and design efforts, and also in parallel with creation of the production facilities and expertise to supply the quantities needed, both for development and production systems. This came about over a period of time through a common desire of Army personnel and Western Electric and Bell Laboratories leaders and engineers to accomplish several objectives in a timely and cost-effective way.

Development responsibility for the diffused-base germanium transistor rested with Ryder and Early in the development area at Murray Hill. Military funding was available to support development of designs for military applications. Manufacture of devices for military systems was centered at the Laureldale shop of Western Electric's Radio Division, a facility that had been acquired and equipped using government funds for this purpose in 1952. Bell Laboratories design support to Laureldale, which was the predecessor of the Reading, Pennsylvania plant, came both from the Allentown Laboratory and from the Murray Hill location. Military supply contracts initiated in 1956 accelerated the availability of the new transistors. These contracts called for delivery in quantity of high-performance devices, i.e., beyond the capabilities of alloy and grown-junction technologies. The contracts paid for final development work, the resolution of small-scale manufacturing processes, and the production of samples for circuit and systems development use, including reliability assessment. The Signal Corps also had funds available to establish production facilities under so-called preparedness contracts.

In the spring of 1957, there were problems of supply, reliability, and cost at Laureldale with the new military devices. W. C. Tinus, Bell Laboratories military vice president, and L. C. Jarvis, the Western Electric shop manager, approached Morton with the proposal that a branch laboratory be set up at Laureldale. Morton responded by forming a task force to attack and solve the problems. S. O. Ekstrand, of Bell Laboratories in Allentown, the task force leader, was supported by knowledgeable personnel on loan to the task force from Murray Hill, Allentown, and Western Electric. The task force was greatly assisted by the work of a committee of device people and military systems engineers, work that specified explicitly the types of devices to be designed and manufactured at Laureldale. The list included diffused germanium and silicon transistors, silicon diodes, and alloy transistors. Dates for completion of the various tasks were also set. The task force operated daily at Laureldale, performing its assignments directly on the manufacturing floor—a new and effective method of procedure for Bell Laboratories and Western Electric—except when lack of a particular facility necessitated a trip elsewhere. The task force effort resulted in industry-coded devices such as the diffused germanium transistor 2N559, the silicon diode 1N673, and later the diffused silicon transistor 2N560.

More specifically, the Laureldale germanium diffused-base codes were: 2N559, a 20-MHz switching transistor; 2N509, an oscillator capable of producing 100 mW at 100 MHz; 2N537, designed for 100 mW at 250 MHz; and 2N694, for small-signal video applications. The 2N694 offered a gain of 15 dB with 25- to 30-MHz bandwidth. These devices, with internal doping similar to Lee's p-n-p, had a 0.001- by 0.002-inch aluminum emitter stripe separated by less than 0.001 inch from a similar rectangular gold stripe, which contacted the diffused-base region. Electrical contacts to the stripes were made by small gold wires that were thermocompression bonded to the stripes, a process developed by O. L. Anderson, H. Christensen, and P. Andreatch.[89,90] [Fig. 1-27] The thermocompression bond was a timely solution to a perplexing problem: how to attach connections to the tiny electrode stripes. In this process, an operator used a micromanipulator and a microscope to align a 0.001-inch gold wire to the stripe. Bonding was accomplished by pressing a hard wedge down on the wire with the application of heat. Subsequently, the wire was drawn to a header post where a second bond was made. This operation in large part accounted for the sight familiar in the early transistor industry: employment on the production floor of many operators seated at bench tables, looking through microscopes and delicately operating micromanipulators. Definition of the

Fig. 1-27. (Left to right) H. Christensen, O. L. Anderson, and P. Andreatch, developers of thermocompression bonding.

collector junction area was accomplished by etching away the top surface of the germanium die outside a circle enclosing the emitter and base strips. This process formed a mesa—a tiny, flat mound standing above the exposed base collector junction and the collector material below—and the general structure became known as a mesa transistor.

As before, a vacuum-tight encapsulation of the transistor was found to be necessary (but in itself not sufficient) to attain high reliability. Proven techniques of encapsulation were borrowed from electron tube technology to accomplish hermetic sealing. As shown in Fig. 1-28, three Kovar*-to-glass seals brought three lead wires through the bottom Kovar header. (Kovar is a special alloy suitable for glass-to-metal seals.) After the transistor was connected to the three lead wires, a can shaped like a top hat was sealed by a weld to the outside rim of the header. A small tubulation through the top of the top hat permitted final vacuum exhaust and other desired processing before sealing the tubulation.

The Kovar piece parts, i.e., the Kovar header and can, were plated with nickel and then with gold to avoid corrosion and subsequent hydrogen ion penetration. The gold-to-gold header-to-can interface would weld under moderate pressure and at a temperature the semiconductor system would tolerate. Later, a copper surface was added to each piece part, and a copper-to-copper cold weld was shown to give satisfactory results. The tubulation was sealed off by mechanically pinching together and cutting the tubing, a cold-welding process in which, again, copper performed satisfactorily.

The wire leads (Kovar) through the glass seals became stubby posts inside the header enclosure. These posts were gold plated, and the collector side of the germanium die (or chip) was alloy bonded to the header, making contact to the collector. This process was improved by introducing a small "preform" of gold-germanium eutectic mixture between the die and the header prior to alloying. One of the leads was bonded to the header to become the collector lead, as seen in Fig. 1-28. The gold wires from the emitter and base stripes were bonded to the two remaining posts, completing the electrical connections to the lead-through wires.

Exhausting the gas from this structure through the tubulation and then sealing the tubulation at a reasonably good level of vacuum did not yield devices of adequate reliability. The vacuum-tight enclosure was necessary to prevent water vapor in the ambient atmosphere from penetrating to the transistor p-n junctions. Simply pulling a vacuum, however, did not eliminate traces of water vapor, stray ions, and other forms of uncleanliness that remained absorbed and/or attached to surfaces inside the sealed can. High-temperature heating (baking) during exhaust was traditionally an effective means of eliminating the offending molecules; however, the ma-

---

* Trademark of Carpenter Technology Corp.

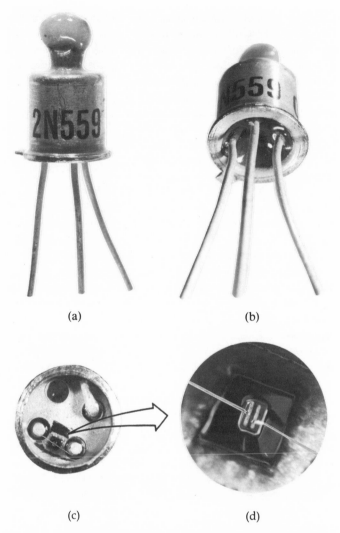

(a)  (b)

(c)  (d)

Fig. 1-28.   The Laureldale 2N559 transistor in a so-called TO-18 package. (a) Front view. (b) Three Kovar-to-glass seals brought the three lead wires through the bottom Kovar header. (c) Top interior view. The wire leads became stubby posts inside the header. One post was bent and welded to the header to become the collector lead. (d) Enlarged view of chip and wires connecting chip to posts.

terials systems of the transistor could not withstand a baking temperature nearly high enough for adequate purification. So reliance was placed on several unsophisticated procedures that, in the combinations employed, gave an empirical solution to the problems. Very reliable transistors resulted. The procedures included: moderate temperature bakes and dry gas flushing;

assembly in "dry boxes," which had rubber gloves for the assembler's hands sealed into openings in the wall of the box; introduction of a getter consisting of unreconstructed Vycor* (a glass frit) to absorb water vapor and thus decrease the likelihood that the troublesome molecules would settle on a sensitive site, for example near the collector junction.

The diffused-base mesa transistor was a very good product, but it was important to find out how good. This, in principle, could be determined from measurement of device failure rate as a function of time and from a knowledge of failure mechanisms. The latter could be determined from autopsies of failed devices by diagnostic experts in the laboratory. The former, determination of failure rates, presented an awkward problem of numbers and time. If the transistors were a sound product with a long-time failure rate of 0.01 percent per 1000 hours, then a life test performed under rated conditions on a sample of 1000 transistors would turn up only about one failure after one year of continuous testing. To improve the statistics and confidence in the result, both longer times and larger samples needed to be introduced into the reliability test plan—making it all the more costly and time consuming.† Clearly, faster and cheaper means than brute force life testing needed to be found to evaluate transistor reliability.

The Laureldale 2N559 transistor was chosen as the principal vehicle in a program to learn more about evaluating transistor reliability. To meet its specifications for application in Nike Zeus circuits, it had to have high reliability. It was also available in sufficient quantity; transistors from individual production lots could be identified, and thus their exact process history established. B. T. Howard and G. A. Dodson solved the numbers problem most elegantly by devising a method of accelerated aging.[91] Called step-stress aging, this process applies stress in uniform increasing steps to a limited sample of devices for a fixed time interval at each step until all devices fail. From the high failure rates obtained at the high values of stress, it is possible to predict the much lower failure rates under normal operating conditions.

With the aid of this technique, D. S. Peck [Fig. 1-29] was able to characterize the 2N559 transistor as delivered from Western Electric manufacture.[92] He found that this transistor had a more than respectable reliability failure rate of nearly 0.001 percent per 1000 hours (10 failures in 1 billion hours—or 10 Failure unITs [FITs]), adequate to meet the stringent reliability

---

* Trademark of Corning Glass Works.

† To be able to establish that a device will perform for a given number of years, in general the device must have existed for an appreciable fraction of that time. For example, a 20-year failure-free life was the design objective for the tubes in the repeaters of the original submarine telephone cable system. When the first transatlantic cable was laid in 1954, the design of the vacuum tubes in the submerged repeaters was two decades old.

Fig. 1-29.   D. S. Peck, who implemented reliability characterization processes and criteria that permitted stringent reliability requirements to be met for transistors.

needs of Nike Zeus. The elusive property of solid-state reliability, somehow foreseen from the early days of the point-contact transistor, was thus achieved and demonstrated for the first time.

By mid-1958, Western Electric germanium transistors of this general type were encircling the earth in Explorer and Vanguard satellites. At the Whippany location, the Bell Laboratories designers of military systems were convinced that enough experience with transistor digital circuits had been gained to allow a confident commitment to a very large system like the Nike Zeus ABM system. Anticipating the lead time required for large production, the Signal Corps in the fall of 1958 asked Bell Laboratories to select and qualify two second sources of supply of Nike Zeus transistors in addition to Western Electric. The selection process was initiated but not completed, as Nike Zeus was destined not to go into production. If second sourcing of the 2N559 had occurred, the intent of Bell Laboratories and Western Electric was that each second source duplicate exactly in its lines the Western Electric production processes and procedures, not just produce a transistor that conformed to the 2N559 performance specifications. The feeling was that true device interchangeability could be achieved only through similar processing.

By the fall of 1958, the success of the Laureldale task force led to visible changes there. The Western Electric facility, originally a silk mill, grew from a few hundred employees to over 2000, and was upgraded from a shop to a plant designation. Also, Bell Laboratories established there an independent branch laboratory under the direction of Ekstrand. This move was followed in a few months by a decision actively to market Laureldale semiconductor products more generally. In contrast to entertainment quality devices, so-called Hi-Rel transistors, distinguished by both high reliability and high performance, were not widely available in the marketplace. So there was an outside need for a product that happened to mesh with a strong desire of Western Electric and Bell Laboratories engineers to build up production rates and thus acquire more production experience sooner. The engineers were convinced, through knowledge of learning curve studies,* that transistor unit cost is related to production experience: those who acquire the production experience will tend to have the lowest cost.

The decision to promote external sales not only satisfied the engineers, it also expressed the sales interests of Western Electric's Radio Division, which operated the Laureldale plant, and coincided with the interests of the Signal Corps. The military had funded the facility with the general purpose of making state-of-the-art products available on a timely basis for use in military systems. External sales of high-quality semiconductor devices would augment the Laureldale product line. It became apparent that in order to cover both the internal and external markets, it would be necessary to establish a sales engineering force. Accordingly, a group of sales engineers was assembled at Laureldale.

To help in stabilizing the Laureldale production operation, the magnetron production facility, together with principal personnel, was transferred from St. Paul, Minnesota. The major production consisted of two magnetrons: the 5780, a tunable X-band 250-kilowatt (kW) tube, and the 5795, a tunable S-band 1000-kW tube. The new Laureldale Laboratory was assigned design responsibility. It also initiated final development of a coaxial-cavity magnetron conceived by J. Feinstein (see Chapter 3, section 3.3). The 100-kW tube designed for the Ku-band was coded 7208 and was introduced into the magnetron production line.

Thus, by the end of the 1950s, Bell Laboratories had solved most of

---

* A learning curve is a plot of unit cost (usually log cost) versus accumulated production (usually log production) of a given product, in which unit cost decreases monotonically with accumulated production. Typically, the slope of the curve is found to be a 20- to 30-percent reduction in cost for each doubling of cumulative product manufactured. This empirical relationship was first noted in the aircraft industry during World War II. Later, it was found to apply to vacuum tube production and was then believed (late 1950s) to apply to transistors. This has since been verified.

the reliability problems in germanium transistor production and paved the way for future introduction of high-quality, inexpensive transistorized products.

### 5.3 Improving Silicon Technology

While zone refining as a technique for purification of germanium worked very well, as indeed it did with many solids, it became a partial failure with silicon. The problem resided in the boat, or crucible, for holding the silicon charge. No material for the boat could be found that did not react with molten silicon. For example, the least reactive container material, very pure fused quartz, reacted with molten silicon, releasing contaminants (mainly oxygen) into the melt. To circumvent this problem, Theuerer invented an ingenious technique called float zoning that eliminated the crucible.[93,94] He employed a vertical rod of silicon clamped at both ends. With an induction heater, he formed a short molten zone into which the molten silicon was confined by surface tension. The zone was then passed along the length of the rod by moving the rod with respect to the heater coil. This method removed impurities in silicon with the important exception of boron, which was not segregated efficiently. (See also *Physical Sciences (1925-1980)*, pp. 582-583, which includes illustrations.) A specially developed chemical process for removing boron corrected this deficiency. These techniques, plus the Pfann process of zone leveling,[45] produced single-crystal silicon remarkably free of unwanted impurities and also uniformly doped with desired impurities. It was a splendid starting material for diffusion.

The surface pitting on silicon, which occurred during high-temperature diffusion experiments, received attention from both the research and development areas. Until it was brought under control, transistors with close emitter-collector spacing could not be made with good yield. The solution, easily explained in terms of simple chemical thermodynamics, was found by C. J. Frosch and L. Derick.[95] They discovered that by covering the silicon with a thin film of silicon dioxide, the diffusion could then be carried out without surface pitting and roughening. The oxide protection was formed in the diffusion furnace by simply exposing the silicon to oxygen at a high temperature. The subsequent diffusion was carried out under conditions such that the oxidizing potential of the carrier gas (generally by the addition of water vapor) favored formation of $SiO_2$ rather than $SiO$.

Frosch found this solution in the course of his effort to develop an open tube diffusion process to replace the cumbersome vacuum-tight sealed systems. Protection of the silicon by an oxide layer was the key to the process. For a long time, his experimental runs were damaged by oxide pitting, giving poor diffusion layer profiles. One day, while he was experimenting with a hydrogen gas atmosphere, the hydrogen accidentially ignited inside the diffusion tube, adding water vapor to the system. This time the surfaces were smooth. With this discovery, pitting was easily explained as due to the reduction of $SiO_2$ to $SiO$, which is volatile.

Frosch and Derick went on to use properties of the thermal oxide for the very important technique of oxide masking, which is now used universally (together with photolithography) for selective, localized diffusions.[95,96] They discovered that $SiO_2$ a few thousand angstroms (Å) thick on the surface would prevent the diffusion of (i.e., mask) certain diffusants while being transparent to others. For example, an $SiO_2$ layer masks diffusion of phosphorus; thus it protects a p-type base layer while permitting formation of emitters by phosphorus diffusion through windows etched in the oxide layer. The technique was a natural batch process. (See also *Physical Sciences (1925-1980)*, pp. 607-608.)

Precise etching of windows was accomplished by photolithography techniques whose application to silicon structures was pioneered by J. Andrus and W. L. Bond.[97,98] Andrus laid on the surface oxide a layer of photosensitive polymer that was also etch resistant (i.e., a photoresist). The polymer was exposed to light from a mercury lamp through a mask that shaded the small emitter window areas. After exposure and removal of the mask, the silicon sample was immersed in a photodeveloper solution. A subsequent wash in water deionized for cleanliness removed the unexposed polymer coating, forming the emitter windows in the photoresist through which the oxide below was removed by an etch. The remaining photoresist coating was then removed by an organic solvent. Thus a precision-controlled opening for the subsequent emitter diffusion was made in the oxide, which becomes the mask for this process.

While the photolithography technique is described above in terms of an emitter fabrication process, it was actually first used by Andrus as a technique to delineate the device structure of the silicon stepping device of Ross, L. A. D'Asaro, and H. H. Loar, as discussed in Chapter 2.[99]

Precise control of the thin transistor base layer—a requirement for both good yield and higher frequency—was best achieved through a special two-step diffusion, a process developed by B. T. Howard.[100] Howard used a predeposition of boron, which formed both a borosilicate glass deposit on the surface and a boron-diffused region a short distance into the silicon. The glassy boron deposit, a surface source of diffusant, was then removed chemically. The boron atoms remaining in the silicon were diffused further by a high-temperature drive-in step. In effect, Howard's predeposition deposited a set number of dopant atoms in the silicon and arranged them, by the later drive-in, into a carefully constructed base layer. Much later, the initial number of atoms could be preset more precisely by ion implantation (see Chapter 2, section 4.4).

About this same time, mid-1957,* the mysterious and worrisome source of lifetime degradation in silicon during high-temperature diffusions was identified.[102] The offending material was gold—in minute amounts. One part in a billion was enough to cause deterioration. The solution had in fact been found earlier but not applied to transistors.[103,104] Gold prefers nickel to silicon, and a nickel plating of the silicon slices could be carried out prior to diffusion to getter the stray gold atoms. A short time later, gold was deliberately introduced to kill lifetime in high-speed (fast-recovery) diodes in which "long" lifetime contributed undesirable delay; for the delay to be as short as several nanoseconds, then the lifetime must also be as short.[105,106] Long-lifetime silicon material, free of gold deathnium, found its real home in p-i-n power diodes, wherein the long lifetime contributed to lower effective internal resistance, by conductivity modulating the high-resistivity region.[107,108]

Thus, the bulk properties of both germanium and silicon could be controlled well. In a straightforward but expensive way, materials scientists could make 10,000- to 20,000-ohm cm ($\Omega$ cm) silicon (i.e., intrinsic silicon at room temperature) with millisecond lifetime by the float zoning method. Primarily, only second-order effects remained as special problems, such as the origin of a remaining 1-percent resistivity change during heat treatment of silicon. More thought on materials was turning to III-V compounds, the intermetallic compound semiconductors formed from the third and fifth column of the periodic table, e.g., indium phosphide and gallium arsenide (GaAs). Following the pioneering work of H. Welker at Siemens in Germany,[109,110] assessment of compound semiconductors was undertaken by a group led by J. A. Burton. GaAs, it turned out, was promising because of its higher energy bandgap (1.4 electron-volts [eV]) compared to silicon (1.12 eV) and because of the high mobility of its conduction electrons (8500 cm$^2$/V-sec) compared to silicon (1500 cm$^2$/V-sec). Also, high-resistivity GaAs material was becoming available for device work through the application of crystal growth and purification techniques already developed for silicon.

---

* This is also the time that Leo Esaki of Sony, in Japan, discovered the tunnel diode.[101] He observed a negative resistance region, at a voltage of approximately 0.1 V, in the forward current-voltage characteristic of a heavily doped germanium p-n junction. He interpreted the characteristic in terms of quantum mechanical tunneling of majority electrons from the valence band through the potential barrier at the junction, analogous to a phenomenon known as field emission at the surface of a conductor when a very high electric field (about 10$^7$ V/cm) is applied. While negative resistance is a well-known and useful effect in electronics, the Esaki effect was recognized as unusual when it was realized that the tunneling time is *very* short compared to conventional transit times. As a negative resistance oscillator, it can be operated in the microwave region and well into millimeter-wave frequencies.

## VI. TRANSISTORS/DIODES FOR BELL SYSTEM APPLICATIONS: FALL 1957

A carefully selected array of different devices had been developed for Bell System applications by the autumn of 1957. A description of these devices illustrates the state of the transistor art at the close of the transistor's first decade. The characteristics listed in some cases were design objectives, not all of which had yet been achieved.

### 6.1 Two-Region Devices—Diodes

#### 6.1.1 Power Rectifiers

Made of silicon by diffusion techniques, the structure was $n^+$-p-$p^+$.* The heavily doped $p^+$ segment reduced forward resistance and made possible a good ohmic contact to the external leads. There were three sizes for maximum forward currents of 0.1 ampere (A), 0.5 A, and 10 A. Reverse saturation current at room temperature was less than 5 microamperes ($\mu$A), except slightly higher for the 10-A code. Peak inverse voltage was at least 200 V (600 had been achieved). Recovery time was about 5 $\mu$s. All devices had a forward drop of less than 1 V at rated currents.

#### 6.1.2 Computer Diodes

These were $p^+$-n-$n^+$ silicon devices made by diffusion. Their rated characteristics were: current, 50 milliamperes (mA); reverse voltage, 45 V; series resistance, 30 $\Omega$; capacitance, less than $4 \times 10^{-12}$ farad (F); reverse saturation current, less than $10^{-9}$ A at 10 V; forward drop, less than 0.7 V at 2.5 mA; recovery time, less than $10^{-8}$ sec. The fast recovery time, essential to the computer diode application, was achieved by the combination of low capacitance (small diameter), low series resistance, and short lifetime. At that time, short lifetime was achieved by quenching the original crystalline material from high temperature or by electron bombardment; both treatments disrupt the regularity of the underlying crystal lattice.

#### 6.1.3 Intermediate-Speed Power Diode

This structure was a diffused silicon $n^+$-p-$p^+$ diode to be used in electronic switching for such things as core drivers. Its characteristics were similar to power diodes with 1-A rating and 0.3-$\mu$s recovery time.

---

* Nomenclature: $n^+$ means much more heavily doped material (i.e., more donor atoms, lower resistivity) than n. Similarly, $p^+$ is more heavily doped with acceptors than p.

### 6.1.4 Voltage Limiters

These were $n^+$-p-$p^+$ diffused silicon devices. They were designed for ten values of voltage between 6.2 and 150 V, rated at 3-W dissipation continuously. A larger-area device to be used as a surge protector could dissipate 1 kW for 0.5 millisecond. For limiters below 6 V, forward-biased silicon diodes were used. These could be stacked to give voltage limits in multiples of 0.7 V, which is a characteristic of silicon junctions.

## 6.2 Three-Region Devices

### 6.2.1 High-Frequency Germanium Transistors

These transistors (3 codes) were p-n-p germanium with a base region diffused to a depth of 1 $\mu$m and an aluminum-alloyed emitter junction. Leads were attached to both the alloyed and diffused layers by the new thermocompression bonding technique. These devices had bandwidths of several hundred megahertz with the "best" specimens observed to oscillate at 1300 MHz and higher. Applications were as oscillators, video amplifiers, and small-signal amplifiers.

### 6.2.2 High-Frequency Silicon Transistor

This device, a diffused-base, diffused-emitter n-p-n silicon transistor, was being developed as a switching transistor. Its high-frequency performance (frequency cutoff) was at least ten times lower than the germanium devices being developed. As a switching transistor, its total switching time (which includes turn-on, storage, and turn-off) was less than 0.1 $\mu$s with a collector current of 30 mA. Anticipated applications were gates, flip-flops, and similar digital functions in switching and computer systems.

## 6.3 Four-Region Devices

### 6.3.1 Two-Terminal Bistable Switch (Shockley Diode)

This device was a diffused silicon p-n-p-n structure with base layers 7.5 to 50 $\mu$m. Typical characteristics were: breakdown voltage, 50 V; turn-on current, nominally 1 mA but controllable from 50 $\mu$A to 100 mA by high-energy electron bombardment; OFF impedance, $10^9$ $\Omega$; capacitance, less than 20 picofarads (pF); dynamic ON impedance, less than 3 $\Omega$; ON voltage drop, somewhat greater than 0.7 V. This device would switch on in nanoseconds and off in tenths of microseconds. Applications included talking path switch, function generator, flip-flops, etc.

### 6.3.2 Three-Terminal Bistable Switch (Thyristor)

With the addition of a base contact to the two-terminal bistable switch, the device can be turned off by introducing a signal current through the base lead. This current, however, has to be nearly equal to the controlled current—quite different from the turn-on condition in which only a small control current is required.

## 6.4 Applications

Within the larger Bell System framework, diffused-base germanium as a technology had matured before big system demands occurred, e.g., electronic switching systems. Therefore, its use was limited, principally in transmission applications, among these the first transistorized submarine cable. The first broadband (6 MHz) submarine cable system (designated SF) was laid in 1968. When the decision was made some years earlier to design its transistorized repeaters, silicon technology was in a state of flux, undergoing rapid evolution. Germanium devices were both better known and better understood, and they more nearly met the time criterion (see footnote, p. 54) of having existed a number of years as a mature technology. Even so, several changes, primarily in the metalization technology, were introduced in 1962 along with dimensional changes to improve the stability and reliability of the devices for the submarine cable application. Also, the collector was brought under control by means of a deep diffusion of zinc to make it more p-type ($p^+$). The operational performance of the SF cable systems has been as impressive as the remarkable performance of its two narrower-band predecessor systems, both based on vacuum tube repeaters. In the early 1970s, the SG cable (30 MHz versus 6 MHz) was designed around a double-diffused silicon transistor developed and manufactured at Western Electric in Reading. This cable was laid in 1976.

## VII. PROGRESS IN SILICON TECHNOLOGY AND DEVICE STRUCTURES

### 7.1 Epitaxial Transistor

The speed, or cutoff frequency, of a diffused-base transistor could be increased by narrowing the collector region. The electrically useful part of the collector was perhaps 10 $\mu$m, whereas the collector thickness in practice was 100 to 200 $\mu$m. This dimension was set by the thickness of the thinnest silicon wafer that could be handled without excessive breakage. The additional collector material contributed series resistance (called "serious" resistance in the lab), which degraded frequency performance. A possible solution to this problem would be to grow a thin layer of high-resistivity single-crystal silicon (i.e., material suitable for good collector characteristics) on a thick substrate of low-resistivity (i.e., low series resistance) material. Such a thin film, or layer, is called an epitaxial layer if its orientation and lattice structure are identical to that of the substrate on which it is grown. (Actually, it is isoepitaxial if the atoms in the film and substrate are identical, but this term is rarely used.) A group led by Ross initiated development to provide an epitaxial silicon layer in which a diffused-base silicon transistor could be made.

The effort of that group could build on previous work by Christensen and Teal, who had demonstrated the feasibility of depositing germanium epitaxially on germanium in the mid-1940s.*[111] For silicon, the process involved, under super-clean conditions, the thermal decomposition of a gas, $SiCl_4$ for example, onto a substrate of single-crystal silicon heated comfortably below its melting point. The resistivity of the layer could be established separately, and in any case it was independent of that of the low-resistivity substrate. The development was successful; the epitaxial layer could be made of device quality. At the June 1960 Solid State Device Research Conference, Theuerer, J. J. Kleimack, Loar, and Christensen described the new, practical epitaxial transistor, in which transistor structures had been formed in an epitaxial layer by diffusion.[112] [Fig. 1-30] The epitaxial technique was anticipated to be broadly and generally applicable, as indeed it has turned out to be.

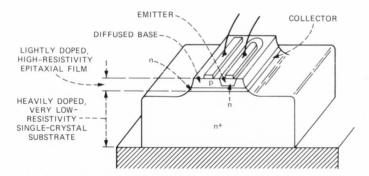

Fig. 1-30.  Epitaxial diffused transistor. A thin layer of high-resistivity silicon, grown on a low-resistivity substrate identical in lattice structure and orientation—an epitaxial layer—forms the material in which a diffused-base transistor is made. The low-resistivity substrate significantly reduced series resistance, avoiding a major problem experienced in nonepitaxial structures.

---

* They later proposed forming a transistor by growing successive layers of opposite conductivity type. This approach had not proved practical, however, primarily because of the difficulty in forming p-n junctions good enough for transistor action.

## 7.2 Oxide Stabilization and the Metal-Oxide-Semiconductor Transistor

From early transistor days, it had been recognized that the properties of a p-n junction, where it reaches the surface, affect both the fabrication and stable operation of silicon and germanium transistors. Furthermore, it was well known that the conductance between the emitter and collector of an n-p-n transistor, with base floating, was increased by the presence of a channel over the p region. A channel was a surface layer of opposite-sign conductivity (n-type on p), an inversion layer, which could arise, for example, from the local electric field associated with ions absorbed on the surface.

It was these surface layers that obscured the early field-effect experiments. Inversion layers had been studied extensively at Bell Laboratories by W. L. Brown, Brattain, C. G. B. Garrett, and H. C. Montgomery[113,114] and others.[115] With the adoption of oxide covering and masking techniques, the properties of the interface between an $SiO_2$ layer and silicon became very important to understand and control. M. M. Atalla headed a group whose responsibility was surface properties and related $SiO_2$ growth and characterization. In their studies, Atalla's group soon discovered oxide passivation, the stabilization of silicon surfaces by clean, thermally grown oxides.[116] They also discovered that working with oxide films was a continuous fight to keep the films and interfaces free of contaminants.

The observation of oxide passivation was an important clue that guided the direction of future bipolar device development. It suggested that p-n junctions emergent at a surface could be protected by a covering oxide and that channels could be avoided. But it also suggested that channels (inversion layers) could be induced by an electric field. And, at the device conference where the epitaxial bipolar transistor was first described, D. Kahng and Atalla reported, in a less heralded paper, the first demonstration of an $Si$-$SiO_2$ metal-oxide-semiconductor (MOS) transistor, the first modern surface-field-effect transistor.[117] [Fig. 1-31]

In this device, Kahng and Atalla induced an inversion layer over a narrow region separating two regions of like conductivity by creating an intense electric field in a thermal $SiO_2$ film grown on the underlying semiconductor. Voltage was applied to a metal-film gate deposited on the $SiO_2$ insulating layer, which was perhaps only 0.2 $\mu$m thick. Conductance was measured between the two like-conductivity regions, source and drain in field-effect parlance.

By contrast, in the very early field-effect experiments of Shockley and Pearson,[7] surface states vitiated observation of the field effect, or reduced its strength many times. The Kahng-Atalla experiment was intended to probe for the presence of surface traps at the $Si$-$SiO_2$ interface as well as to test the quality of a particularly good thermal oxide as an insulator. (Lack of cleanliness would lower its dielectric strength and increase its

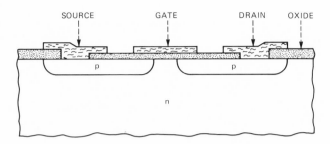

Fig. 1-31.  The metal-oxide-semiconductor (MOS) transistor.

loss.) The particular oxide was that of J. R. Ligenza, who had grown good-quality $SiO_2$ films by high-pressure steam oxidation of silicon.[118] These films had yielded improved device (transistor/diode) stability, and the Kahng-Atalla experiment was designed to quantify this improvement. As Kahng reported, the quality of oxide was good enough to allow demonstration of the unipolar insulated gate field effect, an existence proof that $Si$-$SiO_2$ interfaces can contain manageably small amounts of surface traps.[119] As it has turned out, this may be a unique property of the $Si$-$SiO_2$ system, in which the concentration of surface traps is on the order of $10^{10}$ per $cm^2$, i.e., about one per 10,000 surface silicon atoms. This is perhaps 100 times lower than other known systems. The performance of the novel device, sometimes called insulated gate field-effect transistor (IGFET), conformed to calculations but was far below bipolar standards. Reproducibility of results from different fabrication attempts was poor. This was attributed by Kahng to variability in surface state configurations at the interface and to ionic contamination of the insulating film.

### 7.3 Planar Transistor

Shortly after disclosure of the epitaxial transistor, J. A. Hoerni, a Fairchild Semiconductor scientist, reported on their new planar transistor.[120] [Fig. 1-32] In this flat structure, the p-base region, in addition to the n-emitter region, is localized near the surface by a thin diffusion made through a window, photolithographically defined and etched into the masking oxide. The structure employs, then, two separate applications of the oxide masking-diffusion process. Electrical connection to both the emitter and base at the planar surface was made by aluminum metalization through another set of windows defined by masking.[121] Contact to the collector region from the front (top) surface could also be made, if desired. The technique of forming base and emitter regions by diffusion through windows in the surface oxide caused junction termination at the surface under a silicon dioxide layer, which was left in place. The junctions were thus automatically

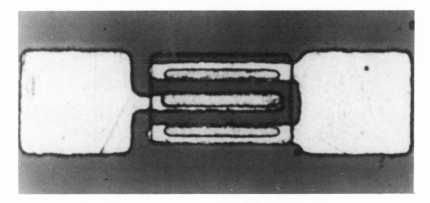

Fig. 1-32.   A planar transistor, a concept that originated at Fairchild.

passivated to the extent known techniques of cleanliness and encapsulation permitted. The planar structure offered the synergistic combination of the separately developed technologies of oxide-masked diffusion, photolithography, aluminum metalization, oxide passivation, and, very naturally, epitaxy, when it was incorporated into the starting material. R. N. Noyce suggested extending the metalization from the electrodes over the $SiO_2$ layer to form larger bonding pads for ease of contacting.[122] He also pointed out that metalization over the oxide could interconnect separated discrete devices and components, an approach that made ICs practical.

The superior utility of the planar structure was not immediately obvious, as was the case with the epitaxial structure. At the time, mesa transistor performance was markedly improved by the addition of epitaxy, and it was superior to nonepitaxial planar performance. An assortment of development problems, which plagued designers of both structures, also tended to obscure the situation. In May 1961, in reporting to a Bell System Presidents' Conference concerned with the impact of important new developments in transistor technology, Hornbeck, then responsible for semiconductor device and electron tube development, gave a Bell Laboratories evaluation of the competing elements. He said: "The planar is flat rather than mesa. It is diffused like the others. It does not appear to be a major step forward. We are still evaluating it. It might have advantages in ease of processing, cost and inherent high reliability. I mention it because the best transistors made tomorrow will take advantage of last year's developments. They may be planar-epitaxial-diffused. They will certainly be epitaxial." He also went on to say: "The situation today is that all transistors being manufactured are made from germanium or silicon. You will note, however, that there is another material—gallium arsenide. In single crystal it looks very much like germanium or silicon. It has the properties that it is inherently faster than germanium with higher temperature capabilities, higher power, than silicon. We can make transistors and diodes from it, but we don't have it under control. It is the material of tomorrow."[123] The mesa structure, however, turned out to be ill suited to the joint employment of the photoresist process together with junction passivation by thermal oxidation.

### 7.4 Reliability Improvements

As reliability testing of silicon transistors progressed in its search for latent junction instability, a new device disease suddenly stood out above other infirmities. Thermocompression bonds failed, apparently randomly, through developing high impedance, and actually parting in some cases. Examination of the bonds disclosed a mottled, purple-colored material on the gold wire and at the interface between the wire and the aluminum metal film below. The disease, accordingly, was called "purple plague." It was widespread in the industry and of major concern. Metallurgical studies showed the purple material to be a gold-aluminum intermetallic compound, $AuAl_2$, containing as much as 6-percent dissolved silicon. The studies also showed the presence of a new family of intermetallic compound phases ($AuAl_2$, $AuAl$, $Au_2Al$, $Au_5Al_2$, and $Au_4Al$), some or all of which might be formed at the interface. Thus, generalized purple plague was a natural product of the metal system and hardly a random defect. Further experimentation showed that the intermetallic compounds, and $AuAl_2$ in particular, had high yield strengths and good electrical conductivity, properties conducive to good leads. The broken leads suggested that another effect must be occurring. This later was diagnosed to be the Kirkendall effect, the diffusion of vacancies through the metallurgical system sometimes aggregating into physical voids. High-temperature aging as well as temperature cycling, both recognized testing procedures, tended to generate the voids and their subsequent motion, which in time caused high resistance and open bonds to appear.

The "solution" for Bell system designs was, first, to increase selectively the amount of gold in the thermocompression bond, e.g., by using a bonding tool developed by Bell Laboratories that did not flatten the entire width of the wire into a pancake or create a narrow neck in the wire, and by subjecting all devices to a 300-degree C, 16-hour bake followed by a centrifuge stress of 20,000 times gravity to eliminate weak bonds. Elsewhere in the industry, aluminum wire was being used to replace gold for bonding to the aluminum metalization, but this shifted the problem from the chip bond to the bond at the gold-plated terminal posts.

The Bell Laboratories-Western Electric double-diffused silicon n-p-n mesa transistor was a triumph of knowledgeable engineering control over contamination as well as chemical and metallurgical complexity. Factors in the control included ultracleanliness in processing, most often requiring specially designed facilities, and in-process testing for quality control. The final assembly operations consisted of resistance welding a top hat can to the header, on which the transistor was mounted, in a nitrogen dry box followed by a vacuum (with a pressure of $3 \times 10^{-5}$ millimeters [mm] of mercury) bake at 325 degrees C, just 31 degrees below the silicon-gold eutectic temperature at which the device would be destoyed metallurgically.

Not all the finished product was good, even though it met the electrical characteristics established by the specification. A requirement of very high reliability—low failure rate in service—had to be met, and in general the product, lot by lot, as delivered from the manufacturing line was not up to this standard. For delivery of a satisfactory product, several things were required. Qualification tests had to be established on a firm engineering basis, i.e., on the existence of a direct correlation between the test result and the performance (failure rate) of the devices in their circuit environment. Also, procedures needed to be found that effectively weeded out from the main population a subset of "weak" devices, those prone to early failure. Realistically, this could be accomplished only by sorting out the principal failure mechanisms affecting the product and attacking each individually.

Some of these were straightforward, easily identified, and dealt with. Leaky seals resulting from a defective resistance weld of the housing allowed moisture to affect the junctions adversely. Although this defect occurred only very infrequently, leak detection testing of all products was instituted as a standard procedure. Particle contamination inside the encapsulation could cause intermittent short circuits, particularly when used in an environment simulated by a shake table. For these applications, typically found in military and space systems, an insulating deposit of $SiO_2$ was laid over the surface after wire bonding, separating any metallic particles arriving at the surface from the underlying junction.

A most disconcerting failure mechanism of both planar and mesa devices was an instability in the electrical characteristics diagnosed as being associated with a minute, uncontrolled surface contamination. The contaminating ions pentrated an $SiO_2$ layer on the surface and, in sufficient numbers, caused a conductivity inversion layer in the bulk silicon beneath. In spite of the best known cleanliness process, this weakness might be characteristic of all the product, albeit in different degrees, and a means had to be found to deal with it in the finished product. High-temperature aging experiments, including step-stress techniques, revealed that this failure mechanism was "well behaved" in that temperature accelerated the failure distribution. (In chemical terms, the process had an activation energy of 1.02 eV.) This became the basis for establishing lot acceptance test criteria and also screening procedures designed to weed out devices prone to early failure (a so-called "freak" population).

Screening consisted of the prebake and centrifuging mentioned before, followed by a high-power burn-in that elevated the device junction temperature to 300 degrees C for two hours. After passing screening, a sample lot was given a high-temperature life test (300 degrees C for 48 hours) and tested for indications of surface-inversion instability. (A failure rate of 15 percent was allowed in the lot acceptance criteria.) If tests indicated an unacceptable proportion of freaks remaining, burn-in was repeated. The activation energy of the process made it possible to relate the high-stress temperature aging to operation in practice at a normal junction temperature: a 20-hour burn-in at 300 degrees C has the effect of nearly 10 million hours at 80 degrees C.

Since bond weakening associated with aluminum-gold intermetallic compounds is also a temperature-accelerated mechanism, the strenuous aging procedures instituted to identify and remove early surface instability failures in turn accelerated the weakening of low-strength bonds. The centrifuging operation, however, screened out weak bonds.

By these means, a frail manufactured product was strengthened to pass stringent requirements. The product delivered to customers was very reliable, characterized by a failure rate of less than 0.001 percent per 1000 hours (10 FITs), but at the same time relatively costly. The high cost was attributable primarily to the fact that some essential manufacturing process understanding was lacking (the source of the damaging ionic contamination had in fact not yet been discovered), with the inevitable consequence of some of the processes being inadequately controlled.

### 7.5 Gettering of Impurities

Preserving minority carrier lifetime in silicon heat treatments was a major challenge in the realization of silicon devices. The initial discovery of G. Bemski and J. D. Struthers[102] on the role of gold as a lifetime killer and their observation of nickel as a getter led to active studies of the recombination properties of other elements and on the methodology of gettering. It also led to an understanding of why certain manufacturing practices were more successsful than others.

For example, starting in the 1950s, silicon diodes were manufactured very successfully with sintered nickel contacts applied prior to the diffusion step. These diodes also involved phosphorous diffusion on the opposite surface from a glassy phosphorous layer as a source.

Later, it was found empirically that the incidence of emitter-to-collector shorts in transistors could be significantly reduced by processing the silicon wafers with a phosphorous glass layer on the back surface. This process was introduced into the manufacture of silicon transistors at the Western Electric Laureldale facility as early as 1958.

The next significant step in gaining understanding of gettering came in 1960 from A. Goetzberger and Shockley[124] of the Shockley Laboratory, who showed that "soft" junctions, (i.e. diodes having poorly defined breakdown characteristics) became "hard" junctions (diodes with sharp breakdown characteristics) after heat treating the wafers in contact with layers of zinc, nickel, or glassy oxides. They showed that prior to the heat treatment, the junction regions contained small metal precipitates that were removed by the gettering process. These metal precipitates may have been gold or other heavy metals, all of which have very high diffusion coefficients in silicon.

Further work by A. R. Bray and R. Lindner, documented in an internal memorandum, confirmed the effectiveness of liquid metal eutectics and glassy oxides in gettering impurities in silicon. Their 1961 memorandum recognized that gettering not only removes metal precipitates, but also dispersed impurities that act as recombination-generation centers, and that the diffusion of emitters from a phosphorous glass source is an example of a natural "hardening" process. Interestingly, they found that the best gold-doped diodes were obtained by heating the silicon in contact with a liquid gold-silicon eutectic. It removes uncontrolled impurities and establishes a controlled concentration of gold.

Later, in an effort to understand surface instabilities, M. Yamin and F. L. Worthing[125] performed electrical conductivity measurements on $SiO_2$ layers that showed that ionic conductivity can occur in $SiO_2$ at temperatures as low as 150 degrees C. This, at the time, was a very surprising observation, since much higher temperatures are normally required before ionic mobility becomes significant. It showed that some ionized impurity was moving in $SiO_2$.

In September of 1964, D. R. Kerr, S. S. Logan, P. J. Burkhardt, and W. A. Pliskin of IBM found that phosphorous glass gettering removed this mysterious impurity.[126]

Later, Fairchild workers, pursuing the same problem, identified the alkali ions, principally sodium,[127] as the source of the $Si-SiO_2$ surface instability. In trace amounts, these mobile charges drift through $SiO_2$ films under the influence of the local electric fields associated with biasing conditions, thus giving rise to unstable device characteristics. The phosphorous glass gettering action is attributable to the fact that sodium is much more soluble in phosphorous glass than it is in the dioxide layer itself.

Thus, phosphorous glass gettering became *the* process that assured stability of MOS transistors in ICs. Soon, some form of gettering was also being incorporated into most bipolar transistor processing. In many cases, no separate gettering step was required such as in n-p-n transistors, which involves a heavy emitter diffusion using phosphorous glass as a source. Furthermore, following the approach that evolved in MOS ICs, phosphorous glass also became an accepted surface protective film improving the stability of the transistors.

## VIII. DISCRETE SEMICONDUCTOR BUSINESS: THE EARLY 1960'S

By 1960, the year the silicon transistor emerged as the product of the future and the year the planar transistor was announced, the semiconductor industry was vigorous and growing. The *Business Week* issue of March 26, 1960 featured a comprehensive report on semiconductors, calling it "the world's fastest growing big business." *Business Week* found 35 "significant producers" manufacturing semiconductors with annual sales of one-half billion dollars, split about equally between transistors and rectifiers/diodes. It estimated that 80 percent of the industry's 40,000 employees were women. It judged that "manufactured goods using semiconductor components—from midget portable radios to giant computers—are the biggest, fastest growing segment" of the $10 billion electronics industry. The article affirmed, "The growth gained terrific impetus from Bell's policy of putting these [design and process patents] virtually in the public domain. Almost 90 percent of the semiconductor items in commercial production came right out of 'Mother Bell's Cookbook.' "[128] In 1960, Bell had some 80 domestic licensees.

The article recognized a favorable, downward trend in price with entertainment-quality transistors at 50 cents, an ordinary computer-grade transistor at $1.70, and a special very high-speed switching transistor for perhaps three times as much.[129] *Business Week* viewed the industry as "an almost classic example of the particular kind of growth that technological innovation can fire up throughout the U.S. economy."[130]

Under the heading "The Peril Overseas," the article called attention to the entrance of Japanese transistors, mostly entertainment quality, into the U.S. markets, rocketing from 11,000 units worth $7000 in 1958 to 1.8 million units valued at $1.1 million in the first nine months of 1959. Other data showed U.S. transistor production passing 100 million units in 1960, having passed in dollar volume that of electron tubes in 1959.[131] Military sales comprised about one half of the total. Looking ahead, *Business Week* anticipated that "the advent of solid circuits will upset some of the familiar marketing patterns in electronics. A computer maker . . . will be able to order . . . . a 'flip-flop circuit' ready-made in a package the size of one of today's transistors. Just this week," the article stated, "the first commercial versions of such devices were offered by Texas Instruments."[132]

## IX. EXAMPLES OF SYSTEM APPLICATIONS

### 9.1 The Telstar I Experiment

Contemporary in time with the *Business Week* cover story, with the concurrence of the National Aeronautics and Space Administration (NASA), AT&T management decided to explore the feasibility of microwave radio-relay communications using a small earth satellite designed as a radio repeater station. Exploratory systems studies in 1955, by J. R. Pierce and colleagues of the radio research group, had suggested a system of artificial satellites in low-earth orbit, linked to ground stations, to provide useful long-range broadband communications. (See a companion volume in this series subtitled *Communications Sciences (1925-1980)*, Chapter 5, section V.) These studies, plus the progress demonstrated by the U.S. Air Force and NASA toward developing reliable rocket-launching systems having respectable payload-to-orbit capacity, led to the conclusion that an experimental test of the ideas was timely. Clearly it was pertinent to the future course of very long-distance telephony.

The system design philosophy was dictated by the launch-limited maximum permissible weight of the orbiting station—Telstar, as it was officially named, or "the bird," as it was called by the developers. This weight was held to something under about 200 pounds by the payload capability of NASA's Thor-Delta launch configuration. The resulting power-transmission capability of the bird led to requirements for ground stations with powerful transmitters, large antennas, and sensitive, very low-noise receivers. System components that met these and other requirements by and large did not exist; they were to be developed by Bell Laboratories for the experiment with a timetable that initially targeted launch before the end of 1961. Guidance into orbit for the Telstar I launch, and for other Delta payloads, was provided by the Bell Laboratories-Western Electric Radio-Inertial Command Guidance System developed to target Titan I, an early U.S. ballistic missile. (For more detailed information, see *National Service in War and Peace (1925-1975)*, Chapter 10.)

A low-altitude orbit of about 1000 miles (or an elliptical one ranging from 500 to 3500 miles) was decided on—for one reason, to keep the signal quality from being degraded by the echo delay associated with the finite transit time of a radio signal from earth to satellite and back. At 1000 miles, this delay is a manageable 0.01 sec.

### 9.1.1 Radiation-Resistant Solar Cells

One consequence of a low-level orbit was that it placed the bird in the adverse space environment of the Van Allen radiation belt. The belt is an indefinite, doughnut-shaped region surrounding the earth's equator and containing energetic charged particles, protons and electrons for the most part, trapped in the magnetic field of the earth. It was discovered in 1958 by Geiger counter sensors placed by J. A. Van Allen aboard Explorer I, the first U.S. satellite. The true severity and extent of the belt, or belts as then envisaged, were not well known. It was clear, however, that the silicon solar cells to be mounted on the outside surface of the bird, as a part of the power supply, needed radiation protection as well as a radiation damage-resistant design. A maximum two-year life for the experiment was visualized, after which the satellite would be disabled.

The requirement for a solar cell design of superior radiation tolerance suggested, first of all, that the heretofore standard p-on-n cell should be replaced by a specially tailored thin-n-on-p structure.[133] With the n-p junction near the front surface, the strongly absorbed short-wavelength (blue) sunlight is utilized most effectively. The more penetrating, longer-wavelength light is absorbed in the p region, creating electron-hole pairs there. The fraction of the minority carriers (electrons) that diffuses to the junction, and thereby contributes to the cell's power generation, is determined by a parameter called the diffusion length, $L_D = \sqrt{D\tau}$, the characteristic distance a minority carrier diffuses with the diffusion coefficient D in a time $\tau$ equal to the minority carrier lifetime, i.e., before it disappears by recombination. With the electron as the important minority carrier, as in a thin-n-on-p cell, the diffusion coefficient is three times that of a hole in the conventional p-on-n cell, the diffusion length is $\sqrt{3}$ longer, and the new cell is more efficient, given the same lifetime. Since radiation damage reduces the minority carrier lifetime in either type, p or n, at about the same rate, the n-on-p cell is more resistant to radiation.

The very superior damage-resistant property of the blue-sensitive thin-n-on-p cell was demonstrated by experiments using a 1 million-eV electron beam.[134] The individual cells, 1 by 2 cm, had an array of thin metal fingers on the surface to collect the current and reduce the internal cell resistance. The surface was given a quarter-wavelength antireflection coating in the blue spectrum to improve the optical match. Each cell also had its own protective cover plate consisting of 0.03 inch of clear sapphire, a material that does not color (darken) when exposed to radiation. To allow for a thermal expansion mismatch, the cover was hinged by a thin platinum ribbon to the ceramic substrate on which the cells were mounted, 12 cells per module. [Fig. 1-33] The entire array, consisting of 3600 cells, was organized into a nominal 14-W, 28-V converter supply to charge the satellite's nickel-cadmium batteries when exposed to sunlight. The Allentown Works of Western Electric took the new cell design in late 1960, and under forced draft began delivering finished modules at the end of January 1961—a typically sound, "unusual" performance by Western.

### 9.1.2 Ionizing Radiation-Tolerant Transistors

In late summer 1961, word reached device development people of circuit performance changes that had occurred in equipment exposed to radiation in an experiment conducted by Bell Laboratories research personnel at the Brookhaven Laboratory of the Atomic Energy Commission. One possible explanation of the events was that the radiation environment had altered

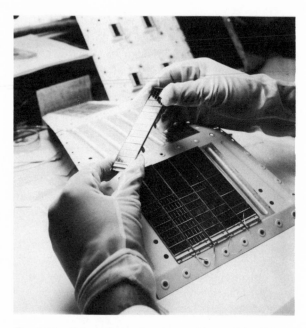

Fig. 1-33.  Solar cell panel and assembly used for Telstar.
The cell used a specially tailored thin-n-on-p structure for
superior radiation damage resistance.

characteristics of some transistors. (See *Physical Sciences (1925-1980)*, Chapter 8, section 1.1.1.) Further experiments were planned to diagnose the source of trouble. By mid-October, additional data from multiple sources left no doubt. High-performance Western Electric silicon mesa transistors in operating circuits (i.e., electrically biased) degraded gradually, by varying amounts, under exposure to a high-energy radiation flux. The results were completely unexpected, since only a negligible change in characteristics was predicted from previous study of radiation damage effects in silicon devices. And these same transistors, by the scores, had been designed into Telstar's circuitry. What to do?

To redesign the circuits around less radiation-sensitive devices, provided such could be found, would seriously delay the project at a further cost of considerable public and private embarrassment; the privately financed Telstar project was virtually competing with publicly (NASA) financed projects to be first to deliver a working experimental communications satellite for launch by NASA. The hope was that an expedient could be devised to salvage the situation from impending disaster.

Extensive autopsies performed on diseased transistors failed to clarify definitively the origin of the damage. Before seal-off, the transistor cans were back-filled with inert, dry nitrogen. One hypothesis was that radiation ionized this gas and that some of the gaseous ions, in variable amounts, sat on the silicon surface on or very near an exposed p-n junction. There were other hypotheses about the causes of the radiation sensitivity. Not all radiation-exposed transistors behaved alike; some were much more seriously affected than others.

After tests on many exposed transistors had been analyzed, the decision was made to set up an extensive screening program designed to select from a large population of devices those least sensitive to radiation exposure. The screening program consisted of exposing transistors in batches to a limited dose of radiation from a cobalt-60 source (1.25 megaelectron-volt gamma rays) and testing individually certain transistor parameters before and after exposure. Empirical criteria were established for selection and rejection.

Because of time limitations, individually screened transistors could be incorporated only in critical parts of the command module of Telstar I. For other circuitry, the transistor codes were characterized as to their radiation sensitivity. The engineering decision was made to use, and be satisfied with, whatever satellite life for Telstar I accompanied the use of the weak components. This turned out to be seven months, under conservative assumptions for the not very well-known radiation level in the Van Allen belt. This period was deemed adequate to establish the practicability of broadband satellite communications by hundreds of tests and demonstrations that were to be performed.

In point of fact, Telstar I was subjected to a more severe radiation environment than that anticipated from the Van Allen belt. Just one day before its July 10, 1962 launch by NASA, the radiation level in Telstar's orbit was enhanced about 100 fold by a giant high-altitude nuclear explosion conducted by the Atomic Energy Commission. Many satellites ceased to function, some after only one orbit through the enhanced radiation field.

And indeed, within about one month of launch, there was some indication that one of the redundant command decoders may have been operating intermittently.[135] Three months after launch, it failed to operate. However, through modified command pulses, certain commands could later be executed. By exercising the command module in a sophisticated manner, use of the decoder was regained and the normal commands could be used again so that the communications experiment could be continued.

A second satellite, Telstar II, was launched on May 7, 1963. In this case, enough time was available to prescreen every radiation-sensitive transistor. The satellite operated flawlessly for two years, at which time a required internal timer deactivated major elements of the satellite circuitry.

Telstar was a pure example of discrete element solid-state circuitry, the milieu of the time. [Fig. 1-34] Its complete component count (without the

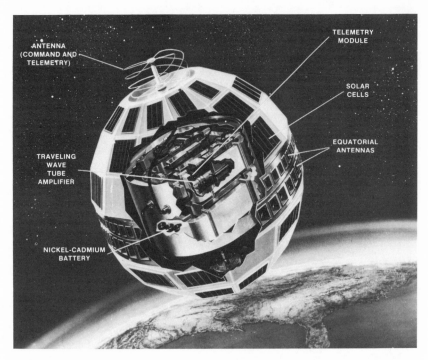

ANTENNA
(COMMAND AND
TELEMETRY)

TELEMETRY
MODULE

SOLAR
CELLS

TRAVELING
WAVE
TUBE
AMPLIFIER

EQUATORIAL
ANTENNAS

NICKEL-CADMIUM
BATTERY

Fig. 1-34.  Telstar I, launched June 10, 1962. New components were developed by Bell
Laboratories to keep the launch weight under 200 lbs.

nickel-cadmium battery) totaled 3600 solar cells, 372 inductors, 1119 transistors, 64 transformers, 1521 diodes, 5 quartz crystals, 1343 capacitors,
1 traveling wave tube, 2949 resistors, and 1 yttrium-iron-garnet limiter.

Telstar could not have met the weight limit if reliance had had to be
placed on electron tubes with their greater bulk and lesser power efficiency;
further, without a diffused silicon technology, the remote power-supply
problem of the satellite would have been, as a practical matter, insur-
mountable. Telstar thus was an example of the new sophisticated products
made possible by discrete transistor technology that can be thought of
as distinguishing a second era of electronics from the earlier vacuum
tube era.

In terms of effort, the contribution to Telstar by the semiconductor
device development group was only a small portion of the Bell Laboratories
total; in other terms, it was an integral part of the heavy project commitment
assigned to the device development area. Under the direction of E. F.
O'Neill, the Telstar system project manager, the additional responsibilities
included: design of the bird structure, its mechanical, thermal and radiation
(absorption and emission) properties, including construction, assembly,

and test of prototype and flyable models; the 2-W, long-life traveling wave output tube (4000 MHz) on the bird, fortified to endure a launch environment 100 times gravity; the 2000-W traveling wave output tube of the transmitter (6000 MHz) of the ground stations; and the very low-noise (4 kelvins) traveling wave ruby maser preamplifier in the receiver of the ground stations. The device development team comprised an aggregation of versatile design competence that was, in its way, unique. It consisted of experienced device/process chemists, innovative mechanical and electrical engineers and physicists, and models laboratories with a reservoir of fabrication capabilities and engineering know-how—all united by a 40-year history of successful technical accomplishments, many of which had delineated the state of the art.

### 9.2 Silicon Device Applications in Digital Systems

Perhaps the most important natural application of transistors has been to digital systems, where the number of devices required per system was large and therefore unsuitable for the employment of vacuum tubes. Two Bell System developments of this nature in the early 1960s were the T-1 carrier system, which employed digital transmission in the form of pulse code modulation (PCM) for short-haul multiplex voice transmission, and the electronic central office (later electronic switching system) based on stored program control, incorporating digital electronic common control for central office switching. The time when these and other new solid-state systems could be introduced into the Bell System depended both on the time required for system development and on the competitive costs of the new systems relative to the costs of older ways of providing the service. For new systems to be economical, it was preferred that no more than 40 percent of the shakedown equipment cost be due to semiconductor components. Low transistor costs could only be achieved in large-scale manufacture as an economy of scale. Thus there was, inevitably, a start-up period in the introduction of new systems during which component costs, viewed simplistically, were uneconomically high. In the case of the digital systems, the start-up costs were spread over several years of production, and diffused silicon transistor manufacture in Western Electric got off the ground. In time, mesa designs were replaced, first by planar versions still using a metal-to-glass hermetic encapsulation. In the 1960s, many millions of the discrete devices were employed in the T-1 carrier system, several switching systems, carrier channel banks, inband signaling equipment, the N carrier system, electronic translators, key telephones, PBXs, and various data sets. Small applications using standard codes in production accounted for greater volume. By 1968, the internal records of Western Electric showed several hundred equipment applications.

### 9.3 Active Devices for TD-3 Radio Relay

Reflecting rapid technological progress, a newly designed microwave radio relay system, designated TD-3, updating the highly successful TD-2 system, was initially installed in 1966. The 20 new solid-state devices developed for TD-3 depict the state of the art of semiconductor solid-state devices at Bell Laboratories-Western Electric around 1965.[136] The devices included a new family of high-frequency, planar, epitaxial n-p-n silicon transistors and assorted diodes. Prominent among the latter was the first application of a III-V compound semiconductor device, an epitaxial GaAs device with a metal-GaAs Schottky barrier junction. This diode had a very low noise figure. It provided the mixer function (actually a single-diode downconverter) of the receiver modulator. Together with a special low-noise transistor, this diode permitted the removal from an earlier receiver design of an RF parametric preamplifier. The transmitter and receiver units used semiconductor devices, except for the traveling wave tube in the transmitter, and operated from a single 24-V battery source. The group of TD-3 devices illustrates the development concept at Bell Laboratories of device design tailored to optimize system performance and reliability.

#### 9.3.1 Transistors

The silicon transistor family developed for TD-3 was characterized by a 1-GHz gain-bandwidth product and an n-p-n planar epitaxial structure. The basic transistor of the family, the 45A shown in Fig. 1-35, was designed for operation at collector currents of 5 to 10 mA. With an increased number of emitter stripes, three other members of the family had, respectively, two, four, and eight times the collector current rating and one-half, one-fourth, and one-eighth the base resistance. These transistors were packaged in a new metal-to-ceramic package (TO-112) designed to minimize parasitic capacitance and lead inductance, and provide low thermal impedance. For use in the FM deviator cavity, the basic transistor was also encapsulated in a conventional TO-18 package designated the 44A, wherein the collector is electrically joined to the metal package, minimizing collector inductance and so maximizing the deviation range. For effective use of the low-noise performance of the GaAs Schottky barrier diode, the following 70-MHz IF preamplifier required a noise figure (2 dB at 12 mA) lower than the best attainable (3.5 dB) within the immediate 45A family. A completely new transistor, designated the 45J, shown in Fig. 1-36, was designed for this application. In it the base resistance was reduced by a factor of four by changing the emitter geometry and some details of the metalization. In particular, the new design used low-resistivity, diffused p-type conductors in place of the usual metal base stripes. This arrangement permitted covering the entire base-emitter interdigitated structure with emitter me-

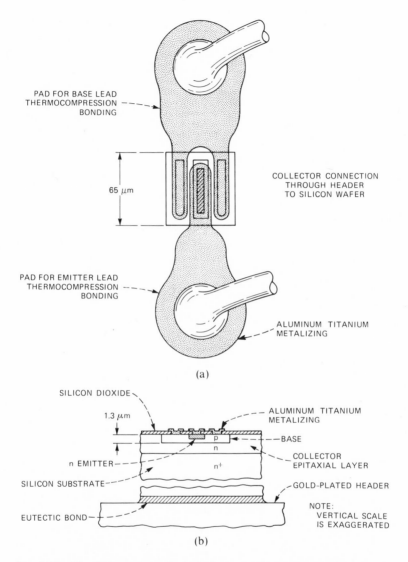

PAD FOR BASE LEAD
THERMOCOMPRESSION
BONDING

65 μm

COLLECTOR CONNECTION
THROUGH HEADER
TO SILICON WAFER

PAD FOR EMITTER LEAD
THERMOCOMPRESSION
BONDING

ALUMINUM TITANIUM
METALIZING

(a)

SILICON DIOXIDE

1.3 μm

ALUMINUM TITANIUM
METALIZING

BASE

n EMITTER

p

n

n⁺

COLLECTOR
EPITAXIAL LAYER

SILICON SUBSTRATE

GOLD-PLATED HEADER

NOTE:
VERTICAL SCALE
IS EXAGGERATED

EUTECTIC BOND

(b)

Fig. 1-35. The 45A transistor, the basic device for the TD-3 radio relay system, an n-p-n planar epitaxial device characterized by a 1-GHz gain-bandwidth product. (a) Top view. (b) Cross section through the center of the device.

talization. Accelerated aging data for the 45A types predicted a transistor failure rate of less than 10 FITs over a 20-year period at junction temperatures of 125 degrees C. For the 45J, which was fabricated with platinum-titanium-gold metalizing, similar data predicted a failure rate of 10 FITs at 250 degrees C junction temperature.

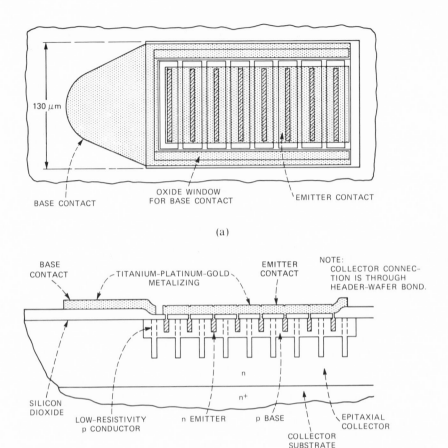

Fig. 1-36. The 45J transistor family, which replaced the 45A, offering a lower noise figure and reduced base resistance by using low-resistivity diffused p-type conductors instead of the usual metal base stripes. (a) Top view. (b) Cross section.

### 9.3.2 Harmonic Generator Varactor Diodes

Beat oscillator power for the microwave transmitter and receiver was supplied by a varactor diode frequency multiplier scheme. The output of a crystal-controlled oscillator at 125 MHz was amplified by a transistor amplifier to a power level of 7 W. A three-stage varactor multiplier chain converted the amplifier output to about 0.5 W at 4 GHz. Three codes of varactor diodes (473A, B, C) were needed for this circuitry. The difference in operating frequency and power level in each multiplier stage called for different diode characteristics to achieve high conversion efficiency, frequency stability, and low noise. The codes were all diffused, planar, epitaxial

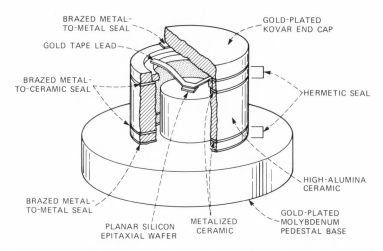

Fig. 1-37.   The V package microwave diode, a miniature ceramic pillbox type with brazed metal-to-ceramic hermetic seals.

silicon diodes in the V package. All were manufactured by the same process, the differences in final characteristics being caused by the choice of epitaxial starting material, the diffusion profile, and a difference in junction area. Epitaxial silicon was essential to these applications, because they required low series resistance, the only lossy element of the varactor, and high breakdown voltage. The diode capacitance was determined by the junction area. The diodes were fabricated by standard planar photolithographic techniques including oxide passivation to stabilize the wafer surface. The V package, shown in cutaway in Fig. 1-37, was a miniature ceramic pillbox type with brazed metal-to-ceramic hermetic seals. The primary characteristics of this package were low shunt capacitance, low series inductance, and good power dissipation. The varactor diodes were suited to harmonic generation at medium power in the frequency range 0.1 to 10 GHz. Reliability was assured by high-temperature screening, encapsulation leak testing, a centrifuge test at 20,000 times gravity, a shock test at 1500 times gravity, and 100-hour power aging of a sample lot.

### 9.3.3 U Package Microwave Diodes

The U package was designed to fill a need for a hermetic microwave package with low parasitic inductance and capacitance. Several codes of varactor diodes intended for different terminal functions were encapsulated in this package. Figure 1-38 shows a U package assembly and subassemblies for the 471A diode. This diode, a diffused, epitaxial silicon mesa structure, was developed to satisfy a requirement for an upconverter diode pair that would give about an 8.5-dB conversion gain in the TD-3 transmitter mod-

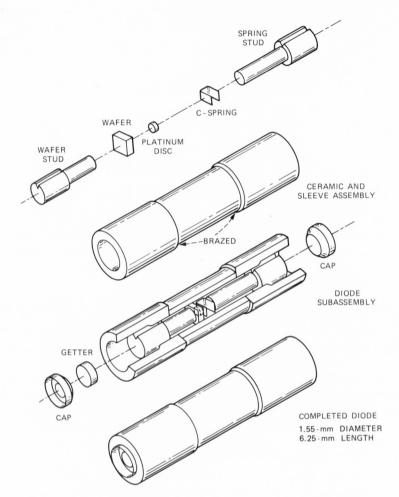

Fig. 1-38. The 471A diode in a U package. This diode, a diffused epitaxial silicon mesa structure, was made using a special chemical etching technique for precise control of the mesa diameter.

ulator. The severe pairing requirements, needed in the hybrid circuit to suppress the carrier, were that the two diodes could differ by no more than 0.1 pF in zero-bias capacitance and no more than 0.5 $\Omega$ in zero-bias series resistance. A special chemical etching technique was evolved that precisely controlled the mesa diameter (i.e., the junction area) and thus the zero-bias capacitance. Diode reliability was achieved by the usual careful attention to cleanliness in manufacture, by employment of a glass frit getter in each sealed-off device to absorb gasses evolved during the closure weld, and by screening tests on finished devices to eliminate weak units. Step-stress power aging data indicated the failure rate in the TD-3 transmitter modulator to be less than 100 FITs at 50 mW per diode with a 95-percent confidence limit.

A tungsten wire S-spring contact was employed in the point-contact version of the U package. In this configuration, two silicon diodes (488A and 493A) were developed for the IF and RF detector circuits, as no suitable hermetically sealed diodes were available. These diodes used the same basic aluminum-doped silicon material. Slightly different processing (one was oxidized at 975 degrees C, the other at 927 degrees C) gave one a higher reverse impedance and the other a lower forward resistance. Reliability tests indicated a negligible failure rate for the TD-3 applications, less than 1 FIT, which were at a significantly lower power level (one-fifth) than that of the 471A upconverter application.

The GaAs Schottky barrier diode (497A) represented the striking advances that had been made in the technology of GaAs. The device structure was thin-n on $n^+$. The 0.2-$\mu$m epitaxial n layer was doped to $5 \times 10^{16}$ net donor atoms per $cm^3$. The $n^+$ substrate wafer was doped to $3 \times 10^{18}$ donors per $cm^3$. These choices, together with a junction diameter near 25 $\mu$m, optimized the significant parameter for the down-converter application, i.e., the forward-biased cutoff frequency, yielding a value nearly 10 times higher than the silicon point-contact diode it replaced and, consequently, a substantially lower system noise figure. Photolithographic techniques were employed to produce the device structure, which consisted of the barrier metal silver evaporated onto the thin-n side of the semiconductor wafer and then overlaid with an electroplated gold contact. A gold-tin solder preform was the low-resistance contact to the back of the wafer. Sealing into the U package assembly configuration was accomplished by a single heating cycle accompanied by a small compressive force. Step-stress aging data on finished devices indicated a failure rate under 10 FITs.

### 9.3.4 Highly Stable Diodes for the IF Deviator

Two electrically very stable diffused silicon diodes provided a stable output from the FM deviator. The 446AC was an 8.2-V regulator diode, and the 457A was a variable capacitance diode. To meet the system requirements of a maximum drift instability of ±100 kilohertz (kHz)* over a three-month period in the 70-MHz IF deviator, the 446AC diode breakdown voltage had to be stable to within $7 \times 10^{-3}$ V (one part in 1000);

---

* In the deviator circuit, an ultrastable 5-V regulated supply using the 446AC regulator diode provides bias for the 457A variable capacitance diodes in the oscillator tank circuits. The baseband signal superimposed on this bias modulates the capacitance of each 457A diode, and consequently the output frequencies of the oscillators.

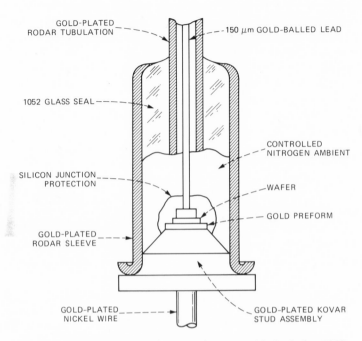

Fig. 1-39. Internal structure of the K package assembly. In the late 1960s, Western Electric was manufacturing tens of millions of them per year for some 700 applications.

also, for the 457A diode, the reverse leakage current had to be stable to within 2 nA and the diode capacitance to within 0.008 pF.

The 446AC was a member of the large K package family of hermetically sealed diodes. Devices of the 446AC code were selected as the most stable from a larger universe of preaged diodes. The preaging consisted of 250 hours of storage at 250 degrees C followed by 750 hours of reverse dc power aging at 25 times the rated condition for use (16 mW). The K package assembly is shown in Fig. 1-39. Basically, it is a glassed Kovar or Rodar* eyelet welded to a Kovar stud, with a tubulation to permit outgassing and dry nitrogen backfilling. The tubulation was pinched off to provide the final seal and also electrical connection to the lead wire inside the tubulation. K package diodes typically were rated at 0.6-W dissipation, with a normal failure rate of about 1 FIT at a junction temperature of 60 degrees C. By the late 1960s, the manufacturing level of K package devices, all at Western Electric in Reading, was tens of millions per year for some 700 applications.

---

* Trademark of Wilbur B. Driver Co.

Manufacture of the 457A semiconductor die was similar to that of the 446AC. An n-type slice was simultaneously diffused with both p and n layers on opposite sides of the slice. This left only a very thin region in the middle where the silicon resistivity was high, resulting in a low series resistance. For the 457A, the key to high stability lay in the final processing. After a final etch to remove any damaged material, an oxide was grown on the exposed silicon for passivation and protection. This made it suitable for encapsulation in the miniature S package, in which the final seal was made at the high temperature of 830 degrees C to reduce or eliminate trapped moisture. Again, an accelerated preaging procedure was used to select the most stable devices, which met the stability requirements.

The S package encapsulation for the 457A was selected because of its low inductance. As shown in Fig. 1-40, the S package assembly incorporated two stud leads that contact the semiconductor die and that are hermetically sealed to a coaxial glass sleeve. The layered metalization of the diode chip,

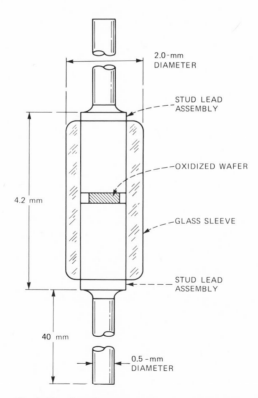

Fig. 1-40.  S package assembly. Close cooperation between Bell Laboratories and Western Electric helped create this low-cost, small, highly reliable product.

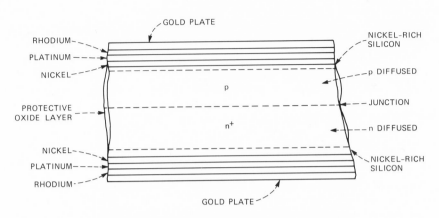

Fig. 1-41.  The high-stability 457A semiconductor die.

illustrated in Fig. 1-41, bonded the chip to the metal studs. The metal of the studs had not only to be a good thermal expansion match to the glass comprising the sleeve, but also, importantly, mismatched slightly to maintain stud contact pressure on the silicon chip.

Development of the S pack had begun in 1961 and actually took place over several years because of many difficulties encountered. The design objective was a low-cost, small, rugged, highly reliable product for use in low-power, high-performance applications. This implied finding solutions to technical problems affecting surface stability, mechanical and thermal compatibility, and device performance. It also implied finding a solution to the problem of manufacturing cost.

Manufacturing processes involving one-at-a-time handling, as in device assembly or testing, ballooned cost beyond acceptable limits. Over time, this obstacle was removed by developing new manufacturing processes employing innovative batch assembly and fabrication techniques augmented by highly mechanized testing and code marking.[137,138] Batch assembly made use of shaker-loaded jigs. In this concept, piece parts, together with fixtures for carrying the assembled parts, are designed compatibly such that vibrating a fixture flooded with piece parts causes the piece parts to fall into separate jig positions in the fixture in the correct order. For the S pack, this meant loading a bottom stud, glass, chip, and top stud, together with glassing cavities and weights, into fixtures holding several hundred assemblies. Several of these glassing fixtures were then placed in the controlled atmosphere of an autoclave furnace where the diodes were sealed, necessarily by the hundreds, in one oven cycle.

The S pack development was particularly demanding of a close working relationship between engineers exercising Bell Laboratories design functions and engineers exercising Western Electric manufacturing responsibilities.

In practice, a nearly seamless interface between the two organizations was operationally necessary to accomplish the group objectives.

### 9.3.5 Silicon Schottky Barrier Diodes

Two new planar, epitaxial silicon Schottky barrier diodes, coded 479A and 479B, were developed for use in the IF amplifier-limiter and discriminator circuits. These applications required a nonlinear diode resistance and low capacitance, low recovery time, and moderate breakdown voltage. Point-contact silicon diodes could meet the first two requirements, but not the third—10- to 20-V reverse breakdown. The point contact was replaced by a 0.001-inch diameter junction built up of a complex metal layer of chromium, palladium, and platinum on epitaxial silicon, a design of Kahng and M. P. Lepselter.[139] Doping level of the epitaxial layer was determined by breakdown voltage, and layer thickness was consistent with a low forward impedance. The metal-silicon contact was formed by a tiny button of palladium on silicon heated to 475 degrees C, forming a kind of alloy. This was then protected by a steam-grown $SiO_2$ layer. Next an overlay of platinum and palladium 0.002 inch in diameter over the button and $SiO_2$ was sealed to the oxide by a thin 300-Å chromium layer. The finished die was mounted to a nail-head lead, isolated electrically from the metal header platform of the TO-18 package. Temperature aging data predicted a failure rate of the diodes of less than 10 FITs for the TD-3 operating conditions.

### 9.3.6 High-Voltage Rectifier Diodes

Stringent voltage breakdown, corona, and mounting requirements were satisfied by molding diode assemblies, four diodes per assembly, in a high-dielectric plastic. Multiple diodes were necessary to attain high-voltage requirements. Localized silicone resin served as an insulator across the glass seal of each diode. The diodes and the internal wiring structure were immersed in alumina-filled epoxy, which also sealed the diodes from the external environment. Alumina filling agent in the epoxy improved the heat dissipation property of the package and thus reduced diode junction operating temperatures. One assembly, coded 463A, was made up of four type 426J, 1-W diodes in series, which had a breakdown voltage of 4800 V and a power rating of 5 W. The other assembly, coded 464A, consisted of four type 426G diodes in a full-wave bridge configuration with a minimum breakdown voltage of 1200 V for each leg. The assembly also had a power rating of 5 W and a maximum forward drop of 2.1 V at 600 mA. Based on aging data, the failure rate assigned to the two codes was less than 100 FITs.

### 9.3.7 P-I-N Diode

For automatic gain control in the IF amplifier, variable loss stages were needed that behaved like pure variable resistances independent of frequency and signal amplitude. The 474A p-i-n silicon diode was designed for this application. The resistance of a p-i-n diode is normally high because of the high-resistivity i layer. Under forward-biased current, however, the resistance of the i layer can be greatly diminished by the presence of excess injected carriers, holes from the p region and electrons from the n region. This is called conductivity modulation,[107] and in this state, a p-i-n diode exhibits a resistance that varies roughly inversely with the forward-biased current. Now suppose the layer width of the i region is made comparable to the length diffusing carriers move in a lifetime (i.e., before recombination). Under this condition the excess mobile carrier density in the i region cannot follow a frequency that is high compared to the inverse lifetime, which in the 474A was about 1 $\mu$s, and the i layer resistance is frequency insensitive at these high frequencies, as in the device realized by Uhlir.[85] Within these restrictions, the design objective was to make the remaining frequency-dependent impedance of the two diffused junctions (p-i and n-i) small compared to the impedance of the i layer. This was accomplished by adjusting the junction doping gradients, the i layer width, and the dc bias current, and by maximizing the lifetime of the i layer. The result achieved was a small frequency dependence of the total diode impedance above 5 MHz and no dependence above 50 MHz. The diode was packaged in a conventional, gettered TO-18 can.

## X. TRANSISTORS IN THE ERA OF INTEGRATED CIRCUITS

By the middle of the 1960s, ICs were already available commercially. The fact that several silicon devices interconnected on a single chip of silicon could be made with acceptable yield is a reflection of the degree of maturity silicon device technology had reached at that time.

ICs started to place ever more stringent requirements on silicon technology and soon the evolution of that technology was driven by these needs, as described in Chapter 2 of this book. Of course, transistors benefited from further technology developments leading to improvements of performance and cost.

The first of these developments—the identification of alkali ions as being responsible for surface instabilities—has already been mentioned. It contributed greatly to device stability.

Motivated by the problem of purple plague and by the high cost of attaching leads to ICs one at a time by thermocompression bonding, Lepselter[140] invented the gold beam lead technology, based on a new contact metallurgy. The final gold layer of the multilayer contacts was formed by a masking operation into gold beams projecting beyond the device borders, as discussed in detail in Chapter 2 of this book. All beam leads of an IC could be thermocompression bonded to gold wiring deposited on ceramic substrates or other suitable circuit boards in a single operation.

Even though beam leads themselves offer little or no advantage for transistors, since only a few contacts are to be made, some transistors were designed towards the end of the 1960s using beam leads so that they could be used in conjunction with beam-leaded ICs. The metalization system itself, however, was introduced for all transistors in place of aluminum to avoid the problem of intermetallic compounds altogether. It permitted gold wires to be bonded to a gold surface both on the device and on the external terminals.

The third IC development that affected discrete devices was the introduction of the sealed-junction concept. It is based on the work of J. V. Dalton,[141] who in 1966 reported that silicon nitride is a formidable barrier to the migration of sodium ions, water vapor, and other impurities, even at high temperatures. After the effectiveness of this material as a junction seal was demonstrated by G. H. Schneer, W. van Gelder, V. E. Hauser, and P. F. Schmidt,[142] the conclusion seemed inescapable that a silicon nitride overcoating could provide the junction seal necessary for high-reliability planar structures. Costly hermetic sealing of transistors in metal-glass packages was not necessary, and the way was paved for the introduction of far less costly nonhermetic packages, specifically the post-molded plastic package, in which silicon chips are first bonded to a lead frame having a large number of device sites. Plastic is then molded around the chip, and in a final "trim-and-form" operation, the device is completed.

The reduced packaging cost, in combination with dramatically increased device yield due to improvements in processing quality, brought the cost of transistors to well below one dollar. Furthermore, large and complicated transistors became possible, permitting the design and economic manufacture of high-power devices.

Even though the development of ICs quickened in the second half of the 1960s, new transistor designs continued to be needed for such functions as interfacing ICs with the external world, as "glue" between ICs or in power supplies, where large power-handling capability is important. The need for transistors in system applications is reflected in the number of designs transmitted by Bell Laboratories to Western Electric; Fig. 1-42 shows the number of designs transmitted per year. It breaks down the devices by hermetically sealed transistors, nonhermetic transistors, and bare transistor chips (either beam-leaded transistors or transistor chips that are wire bonded into other assemblies). While the peak of the design activity is in the past, there is a continuing need for special transistor designs. Even at the time of the publication of this book, the transistor business continues to be significant.

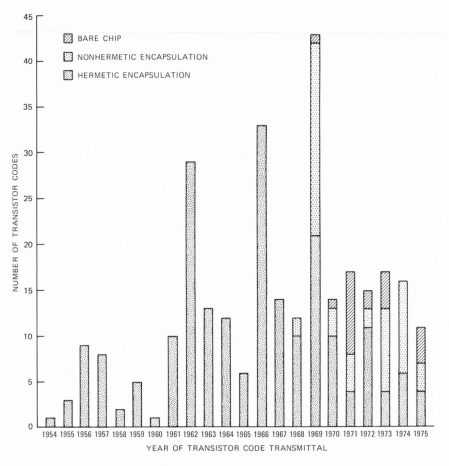

Fig. 1-42.   Number of transistor designs transmitted by Bell Laboratories to Western Electric, from 1954 to 1975.

## XI. MILLIMETER AND MICROWAVE AVALANCHE DIODES

### 11.1 Device Development

In 1964, an exploratory device development group led by B. C. De Loach, Jr. at the Murray Hill Bell Laboratories site was engaged in the early stages of a program aimed at developing a solid-state source of millimeter-wave power. At that time, the circular waveguide, millimeter-wave transmission system gave promise of practical, very wide-band, digital communication over long distances. (See *Communications Sciences (1925-1980)*, Chapter 6, section 4.2.) Electron tubes, both traveling wave tubes and klystrons, could provide millimeter-wave power for this application, but they were also encumbered by certain inherent cost and lifetime limitations. A reliable solid-state device alternative would be welcomed. After considering several possibilities, the De Loach group decided to concentrate its efforts on the Gunn diode as a promising approach. The Gunn effect, a negative resistance associated with the band structure of GaAs and other complex energy band semiconductors, was first reported by J. B. Gunn of IBM at the June 1963 Solid State Device Research Conference in Lansing, Michigan.[143] The Gunn diode has developed into a versatile source of high-frequency power, offering lower noise but less power than avalanche diodes.

While "wringing out" microwave circuitry assembled to measure properties of Gunn diodes, R. L. Johnston inserted a silicon diode, for test purposes, into the microwave cavity. When the diode was pulsed into reverse avalanche breakdown, microwave oscillations were observed.[144] Furthermore, almost any other silicon diode readily at hand produced similar oscillations. These first microwave oscillations were obtained with diodes having simple, abrupt p-n junctions and not the complex, tailored $p^+$-n-i-$n^+$ structure, treated in the Read theory.[84] T. Misawa showed that the Read mechanisms can be obtained from a junction diode (or metal-semiconductor Schottky-barrier contact) with almost any doping profile.[145] Ryder, De Loach's department head at the time, named the generic set of devices, in which avalanche impact ionization and carrier transit time delays produce negative resistance, IMPATT diodes. The acronym stands for IMPact Avalanche and Transit Time.

Further experimentation by De Loach and Johnston [Fig. 1-43] produced continuous wave (CW) operation in December 1964.[146,147] This was followed by the success of the research group in achieving CW oscillations with a Read diode, $n^+$-p-i-$p^+$ in structure, fabricated a year earlier. The research group had realized that the failure of its earlier experiments must have been due to circuit coupling problems rather than device design.

In the years immediately following the initial experimental breakthrough, IMPATT diodes were operated over the frequency spectrum from 300 MHz to 300 GHz, i.e., from 1 meter to 1 millimeter wavelength. Efficiency was characteristically less than 15 percent. The devices were constructed from GaAs and germanium, as well as silicon. By 1971, an improvement of two to three orders of magnitude had been achieved in power output, as well as somewhat more than one order of magnitude in upper frequency limit.[148] IMPATT diodes could generate higher CW power at millimeter-wave frequencies (greater than 1-W CW at 50 GHz) than any other solid-state device.

A limitation of IMPATT diodes is that they are very noisy, an inevitable consequence of the avalanche process. During each cycle, starting from one or two initial pair-production ionizing events, a large number of electron-hole pairs are built up by impact ionization in the high electric field of the breakdown region. Particularly at the onset, the random nature of

Fig. 1-43.  B. C. De Loach (left) and R. L. Johnston (right) monitoring an oscilloscope while tuning microwave circuitry by moving the pistons in two intersecting brass tubes to optimize power output from an IMPATT diode.

the ionization process causes fluctuations from cycle to cycle in the time to build up the large number of carriers in each avalanche. These time fluctuations show up as noise, bringing amplifier noise figures to 20 to 40 dB above thermal noise. GaAs turns out to be substantially lower noise (typically 10 dB) than silicon.[141]

### 11.2 Applications in the Bell System

Limited numbers of IMPATT diodes were employed in the repeaters of the field trial of the WT-4 millimeter-wave system, which was held in 1975. This cylindrical waveguide system operated over the frequency range 40 to 110 GHz. A 1-W silicon IMPATT amplifier at 6 GHz found use beginning in 1973 in the initial installations of the TM-2 system, a short-haul radio relay application.[149] The largest application of IMPATT diodes was in a 3-W amplifier at 11 GHz in the TN-1 radio relay system, where thousands were employed starting in 1975. An early diode burnout problem

was encountered in the TN-1 application. This turned out to be a mani-
festation of a general design problem with IMPATT amplifiers, an extreme
sensitivity of the operation to circuit design. Detuning and other unwanted
side effects must be eliminated by careful attention to the circuit parameters
and oscillation amplitude.[150] In a carefully designed, benign circuit (and
mechanical) environment, long diode life has been achieved (measured in
years), and predictions were made of $10^7$ hours mean time between failure
for some designs.

Because the layer thicknesses of some regions of high-frequency IMPATT
diodes are both very thin (one or a few micrometers) and closely controlled,
the most refined techniques were employed in their fabrication, among
them ion implantation and molecular beam epitaxy. These processes can
control layer thickness and doping down to atomic dimensions.

To avoid burnout caused by internally generated heat, designers had
the finished diodes mounted in intimate contact with an excellent heat
sink, copper or diamond, the heat-generating junction positioned nearest
the sink, using, for example, the V package in Fig. 1-37. As illustrated in
Fig. 1-44, the small diode package is almost lost in the complicated plumbing
of the microwave circuit. Indeed, the design complications of IMPATT
oscillators and amplifiers reside primarily in microwave engineering rather
than semiconductor device design.

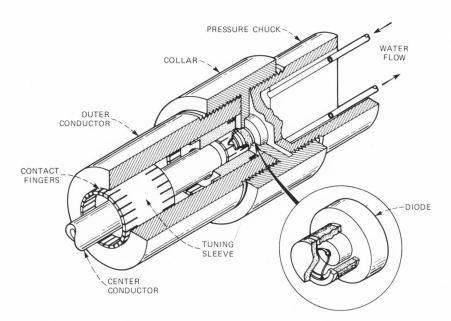

Fig. 1-44.   Microwave oscillator using the IMPATT diode package.

## XII. IMPACT OF THE INVENTION

The electron tube and the transistor each inaugurated an era in which new technology brought something strikingly new to our society and to our social well-being. The most important contribution of the discrete transistor probably has been a practical, high-speed digital computer. In addition to computers, the transistor has made possible that extraordinary human undertaking—landing man on the moon and returning him safely to earth—as well as the communications satellite. Together, these accomplishments of semiconductor technology mark a second era in electronics technology—that associated with the invention of the transistor. The first era, of course, was inaugurated by the electron tube.

Today we are in the midst of a third era—that associated with the silicon integrated circuit (SIC). While the hardware of this era is miniaturized (the functions of a 1960-era central processor like the IBM 7040 can be performed by an SIC microprocessor contained in a package of a few square centimeters), the identifying feature of the new era, and its driving force, may well be the dramatic reduction of many orders of magnitude in the cost of electronic functions. A consequence of this in our society is that, by performing thousands of times more electronic functions, we are gradually reorganizing the methods and procedures by which we conduct our daily affairs.

## REFERENCES

1. J. H. Scaff, H. C. Theuerer, and E. E. Schumacher, "P-type and N-type Silicon and the Formation of the Photovoltaic Barrier in Silicon Ingots," *Trans. A.I.M.E.* **185** (June 1949), pp. 383-388.
2. J. H. Scaff and R. S. Ohl, "Development of Silicon Crystal Rectifiers for Microwave Radar Receivers," *Bell Syst. Tech. J.* **26** (January 1947), pp. 1-30.
3. W. Shockley, *Electrons and Holes in Semiconductors* (Princeton, New Jersey: D. Van Nostrand Co., 1950).
4. O. Heil, British Patent No. 439,457; convention filing date (Germany) March 2, 1934; issued December 6, 1935.
5. J. E. Lilienfeld, U.S. Patent No. 1,745,175, filed October 8, 1928, issued January 28, 1930; U.S. Patent No. 1,877,140, filed December 8, 1928, issued January 13, 1932; U.S. Patent No. 1,900,018, filed March 28, 1928, issued March 7, 1933.
6. J. Bardeen, "Surface States and Rectification at a Metal Semi-Conductor Contact," *Phys. Rev.* **71** (May 15, 1947), pp. 717-727.
7. W. Shockley and G. L. Pearson, "Modulation of Conductance of Thin Films of Semi-Conductors by Surface Charges," *Phys. Rev.* **74** (July 15, 1948), pp. 232-233.
8. J. Bardeen and W. H. Brattain, "The Transistor, A Semi-Conductor Triode," *Phys. Rev.* **74** (July 15, 1948), pp. 230-231.
9. G. L. Pearson and W. H. Brattain, "History of Semiconductor Research," *Proc. IRE* **43** (December 1955), p. 1801.
10. W. H. Brattain, Lab Notebook 18194 (December 16, 1947), p. 193. [AT&T Bell Laboratories Archives Collection.]
11. W. H. Brattain, Lab Notebook 21780 (December 24, 1947), pp. 6-8. [AT&T Bell Laboratories Archives Collection.]

12. W. Shockley, "The Theory of *p-n* Junctions in Semiconductors and *p-n* Junction Transistors," *Bell Syst. Tech. J.* **28** (July 1949), pp. 435-489.
13. W. Shockley, "The Path to the Conception of the Junction Transistor," *IEEE Trans. Electron Dev.* **23** (July 1976), pp. 597-620.
14. W. H. Brattain and J. Bardeen, "Nature of the Forward Current in Germanium Point Contacts," *Phys. Rev.* **74** (July 15, 1948), pp. 231-232.
15. J. N. Shive, "Double-Surface Transistor," *Phys. Rev.* **75** (February 15, 1949), pp. 689-690.
16. J. Bardeen and W. H. Brattain, "Physical Principles Involved in Transistor Action," *Phys. Rev.* **75** (April 15, 1949), pp. 1208-1225.
17. *New York Herald Tribune* (July 1, 1948).
18. "The Transistor—A Crystal Triode," *Electronics* **21** (September 1948), p. 68.
19. J. R. Haynes and W. Shockley, "Investigation of Hole Injection in Transistor Action," *Phys. Rev.* **75** (February 15, 1949), p. 691.
20. W. Shockley, G. L. Pearson, and J. R. Haynes, "Hole Injection in Germanium—Quantitative Studies and Filamentary Transistors," *Bell Syst. Tech. J.* **28** (July 1949), pp. 344-366.
21. See reference 3, p. 59.
22. H. Suhl and W. Shockley, "Concentrating Holes and Electrons by Magnetic Fields," *Phys. Rev.* **75** (May 15, 1949), pp. 1617-1618.
23. J. Czochralski, "A New Method for the Measurement of the Velocity of Crystallization of Metals," *Z. Phys. Chem.* **92** (April 1917), pp. 219-221.
24. G. K. Teal and J. B. Little, "Growth of Germanium Single Crystals," *Phys. Rev.* **78** (June 1950), p. 647.
25. W. G. Pfann, U.S. Patent No. 2,577,803; filed December 29, 1948; issued December 11, 1951.
26. R. M. Ryder and R. J. Kircher, "Some Circuit Aspects of the Transistor," *Bell Syst. Tech. J.* **28** (July 1949), pp. 367-400.
27. F. B. Llewellyn and L. C. Peterson, "Vacuum Tube Networks," *Proc. IRE* **32** (March 1944), pp. 144-166.
28. J. N. Shive, "A New Germanium Photo-Resistance Cell," *Phys. Rev.* **76** (August 15, 1949), p. 575.
29. F. S. Goucher, "The Quantum Yield of Electron Hole Pairs in Germanium," *Phys. Rev.* **78** (June 1, 1950), p. 646.
30. Signal Corps Contract No. W36-039-sc-44497, 7th Quarterly Progress Report (December 1, 1950 to February 28, 1951), Section 2. [AT&T Bell Laboratories Archives Collection: *Semiconductors.*]
31. See reference 30, Sections 4 and 5.
32. G. K. Teal, M. Sparks, and E. Buehler, "Growth of Germanium Single Crystals Containing *p-n* Junctions," *Phys. Rev.* **81** (February 15, 1951), p. 637.
33. W. Shockley, M. Sparks, and G. K. Teal, "*p-n* Junction Transistors," *Phys. Rev.* **83** (July 1, 1951), pp. 151-162.
34. G. K. Teal, U. S. Patent No. 2,727,840; filed June 15, 1950; issued December 20, 1955.
35. R. L. Wallace, Jr. and W. J. Pietenpol, "Some Circuit Properties and Applications of *n-p-n* Transistors," *Bell Syst. Tech. J.* **30** (July 1951), pp. 530-563.
36. K. B. McAfee, E. J. Ryder, W. Shockley, and M. Sparks, "Observations of Zener Current in Germanium *p-n* Junctions," *Phys. Rev.* **83** (August 1, 1951), pp. 650-651.
37. C. Zener, "A Theory of the Electrical Breakdown of Solid Detectors," *Proc. Royal Soc. London* **145** (July 1934), pp. 523-529.
38. K. G. McKay and K. B. McAfee, "Electron Multiplication in Silicon and Germanium," *Phys. Rev.* **91** (September 1, 1953), pp. 1079-1084.
39. A. G. Chynoweth and K. G. McKay, "Internal Field Emission in Silicon *p-n* Junctions," *Phys. Rev.* **106** (May 1, 1957), pp. 418-426.

40. See reference 33, p. 153.
41. W. G. Pfann, "Principles of Zone Melting," *Trans. A.I.M.E.* **194** (July 1952), pp. 747-753.
42. W. G. Pfann, U.S. Patent No. 2,739,088; filed November 16, 1951; issued March 20, 1956.
43. R. A. Logan and M. Schwartz, "Restoration of Resistivity and Lifetime in Heat Treated Germanium," *J. Appl. Phys.* **26** (November 1955), pp. 1287-1289.
44. R. A. Logan and M. Sparks, U.S. Patent No. 2,698,780; filed February 3, 1953; issued January 4, 1955.
45. W. G. Pfann and K. M. Olsen, "Purification and Prevention of Segregation in Single Crystals of Germanium," *Phys. Rev.* **89** (January 1, 1953), pp. 322-323.
46. Bell Telephone Laboratories, *The Transistor* (November 15, 1951). [AT&T Bell Laboratories Archives Collection: *Historical Books.*]
47. Introduction to program agenda, *Transistor Technology* I and II (Western Electric Co. Patent Licensing Division, September 1952). [AT&T Bell Laboratories Archives Collection: *Historical Books.*]
48. The Transistor Teachers Summer School, "Experimental Verification of the Relationship between Diffusion Constant and Mobility of Electrons and Holes," *Phys. Rev.* **88** (December 15, 1952), pp. 1368-1369.
49. J. E. Saby, "Fused Impurity P-N-P Junction Transistors," *Proc. IRE* **40** (November 1952), pp. 1358-1360; also a paper presented at Solid State Device Research Conference, Durham, New Hampshire (June 1951).
50. J. J. Ebers, "Four-Terminal *P-N-P-N* Transistors," *Proc. IRE* **40** (November 1952), pp. 1361-1364.
51. J. J. Ebers, U.S. Patent No. 2,655,610; filed July 22, 1952; issued October 13, 1953.
52. W. Shockley, U.S. Patent No. 2,655,609; filed July 22, 1952; issued November 13, 1953.
53. W. Shockley, "Transistor Electronics: Imperfections, Unipolar and Analog Transistors," *Proc. IRE* **40** (November 1952), pp. 1289-1313.
54. W. Shockley, "A Unipolar 'Field-Effect' Transistor," *Proc. IRE* **40** (November 1952), pp. 1365-1377.
55. G. C. Dacey and I. M. Ross, "Unipolar Field-Effect Transistor," *Proc. IRE* **41** (August 1953), pp. 970-979.
56. G. C. Dacey and I. M. Ross, U.S. Patent No. 2,778,956; filed October 31, 1952; issued January 22, 1957.
57. See reference 55, p. 978.
58. G. L. Pearson, "A High Impedance Field-Effect Silicon Transistor," *Phys. Rev.* **90** (April 15, 1953), p. 336; also a paper presented at American Physical Society Meeting, Cambridge, Massachusetts (January 1953).
59. G. C. Dacey and I. M. Ross, "Field Effect Transistor," *Bell Syst. Tech. J.* **34** (November 1955), pp. 1149-1189.
60. S. Darlington, U.S. Patent No. 2,663,806; filed May 9, 1952; issued December 22, 1953.
61. R. L. Wallace, Jr., L. G. Schimpf, and E. Dickten, "A Junction Transistor Tetrode for High-Frequency Use," *Proc. IRE* **40** (November 1952), pp. 1395-1400.
62. J. M. Early, "Effects of Space-Charge Layer Widening in Junction Transistors," *Proc. IRE* **40** (November 1952), pp. 1401-1406.
63. J. M. Early, "P-N-I-P and N-P-I-N Junction Transistor Triodes," *Bell Syst. Tech. J.* **33** (May 1954), p. 517.
64. G. K. Teal and E. Buehler, "Growth of Silicon Single Crystals and of Single Crystal Silicon *p-n* Junctions," *Phys. Rev.* **87** (July 1, 1952), p. 190.
65. G. L. Pearson and B. Sawyer, "Silicon *P-N* Junction Alloy Diodes," *Proc. IRE* **40** (November 1952), pp. 1348-1351.
66. J. J. Ebers and J. L. Moll, "Large-Signal Behavior of Junction Transistors," *Proc. IRE* **42** (December 1954), pp. 1761-1772.

67. J. L. Moll, "Large-Signal Transient Response of Junction Transistors," *Proc. IRE* **42** (December 1954), pp. 1773-1783.
68. J. M. Early, "Design Theory of Junction Transistors," *Bell Syst. Tech. J.* **32** (November 1953), pp. 1271-1312.
69. S. L. Miller and J. J. Ebers, "Alloyed Junction Avalanche Transistors," *Bell Syst. Tech. J.* **34** (September 1955), pp. 883-902.
70. J. G. Linvill, "Transistor Negative-Impedance Converters," *Proc. IRE* **41** (June 1953), pp. 725-729.
71. J. A. Hornbeck and J. R. Haynes, "Trapping of Minority Carriers in Silicon I. P-Type Silicon," *Phys. Rev.* **97** (January 1955), pp. 311-321.
72. J. R. Haynes and J. A. Hornbeck, "Temporary Traps in Silicon and Germanium," *Phys. Rev.* **90** (April 1, 1953), pp. 152-153.
73. J. H. Scaff and H. C. Theuerer, U.S. Patent No. 2,567,970; filed December 24, 1947; issued September 18, 1951.
74. C. S. Fuller, "Diffusion of Donor and Acceptor Elements into Germanium," *Phys. Rev.* **86** (April 1, 1952), pp. 136-137.
75. C. S. Fuller and J. A. Ditzenberger, "Diffusion of Boron and Phosphorus into Silicon," *J. Appl. Phys.* **25** (November 1954), pp. 1439-1440.
76. G. L. Pearson and C. S. Fuller, "Silicon P-N Junction Power Rectifiers and Lighting Protectors," *Proc. IRE* **42** (April 1954), p. 760.
77. D. M. Chapin, C. S. Fuller, and G. L. Pearson, "A New Silicon p-n Junction Photocell for Converting Solar Radiation into Electrical Power," *J. Appl. Phys.* **25** (May 1954), pp. 676-677.
78. C. A. Lee, "A High-Frequency Diffused Base Germanium Transistor," *Bell Syst. Tech. J.* **35** (January 1956), pp. 23-34.
79. G. C. Dacey, C. A. Lee, and W. Shockley, U.S. Patent No. 3,028,655; filed March 23, 1955; issued April 10, 1962.
80. M. Tanenbaum and D. E. Thomas, "Diffused Emitter and Base Silicon Transistors," *Bell Syst. Tech. J.* **35** (January 1956), pp. 1-22.
81. H. Reiss, C. S. Fuller, and F. J. Morin, "Chemical Interactions Among Defects in Germanium and Silicon," *Bell Syst. Tech. J.* **35** (May 1956), pp. 535-636.
82. J. L. Moll, M. Tanenbaum, J. M. Goldey, and N. Holonyak, "P-N-P-N Transistor Switches," *Proc. IRE* **44** (September 1956), pp. 1174-1182.
83. W. Shockley, "Negative Resistance Arising from Transit Time in Semiconductor Diodes," *Bell Syst. Tech. J.* **33** (July 1954), pp. 799-826.
84. W. T. Read, Jr., "A Proposed High-Frequency, Negative-Resistance Diode," *Bell Syst. Tech. J.* **37** (March 1958), pp. 401-446.
85. A. Uhlir, Jr., "Two-Terminal P-N Junction Devices for Frequency Conversion and Computation," *Proc. IRE* **44** (September 1956), pp. 1183-1191.
86. M. E. Hines and H. E. Elder, "Amplification with Nonlinear-Reactance Modulators," paper presented at Solid State Device Research Conference, Berkeley, California (June 1957).
87. J. M. Manley and H. E. Rowe, "Some General Properties of Nonlinear Elements—I. General Energy Relations," *Proc. IRE* **44** (July 1956), pp. 904-913.
88. M. B. Prince, "Diffused p-n Junction Silicon Rectifiers," *Bell Syst. Tech. J.* **35** (May 1956), pp. 661-684.
89. O. L. Anderson, H. Christensen, and P. Andreatch, "Technique for Connecting Electrical Leads to Semiconductors," *J. Appl. Phys.* **28** (August 1957), p. 923.
90. O. L. Anderson and H. Christensen, U.S. Patent No. 3,006,067; filed October 31, 1956; issued October 31, 1961.
91. G. A. Dodson and B. T. Howard, "High-Stress Aging to Failure of Semiconductor Devices," *Proc. 7th Symp. on Reliability and Quality Contr. in Electron.*, Philadelphia, Pennsylvania (January 1961), pp. 262-272.

92. D. S. Peck, "Semiconductor Reliability Predictions from Life Distribution Data," in *Semiconductor Reliability*, ed. J. E. Shwop and H. J. Sullivan (Elizabeth, New Jersey: Engineering Publishers, 1961), pp. 51-67.

93. H. C. Theuerer, "Removal of Boron from Silicon by Hydrogen Water Vapor Treatment," *Trans. A.I.M.E* **206** (1956), pp. 1316-1319.

94. H. C. Theuerer, U.S. Patent No. 3,060,123; filed December 17, 1952; issued October 23, 1962.

95. L. Derick and C. J. Frosch, U.S. Patent No. 2,802,760; filed December 2, 1955; issued August 13, 1957.

96. C. J. Frosch and L. Derick, "Surface Protection and Selective Masking during Diffusion in Silicon," *J. Electrochem. Soc.* **104** (September 1957), pp. 547-552.

97. J. Andrus and W. L. Bond, "Photograving in Transistor Fabrication," in *Transistor Technology*, Vol. III, ed. F. J. Biondi (Princeton, New Jersey: D. Van Nostrand Co., 1958), pp. 151-162.

98. J. Andrus, U.S. Patent No. 3,122,817; filed August 15, 1957; issued March 3, 1964.

99. I. M. Ross, L. A. D'Asaro, and H. H. Loar, paper presented at Solid State Device Research Conference, Lafayette, Indiana (June 1956).

100. B. T. Howard, U.S. Patent No. 3,066,052; filed June 9, 1958; issued November 27, 1962; also a paper presented at Electrochemical Society Meeting, Buffalo, New York (October 1957).

101. L. Esaki, "New Phenomenon in Narrow Germanium *p-n* Junctions," *Phys. Rev.* **109** (January 15, 1958), pp. 603-604.

102. G. Bemski and J. D. Struthers, "Gold in Silicon," *J. Electrochem. Soc.* **105** (October 1958), pp. 588-591; also a paper presented at Electrochemical Society Meeting, Buffalo, New York (October 1957).

103. G. Bemski, U.S. Patent No. 2,827,436; filed January 16, 1956; issued March 18, 1958.

104. S. J. Silverman and J. B. Singleton, "Technique for Preserving Lifetime in Diffused Silicon," *J. Electrochem. Soc.* **105** (October 1958), pp. 591-594; also a paper presented at Electrochemical Society Meeting, Buffalo, New York (October 1957).

105. D. F. Ciccolella, J. H. Forster, and R. R. Rulison, U.S. Patent No. 3,067,485; filed August 31, 1958; issued December 11, 1962.

106. A. E. Bakanowski and J. H. Forster, "Electrical Properties of Gold-Doped Diffused Silicon Computer Diodes," *Bell Syst. Tech. J.* **38** (January 1960), pp. 87-105.

107. H. S. Veloric and M. B. Prince, "High-Voltage Conductivity-Modulated Silicon Rectifier," *Bell Syst. Tech. J.* **36** (July 1957), pp. 975-1004.

108. M. B. Prince, U.S. Patent No. 2,790,940; filed April 22, 1955; issued April 30, 1957.

109. H. Welker, "Über Neue Halbleitende Verbindungen," *Z. Naturforsch.* **7a** (1952), pp. 744-749.

110. H. Welker, "Über Neue Halbleitende Verbindungen II," *Z. Naturforsch.* **8a** (1953), pp. 248-251.

111. H. Christensen and G. K. Teal, U.S. Patent No. 2,692,839; filed April 7, 1951; issued October 26, 1954.

112. H. C. Theuerer, J. J. Kleimack, H. H. Loar, and H. Christensen, "Epitaxial Diffused Transistors," *Proc. IRE* **48** (September 1960), pp. 1642-1643.

113. W. L. Brown, W. H. Brattain, C. G. B. Garrett, and H. C. Montgomery, "Field Effect and Photo Effect Experiments on Germanium Surfaces, 1. Equilibrium Conditions Within the Semiconductor," in *Semiconductor Surface Physics*, ed. R. H. Kingston (Philadelphia: University of Pennsylvania Press, 1957), pp. 111-125.

114. C. G. B. Garrett, W. H. Brattain, W. L. Brown, and H. C. Montgomery, "Field Effect and Photo Effect Experiments on Germanium Surfaces, 2. Non-Equilibrium Conditions Within the Semiconductor," in *Semiconductor Surface Physics*, ed. R. H. Kingston (Philadelphia: University of Pennsylvania Press, 1957), pp. 126-138.

115. R. H. Kingston, ed., *Semiconductor Surface Physics* (Philadelphia: University of Pennsylvania Press, 1957).

116. M. M. Atalla, E. Tannenbaum, and E. J. Scheibner, "Stabilization of Silicon Surfaces by Thermally Grown Oxides," *Bell Syst. Tech. J.* **38** (May 1959), pp. 749-783.

117. M. M. Atalla, U.S. Patent No. 3,206,670; filed March 8, 1960; issued September 14, 1965.
    D. Kahng, U.S. Patent No. 3,102,230; filed May 31, 1960; issued August 27, 1963.
    D. Kahng and M. M. Atalla, "Silicon-Silicon Dioxide Field Induced Surface Devices," paper presented at Solid State Device Research Conference, Pittsburgh, Pennsylvania (June 1960).

118. J. R. Ligenza and W. G. Spitzer, "The Mechanisms for Silicon Oxidation in Steam and Oxygen," *J. Phys. Chem. Solids* **14** (July 1960), pp. 131-136.

119. D. Kahng, "A Historical Perspective on the Development of MOS Transistors and Related Devices," *IEEE Trans. Electron Dev.* **ED-23** (July 1976), pp. 655-657.

120. J. A. Hoerni, "Planar Silicon Diodes and Transistors," *IRE Trans. Electron Dev.* **8** (March 1961), p. 178; also a paper presented at Professional Group on Electron Devices Meeting, Washington, D. C. (October 1960).
    J. A. Hoerni, U.S. Patent No. 3,025,589; filed May 1, 1959; issued March 20, 1962;
    J. A. Hoerni, U.S. Patent No. 3,064,167; filed May 19, 1960; issued November 13, 1962.

121. G. E. Moore and R. N. Noyce, U.S. Patent No. 3,108,359; filed June 30, 1959; issued October 29, 1963.

122. R. N. Noyce, U.S. Patent No. 2,981,877; filed July 30, 1959; issued April 25, 1961.

123. J. A. Hornbeck, "Devices for Digital Circuits," paper presented at Bell Systems Presidents' Conference, The Homestead, Virginia (May 2, 1961). [AT&T Bell Laboratories Archives Collection.]

124. A. Goetzberger and W. Shockley, "Metal Precipitates in Silicon p-n Junctions," *J. Appl. Phys.* **31** (October 1960), pp. 1821-1824.

125. M. Yamin and F. L. Worthing, "Charge Storage and Dielectric Properties of $SiO_2$ Films," *Extended Abstracts, Electron. Div., Electrochemical Soc. Meeting* **13**, Toronto, Canada (May 3-7, 1964), pp. 182-184.

126. D. R. Kerr, J. S. Logan, P. J. Burkhardt, and W. A. Pliskin, "Stabilization of $SiO_2$ Passivation Layers with $P_2O_5$," *IBM J. Res. Develop.* **8** (September 1964), pp. 376-384.

127. E. Snow, A. S. Grove, B. E. Deal, and C. T. Sah, "Ion Transport Phenomena in Insulating Films," *J. Appl. Phys.* **36** (May 1965), pp. 1664-1673.

128. "Semiconductors," *Business Week* (March 26, 1960), pp. 74–121.

129. See reference 128, p. 103.

130. See reference 128, p. 74

131. See reference 128, p. 113.

132. See reference 128, p. 83.

133. K. D. Smith, H. K. Gummel, J. D. Bode, D. B. Cuttriss, R. J. Nielsen, and W. Rosenzweig, "The Solar Cells and Their Mounting," *Bell Syst. Tech. J.* **42** (July 1963), pp. 1765-1816.

134. W. Rosenzweig, H. K. Gummel, and F. M. Smits, "Solar Cell Degradation Under 1-Mev Electron Bombardment," *Bell Syst. Tech. J.* **42** (March 1963), pp. 399-414.

135. J. S. Mayo, H. Mann, F. J. Witt, D. S. Peck, H. K. Gummel, and W. L. Brown, "The Command System Malfunction of the *Telstar* Satellite," *Bell Syst. Tech. J.* **42** (July 1963), pp. 1631-1657.

136. H. E. Elder, "Active Solid-State Devices," *Bell Syst. Tech. J.* **47** (September 1968), pp. 1323-1377.

137. R. L. Kaufman, W. R. Harlin, and W. J. Frankfort, "The S-Pack Diode, Parts I-III," *W. Elec. Eng.* **10** (July 1966), pp. 2-14.

138. J. E. Beroset, H. A. Griesemer, D. M. Large, and K. C. Whitefield, "Magnetic Suspension Parts Handling," *W. Elec. Eng.* **11** (July 1967), pp. 36-40.

139. D. Kahng and M. P. Lepselter, "Planar Epitaxial Silicon Schottky Barrier Diodes," *Bell Syst. Tech. J.* **44** (September 1965), pp. 1525-1528.

140. M. P. Lepselter, "Beam-Lead Technology," *Bell Syst. Tech. J.* **45** (February 1966), pp. 233-253.

141. J. V. Dalton and J. Drobek, "Structure and Sodium Migration in Silicon Nitride Films," *J. Electrochem. Soc.* **115** (August 1968), pp. 865-868.

142. G. H. Schneer, W. van Gelder, V. E. Hauser, and P. F. Schmidt, "A Metal-Insulator-Silicon Junction Seal," *IEEE Trans. Electron Dev.* **ED-15** (May 1968), pp. 290-293.

143. J. B. Gunn, "Microwave Oscillations of Current in III-V Semiconductors," *Solid State Commun.* **1** (September 1963), pp. 88-91.

144. R. L. Johnston, B. C. De Loach, Jr., and B. G. Cohen, "A Silicon Diode Microwave Oscillator," *Bell Syst. Tech. J.* **44** (February 1965), pp. 369-372.

145. T. Misawa, "Negative Resistance in p-n Junctions Under Available Breakdown Conditions, Part I," *IEEE Trans. Electron Dev.* **ED-13** (January 1966), pp. 137-143.

146. B. C. De Loach, Jr. and R. L. Johnston, "Avalanche Transit-Time Microwave Oscillators and Amplifiers," *IEEE Trans. Electron Dev.* **ED-13** (January 1966), pp. 181-186.

147. C. A. Lee, R. L. Batdorf, W. Wiegmann, and G. Kaminsky, "The Read Diode—An Avalanching, Transit-Time, Negative-Resistance Oscillator," *Appl. Phys. Lett.* **6** (March 1966), pp. 89-91.

148. S. M. Sze and R. M. Ryder, "Microwave Avalanche Diodes," *Proc. IEEE* **59** (August 1971), pp. 1140-1154.

149. J. E. Morris and J. W. Gewartowski, "A 1-Watt, 6-Gigahertz IMPATT Amplifier for Short-Haul Radio Applications," *Bell Syst. Tech. J.* **54** (April 1975), pp. 721-733.

150. C. A. Brackett, "The Elimination of Tuning-Induced Burnout and Bias-Circuit Oscillations in IMPATT Oscillators," *Bell Syst. Tech. J.* **52** (March 1973), pp. 271-306.

# Chapter 2

# Silicon Integrated Circuits

*Much of the early technology of silicon integrated circuits grew out of the physics, chemistry, and metallurgy of discrete transistors, including a number of key Bell Laboratories developments: oxide-masked diffusion, photolithographic methods, thermocompression bonding, and epitaxial deposition. Important Bell Laboratories innovations for integrated circuits encompassed beam lead sealed-junction devices, ion implantation, complementary bipolar integrated circuits, the first metal-oxide-semiconductor (MOS) field-effect transistor, charge-coupled devices, and the self-aligned silicon gate MOS transistor. Also important to the complexity and power of modern integrated circuits are Bell Laboratories advances in the areas of mask making with electron beam exposure systems and computer aids to design.*

## I. INTRODUCTION

This chapter traces the development of silicon integrated circuits (SICs) in the Bell System since the mid-1950s. It is really three stories: technology development, innovation in device structures, and applications. We begin with early technology development and have intertwined the stories to show the constant and continuing interdependence of technology, device and systems needs, and opportunity. The developments encompassed Bell Laboratories people from many organizations in research, device development, and systems development, and Western Electric personnel at its Engineering Research Center in Princeton, New Jersey and at the manufacturing locations in Allentown and Reading, Pennsylvania.

The development of integrated circuits (ICs) in the Bell System was aimed not only at improved performance, but at lower cost and higher reliability. The impact of ICs on telecommunications, the semiconductor and electronics industries, and on society as a whole, was profound. It started with the introduction of small-scale and medium-scale integrated (SSI and MSI) circuits during the 1960s. Even more profound changes

---

Principal authors: J. M. Goldey, J. H. Forster, and B. T. Murphy

occurred as the full implications unfolded for low-cost, large-scale integrated (LSI) and very large-scale integrated (VLSI) circuits, such as microprocessors, memories, and custom circuits tailored for specific applications.

## II. EARLY TECHNOLOGY DEVELOPMENTS

SIC technology is a complex composite of chemical, electronic, mechanical, metallurgical, and optical technologies. It is, nevertheless, not a new technology unto itself, but it is part of an evolving silicon semiconductor device technology, which initially supported transistors and diodes only. The history of this technology up to the early 1960s, when discrete transistors reached a state of maturity, is described in detail in Chapter 1. Since then, SICs have dominated the evolution of the technology, as covered in this chapter. However, it would be difficult, if not impossible, to describe the history of SIC technology without summarizing certain parts of the preceding technology, since most of the basic innovations used to achieve complex SICs were initially conceived as a means for improving transistors.

While many aspects of SIC technology are based on Bell Laboratories research and development, this is not to imply that other organizations did not make significant contributions, for indeed they did. Notable among these companies were Texas Instruments and Fairchild Industries, and somewhat later, Intel. It was J. S. Kilby at Texas Instruments who first recognized that silicon technology could produce in one wafer of silicon not only transistors and diodes, but also resistors and capacitors, and that by interconnecting these elements, circuit functions could be realized.[1] R. N. Noyce at Fairchild contributed the idea of using deposited aluminum films on a planar structure to form the interconnections.[2] These early developments were chronicled in greater detail in the *IEEE Spectrum* in 1976—a truly interesting story.[3] The work led to the first commercial offering of production ICs in 1961 by both Texas Instruments and Fairchild. The initial applications were almost exclusively for military purposes; the military funded much of the early IC research and development work. Most Bell Laboratories work, like that of Fairchild, however, was not government funded.

In this chapter, we restrict ourselves, with a few exceptions, to a review of Bell Laboratories work, and begin with an overview of certain relevant features of transistor technology.

Figure 2-1 shows a cross section of a simple bipolar SIC consisting of a diode, transistor, and diffused resistor, interconnected by thin metal films deposited on an insulating layer over the silicon. Note in the drawing that there are multiple layers of both n-doped and p-doped material, and that interconnection is provided by well-defined paths of metal connected to the silicon through holes defined in the insulator.

A key process in SIC technology is the introduction of dopant impurities into the silicon—the right amount in the right place—to form active and

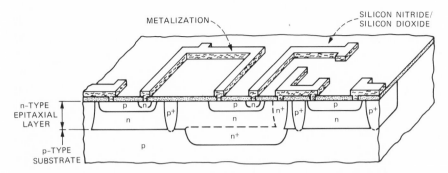

Fig. 2-1. Silicon bipolar IC schematic with a diode at left, transistor at the center, and diffused resistor at right. There are multiple layers of both n-doped and p-doped material, and interconnections are provided by well-defined paths of metal connected to the silicon through holes defined in the SiN over SiO₂ insulating film. Electrical isolation between the circuit elements is achieved by p-type diffused regions extending from the surface to the p-type substrate.

passive components. Basic to this operation is the oxide-masked diffusion process. The pioneering work on diffusion for silicon devices was carried out by C. S. Fuller and J. A. Ditzenberger.[4,5] In this process, impurities, generally in vapor form, surrounding the silicon at high temperature, enter and then move into the semiconductor at a rate fast enough to be practical, yet slow enough to be precisely controllable. The amount of impurity and the distance diffused can be varied over several orders of magnitude by adjusting the time, temperature, or vapor concentration. For example, diffusion depths are precisely controllable over a range from a small fraction of 1 micrometer ($\mu$m) to 20 $\mu$m or greater; surface concentrations are available from $10^{16}$ to $10^{20}$ atoms per cm$^3$, with the total number of diffused atoms between $10^{11}$ and $10^{16}$ atoms per cm$^2$ of silicon area. Because of the good control available, it is easy to change the conductivity type of a piece of silicon several times by using increasingly higher concentrations of different impurities and thus fabricate transistors and other components.

Oxide masking, first studied by C. J. Frosch and L. Derick, is a process by which the diffusion of impurities is limited to certain portions of the silicon slice, thus forming regions for specific functions, e.g., transistor base, resistor, etc.[6] Where oxide is present, no diffusion occurs, and where it is absent, diffusion of dopants into the silicon proceeds freely. Oxide masking in combination with photolithography offers the opportunity to control areas to very fine tolerances. For example, by 1975, oxide windows as small as 7.5 $\mu$m wide were common, and much smaller windows followed thereafter.

It was recognized from the earliest days that since diffusion is a batch process (i.e., many slices, each with many devices, can be diffused simultaneously), it would be inherently a low-cost technology, a key feature for the economic manufacture of ICs.

Initially, control of oxide geometry was achieved by such crude methods as using dots of black wax to protect the oxide from its etchants. The foundations of modern IC photolithographic technology were laid by J. Andrus, who first used photosensitive etch-resistant polymers (photoresists) to define patterns in the masking oxide.[7] Although many advances have been made in the mask-making art (see section VII), and significant improvements in photoresist materials and alignment apparatus have been achieved, the basic photoresist processes used throughout the world in the fabrication of ICs are essentially those of Andrus.

The first application of photo-defined oxide-masked diffusion to devices was in the semiconductor stepping device, developed in the mid-1950s by I. M. Ross, L. A. D'Asaro, and H. H. Loar.[8] [Fig. 2-2] A photo of this device is shown in Fig. 2-3(a) and a schematic in Fig. 2-3(b). It consists of four coupled special-purpose p-n-p-n transistors that perform the memory and space transfer functions of a four-element stepping circuit, and represents, in the eyes of many, the world's first SIC.

Aluminum contacts to silicon are used extensively in SICs. As discussed in Chapter 1, aluminum metalization was used early in the development of transistors. While aluminum, which is an acceptor in silicon, is a natural choice for contacting p-type material, S. L. Matlow and E. L. Ralph of Hoffman Electronics Corporation demonstrated that aluminum can also form ohmic contacts to $n^+$ layers on silicon.[9] G. E. Moore and Noyce at Fairchild were the first to use aluminum to contact both n and p regions of a diffused silicon transistor.[10]

Fig. 2-2.    I. M. Ross (left) and L. A. D'Asaro (right), who, along with H. H. Loar, developed the semiconductor stepping transistor.

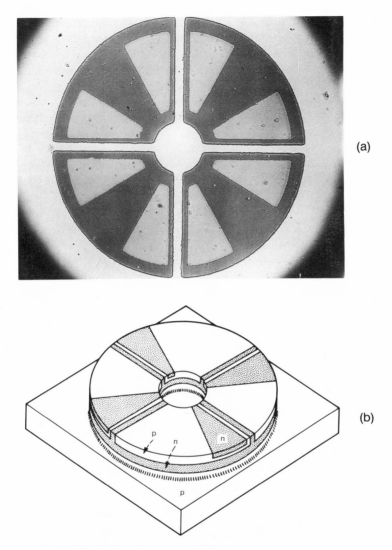

Fig. 2-3. The silicon stepping transistor, the first device to use photo-defined oxide-masked diffusion. (a) The light regions are n-type and the dark are p-type. Both were formed by oxide-masked diffusion. [D'Asaro, *IRE WESCON Conv. Rec., Pt. 3—Electron Div.* (1959): 42] (b) Schematic of the silicon stepping transistor, consisting of four p-n-p-n diodes integrated in a single structure. With electrodes attached, this device operated as a four-stage counter up to a rate of $2 \times 10^6$ counts per second.

The epitaxial process in which thin layers of single-crystal silicon are deposited on single-crystal silicon substrates is of particular importance for bipolar ICs. Epitaxial work in silicon carried out in the late 1950s by

H. C. Theuerer, J. J. Kleimack, Loar, and H. Christensen under the guidance of Ross led to the announcement of the epitaxial transistor at the Solid State Device Research Conference in June 1960.[11] Since then, this technology has been used in a broad spectrum of semiconductor devices.

Another significant step toward ICs was made by J. A. Hoerni at Fairchild with the introduction of planar technology, in which contacts to all electrically active regions can be made on the top surface.[12] In a third significant step by Fairchild people, and as a direct consequence of the preceding developments, Noyce extended the aluminum metalization over the oxide for ease of contacting devices of small geometry.[2] This metal-over-oxide approach was later used as a means of interconnecting devices in SICs.

In most ICs, connection to the chip is made by a gold or aluminum wire leading from an aluminum bonding pad on the silicon chip to an external lead on a header. The process of thermocompression bonding, developed at Bell Laboratories by Christensen, O. L. Anderson, and P. Andreatch, made this approach practical.[13]

In 1964, J. A. Morton, then vice president of Electronic Components Development, described the importance of the technologies discussed so far in the following words: "With its [i.e., epitaxy's] introduction in the late 1950s, however, it provided devices of improved performance that diffusion and oxide masking could not provide alone, and did this by augmenting the existing manufacturing processes rather than making them obsolete. The combination of epitaxy, oxide masking, and diffusion allows the near realization of the inherent capabilities of semiconductor materials and provides an elegant, balanced technology for integrated circuits."

Although the technology base was established by the early 1960s, the technology continued to advance during the IC era. By then, a vigorous semiconductor industry had developed, and while significant contributions came from the industry at large, important developments continued to come from Bell Laboratories as well.

### III. OVERVIEW OF SILICON INTEGRATED CIRCUIT EVOLUTION

One of the earliest Bell Laboratories SIC developments was a family of logic circuits for use in the Nike-X system, on which work began in 1962. (For more information on the Nike-X system, see a companion volume in this series subtitled *National Service in War and Peace (1925-1975)*, Chapter 7.) Multichip SICs, in which transistors and resistors were fabricated on individual chips, albeit several components per chip, were bonded on a common substrate and interconnected by thermocompression-bonded wires and then hermetically sealed. The silicon device technology of that time, described above, was used to provide custom circuits that enabled the Nike system to meet demanding objectives. A photo of one of these early circuits is shown in Fig. 2-4.

Fig. 2-4.  Early developmental version of Nike-X multichip IC. Two types of multiple-resistor chip and one type of multiple-transistor chip were used in its construction.

Customized ICs for specific system applications are characteristic of Bell System SIC developments. Customization required a strong and continued interaction between members of the systems development and the electronics technology development areas. Such interactions served to shape the development of systems hardware in response to the promise of the developing SIC technology and to tailor specific SIC designs to optimize performance, reliability, and/or cost of the equipment. However, despite the custom nature of many SIC developments, a considerable number of these found general Bell System applications, and often the experience gained in specific developments was applied to produce device families for general Bell System use. Thus, SICs became important in a variety of Bell System developments and an important economic factor in Western Electric production.

A condensed summary of the way in which SICs were developed for some of the major Bell System applications is provided in the chronological flow chart for SIC development in Fig. 2-5, which shows major milestones in both applications and new device and circuit configurations. The chart illustrates how IC development proceeded at Bell Laboratories. Significant technological events and applications are also shown for newer product

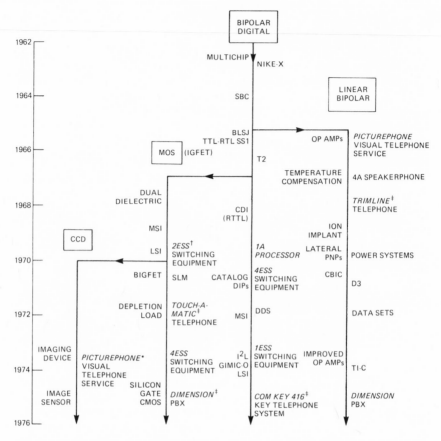

Fig. 2-5.   SIC development in the Bell System. To the left of each branch are shown significant technological and circuit developments with associated applications shown on the right. Bipolar digital circuits came first, based on diffused silicon transistor technology; they led to LSI, an example of which is the COM KEY 416 key telephone system. Two tributaries of the mainstream are linear bipolar and MOS ICs. The last branch away from the MOS line is the CCD line. (* Registered service mark of AT&T Co. † Trademark of AT&T Technologies, Inc. ‡ Registered trademark of AT&T Co.)

lines. In the interests of brevity, however, not all technological events or applications have been included in this figure.

After the early application of the digital bipolar silicon components in Nike-X, the buried collector, junction-isolated monolithic structure (also known as standard buried collector and referred to hereafter as SBC; see section 4.1) rapidly became the most frequently used bipolar structure. A series of developments and applications followed, and early in 1974 integrated injection logic ($I^2L$; see section 4.3) and guard-ring isolated monolithic integrated circuit (GIMIC) technologies began to be applied to bipolar LSI (see section 4.4). Soon after the early development of bipolar digital circuits, the SBC technology was applied to linear SIC development (linear bipolar branch) and continued through many applications and modifications until a completely new, complementary bipolar integrated circuit (CBIC) structure was introduced in 1971 (see section 4.6). The second branch from the bipolar digital line occurred about 1967 when metal-oxide-semiconductor (MOS) development was initiated (see section VI), which later became the leading LSI/VLSI technology used in the Bell System.

The last branch on the SIC flowchart, the charge-coupled device (CCD), was a Bell Laboratories invention, conceived in late 1969 (see section 6.5). The CCD extended the available silicon technology to produce new circuit functions by introducing a concept of voltage-controlled charge transfer along semiconductor surfaces. In its simplest form, a CCD is an array of MOS capacitors between which charge can be transferred by manipulation of electrode potentials. Appropriate charge generators and detectors are also needed.

## IV. BIPOLAR INTEGRATED CIRCUITS

### 4.1 Bipolar Device Structure

Following the transfer of Nike-X circuit development from the Murray Hill, New Jersey location of Bell Laboratories to the Allentown location, the potential for ICs was quickly recognized, and design interactions between system and development areas proceeded rapidly, concurrent with the technology development. The first ICs to be widely used were SBC bipolar circuits with junction isolation.[14] A cross section of the basic SBC transistor shows the buried collector, so called because the low resistivity n-type collector region is first formed by diffusion in a p-type substrate, then buried under a relatively high-resistivity n-type layer grown epitaxially over the entire surface. [Fig. 2-1] P-type diffusions through the epitaxial layer form pairs of back-to-back diodes that electrically isolate collectors from one another. The base and emitter junctions are then formed in the epitaxial layer by successive diffusion processes. This structure allows the collector to be brought out at the top surface where it can be contacted by the same deposited metal layer with which emitter and base are contacted.

One undesirable feature of bipolar transistors, long a limitation on their use in switching circuits, is their excess switching delay because of minority carrier storage, which results when the collector junction becomes forward biased in the ON condition. This delay can be reduced or eliminated if a diode with little or no charge storage is connected in parallel to the collector junction. J. R. Biard at Texas Instruments was the first to propose the use of a metal-semiconductor diode (Schottky diode) as a shunting diode,[15] after D. Kahng and D'Asaro had demonstrated that such diodes have practially no minority carrier storage.[16,17] Using an improved Schottky diode structure,[18] fully compatible with the evolving transistor technology, E. R. Chenette, J. J. Kleimack, R. Edwards, and R. A. Pedersen[19] demonstrated a drastic reduction in switching delay by integrating these diodes into the transistor structure, a technique now widely used in bipolar ICs.

### 4.2 Beam Lead Sealed-Junction (BLSJ) Technology

It was mentioned above that the basic technology provided batch fabrication of the transistors, resistors, and diodes of an IC. But it was costly to attach leads one at a time by thermocompression bonding individual wires using the bonders then available. Beam lead technology, invented by M. P. Lepselter [Fig. 2-6], offered a new contact metallurgy and a structure that provided discrete devices and IC chips with leads attached, as shown in Fig. 2-7.[20] The gold beam leads are formed during slice fabrication and are thus an integral part of the chip. The free ends of the beams can be simultaneously attached to a suitable metalization on an appropriate substrate, thus eliminating the one-at-a-time bonding of individual wires.

If reliable semiconductor devices and ICs were ever to be freed from the high costs associated with packaging in expensive, hermetically sealed

Fig. 2-6.   M. P. Lepselter, inventor of beam lead technology, which carries batch processing to lead formation and IC attachment.

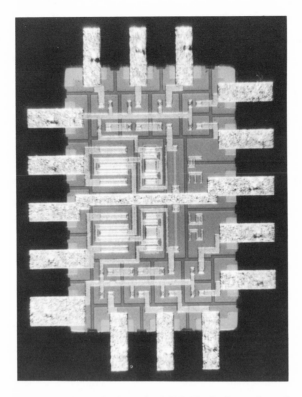

Fig. 2-7. Photo of a beam lead SIC. The leads are formed during the slice processing and are subsequently simultaneously bonded to a metalized ceramic substrate.

enclosures, it was clear that junctions and surfaces would have to be protected from contaminants by other means. As covered in more detail in section VI of this chapter, Fairchild workers, pursuing the improvement of MOS technology, first identified alkali ions, principally sodium, as the source of the Si-SiO$_2$ surface instability.[21] In trace amounts, these mobile charges drift through SiO$_2$ films under the influence of the local electric fields associated with biasing conditions, causing unstable device characteristics. It became clear that a junction protection must be impervious to sodium.

Beam lead technology provided a partial solution to this problem by extending the contacts over the junctions where they came to the surface. However, the complex topography of ICs was not geometrically compatible with this form of protection in many cases. J. V. Dalton, in a news item presented at the Electrochemical Society meeting in May 1966, reported that a film of SiN is an effective barrier to the migration of sodium ions, even at high temperature.[22] An unusually inert compound, SiN has turned out to be impervious to water vapor and to nearly all other harmful contaminants. G. H. Schneer, W. van Gelder, V. E. Hauser, and P. F. Schmidt, working at the Allentown Laboratory, demonstrated the elegance of SiN as a junction seal against sodium contamination by applying it to a transistor structure under development.[23] They deposited a 1000-angstrom film of amorphous SiN as an overcoat on slices containing planar n-p-n transistors after final oxidation but before etching contact windows. Thus the junctions were passivated by a clean Si-SiO$_2$ interface. New and conventional batch chemical techniques were employed to open contact windows through the composite SiN-SiO$_2$ layers. This sealed-junction technology was rapidly combined with beam leads to form the beam lead sealed-junction (BLSJ) technology, which was used in a wide variety of Bell System ICs. These developments extended batch fabrication concepts (originating with diffusion for junction formation) to the lead attachment and encapsulation processes. BLSJ devices were the first high-reliability devices manufactured that did not require a hermetic enclosure.

At the time these developments took place, the tantalum thin-film technology evolved as described in detail in Chapter 9 of this volume. That technology offered interconnections, resistors, and capacitors (as needed) on a ceramic substrate. Beam lead silicon chips could readily be attached by thermocompression bonding to gold metalization appropriately deposited on the ceramic substrate at interconnection points. Figure 2-8 shows an example of such a circuit interconnecting several beam lead ICs.

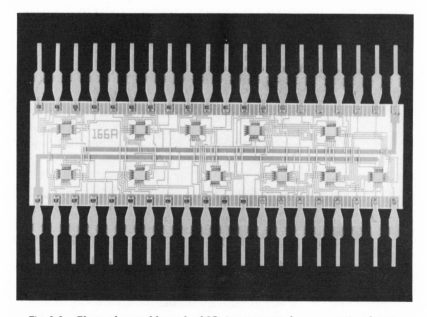

Fig. 2-8.   Photo of several beam lead ICs interconnected on a ceramic substrate.

This system combined a number of apparent advantages, particularly for the level of integration of the late 1960s. Active devices could be pretested and selected before attachment to the substrate. Resistors and capacitors could be made over a wide range of values and, when necessary, to very close tolerances. Many silicon chips could be thermocompression bonded in a batch process—as many as 50 chips per substrate. This in turn, however, called for complex gold interconnection patterns on the ceramic substrate with multiple gold crossovers, introducing a fair degree of complexity into the film technology.

Since gold beam leads for attachment were an essential feature of the concept, and since BLSJ devices were highly reliable, the standard aluminum metalization of prior planar transistor designs was supplanted by multilayer metalizations, always ending with gold. Gold then became the standard interconnecting metalization on bipolar silicon chips in Bell Laboratories designs of SICs. Later, this turned out to be a diversion from the central development path for MOS and other complex extensions of planar IC technology. Aluminum has turned out to be simpler and better for LSI and VLSI circuits.

### 4.3 Logic Circuits

In 1965 and 1966, designs combining SBC and BLSJ technologies produced a family of transistor-transistor logic (TTL) circuits for general-purpose use. Some of these found their first applications in the M1-2 multiplexer and the D2 channel bank as reported by J. S. Mayo and M. G. Stickler.[24] Also at this time, the same technology was applied to the development of a family of resistor-transistor logic circuits that were used originally in the 203 data sets and later in the D3 channel bank and other systems. Both of these circuit families were later used in many other applications. In 1966, a redesign of the Nike-X circuits was begun using BLSJ technology. Most of the BLSJ circuits remained multichip in form, while a few were redesigned as SICs.

The applications of bipolar logic in switching systems were particularly extensive. Continuous interaction between the switching and device areas on IC design and application began with the consideration of multichip circuits for the then developing 1ESS* switch in 1962. Although multichip circuits did not prove economic in that application, later a family of BLSJ logic circuits (23-type ICs) was designed to be equivalent to the discrete low-level logic (LLL) gates in the 1ESS electronic switch. Although timing and economics did not permit their application in 1ESS switches, these devices were later used in the traffic service position system (TSPS). The TTL chips discussed above, first used in transmission systems, were sub-

---

* Trademark of AT&T Technologies, Inc.

sequently used in the core memory introduced as call store in the 1ESS electronic switch in 1968. Other custom bipolar designs completed in 1969 were used in the 2ESS switch peripheral decoder. (These and other aspects of switching systems are covered in *Switching Technology (1925-1975)*, a companion volume of this series.)

In mid-1966, it was evident that the available SIC speed performance could make possible electronic switching system central processors with greatly improved call-handling capability, and joint efforts were continued to improve system and device designs for maximum performance. The result of this work was the 1A logic family, a series of high-speed logic chips first used in the 1A processor and later used in the 4ESS electronic switch, 1ESS switch trunk link network, and No. 3A central control used in 3ESS and 2B ESS switches. These circuits were made in the collector diffusion isolation (CDI) structure, invented by B. T. Murphy.[25] Their functional density was superior to circuits commercially available at that time, and they were designed to improve system performance at minimum cost. CDI technology, like the earlier SBC structure, uses junction isolation and a buried collector, but the epitaxial layer is p type rather than n type, and the isolation is therefore obtained by an n-type diffusion that also serves to contact the buried collector. As shown in Fig. 2-9, a separate p-type isolation is not required. Because of this, and since no localized base diffusion is needed, the CDI structure is simpler and more compact than the SBC structure, allowing more circuit functions to be packed on a chip of a given size.

Up to 50 of these circuits were interconnected on a single piece of ceramic in a physical configuration known as an FA pack (see Chapter 9,

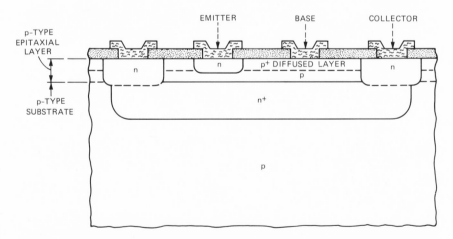

Fig. 2-9. Cross section of the CDI structure, which uses a grown p-type epitaxial base layer followed by a nonlocalized p diffusion. Isolation between circuit elements is provided by the junctions formed between the n collector plugs and the base.

Fig. 9-9). The switching system design organization developed a computer-aided system with which many hundreds of FA packs were designed. The low interconnection capacitance minimized crosstalk between different circuit nodes so that the system operated with excellent noise margins at 3-volt (V) power supply, so that the required speed could be achieved at significantly reduced power consumption, a performance considerably better than that for conventional 5-V circuitry of that time. Only many years later did the low-power, Schottky diode transistor-transistor logic (LS-TTL) family reach that performance.

As the level of integration grew, FA packs needed in large quantities were cost reduced with new MSI or LSI chips designed in I$^2$L.[26,27] One such application, a cost-reduced FA-772 circuit pack, used in the 1ESS switch, went into manufacture in the mid-1970s. In the new circuit pack, 24 CDI and 12 SBC chips were replaced with 2 I$^2$L and 2 SBC chips.

At the same time, bipolar LSI came of age in many applications. Bipolar read-only memory (ROM) chips (1024 bits) were introduced in the 209A data set early in 1974. Following the FA-772, a 250-gate I$^2$L chip went into manufacture for use in the COM KEY* 416 telephone system, and other I$^2$L chips followed in the 400H key telephone unit and other applications.

### 4.4 Ion Implantation

As stated above, oxide-masked diffusion, for introducing dopants into the semiconductor in a controlled manner, is one of the most important processes of IC technology. Ion implantation provided a significant extension of that technology. In the ion implantation process, an energetic (30- to 400-kiloelectron-volt) beam of ions is directed at a silicon slice so that the ions are embedded in the lattice. They can later be annealed and redistributed by diffusion to form a highly controlled doped semiconductor region. Masking, as in diffusion, can be achieved by the use of oxides or by photoresist alone. The strength of the ion implantation process is its precise control of the total number of dopant ions implanted simply by measuring the total charge deposited per unit area. At any given doping level, control is at least an order of magnitude better with ion implantation than with conventional diffusion, and controlled dosages down to $10^{11}$ atoms per cm$^2$ are readily achieved with ion implantation, compared to about $10^{13}$ atoms per cm$^2$ for diffusion. Ion implantation is an extremely effective process at very high as well as at low dosage levels. It produces lower-resistance buried collector layers at higher yield than does diffusion, and this was one of its first major applications in ICs.

Actually, the origin of ion implantation dates back to 1950 when, in

---

* Trademark of AT&T Co.

connection with treating silicon by ion bombardment, R. S. Ohl[28] recognized that ions of donors or acceptors could be introduced to affect the properties of a surface layer. A patent on ion implantation for semiconductor device fabrication as it is known today was filed by W. Shockley in 1954,[29] well ahead of the initiation of active development at Bell Laboratories in the mid- to late 1960s by A. U. MacRae.[30] This development benefited greatly from work in the research area under the leadership of W. L. Brown, in which high-energy ion beams were used to study channeling in silicon.[31]

An important aspect of any technology is the equipment that makes it work. Investigators at the Engineering Research Center of Western Electric recognized this fact and set out to design and build an ion implantation machine with full industrial capability. Their PR-30 machine, designed by a team led by W. Samaroo and J. L. Blank, was the first designed and built for high-volume manufacture capable of providing both high and low dosages.[32] It placed Western Electric in a position as an early, large, and diverse user of ion implantation technology.

New structures are also achievable with ion implantation technology, and the GIMIC structure described by P. T. Panousis and R. L. Pritchett is an important example.[33] This structure is similar to a triple-diffused structure, which requires no epitaxy. It is practical only with the doping control offered by ion implantation. Because of its simplicity, it leads to lower wafer fabrication cost.

GIMIC LSI chips were designed for use in D4 channel banks, 4ESS switches, DIMENSION* private branch exchanges (PBXs), and other systems.

### 4.5 Catalog Transistor-Transistor Logic (TTL)

In 1970, rapid improvements in IC technology and the shortening design cycle for systems led to the need for off-the-shelf, general-use ICs for digital systems in dual in-line packages (DIPs). Fourteen TTL codes were designed for digital data systems in 1971. In 1972, TTL codes were stocked for off-the-shelf availability and described in a Western Electric SIC catalog.

Catalog TTL provided ICs with a limited range of functions. For higher levels of integration, TTL gate arrays were designed and made available in 1974. Since only the interconnection levels are involved, gate arrays can be quickly and inexpensively customized to perform a specific circuit function by using an appropriate interconnection pattern. A single gate array typically replaced four or more SSI circuits to provide extra system economy.

At the high-speed end of the performance range, a new performance limit was established with the emitter coupled logic circuits first made in

---

* Trademark of AT&T Co.

1968[34,35] using an air isolation process originally described by Lepselter, H. A. Waggener, and R. E. Davis that virtually eliminated isolation capacitance between transistors in the circuit.[36]

### 4.6 Bipolar Linear Circuits

The application of BLSJ technology to linear circuits followed its application to digital circuits almost immediately. Development work on monolithic IC technology at the Allentown location during 1965 included operational amplifiers, preamplifiers, and custom circuits for the Touch-Tone calling dial. Since a major portion of the development effort at Allentown was directed toward digital ICs, linear development effort was expanded in 1966 at the Reading location, with the development of two families of custom ICs for PICTUREPHONE* visual telephone service and for an improved speakerphone design. This activity represented the first major system development effort in the Bell System using a family of linear ICs.

Activity in the development of linear ICs grew rapidly through the latter half of the 1960s. Temperature-compensated voltage regulators were developed in 1966. The first use of DIPs in the Bell System was in 1967 for PICTUREPHONE visual telephone service devices. In 1968, a joint effort with a group developing power systems was undertaken to provide a standard family of devices. Further innovations in 1969 included the use of ion implantation for bipolar-compatible, junction field-effect transistors and high-value resistors, melt-back trimming to fine tune resistor values, and the inclusion of lateral p-n-p transistors.

At this time, it was recognized that development of only custom linear ICs did not necessarily best serve the total needs of the Bell System, and that if general-purpose amplifiers, oscillators, timers, etc., could be developed to fit broad areas of application, economy in development effort and increased efficiency could be realized.

To achieve this objective, a multiorganizational task force, chaired by J. B. Singleton and C. F. Simone, was formed in 1969. It included representatives from several Bell Laboratories areas and Western Electric, and encompassed all aspects of linear ICs, including circuit and device design, designer training applications, and product planning. The task force was responsible for development of a large family of high-performance general-purpose linear ICs that provided an excellent fit to Bell System needs. Western Electric shipped a large number of ICs of task force genesis for use in a variety of Bell System projects, such as the D3 channel bank, data sets, speakerphone, PICTUREPHONE visual telephone service, voiceband interface unit (VIU), and DIMENSION PBX. This broad family of linear ICs was made possible by technological advancements such as

---

* Registered service mark of AT&T Co.

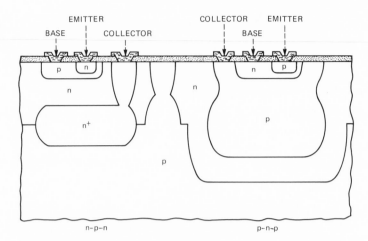

Fig. 2-10.   The complementary bipolar integrated circuit (CBIC) structure, the first to provide high-performance vertical n-p-n and p-n-p transistors on the same substrate. Up to six ion implantation operations are used in its fabrication.

the development of lateral p-n-p transistors, 60-V capability, and high-value ion-implanted resistors.

The most significant technological advance made by the task force was the CBIC structure.[37] This structure is shown in Fig. 2-10. With this technology, which makes substantial use of ion implantation, high-performance isolated vertical p-n-p transistors can be made on the same chip as high-performance n-p-n transistors, with major improvements in frequency-power products and simplification of stabilization circuitry for linear SICs. A second generation of improved performance operational amplifiers was developed, and custom circuits such as a pulse regenerator for the T1-C carrier system and a high-speed analog switch for the DIMENSION PBX used this technology.

In addition to the general-purpose catalog family of linear ICs, a number of important custom linear SICs were developed. These SICs included detectors, comparators, oscillators, converters, variolossers, voltage regulators, electronic hybrids, data couplers, pulse regenerators, analog switches, and digital/analog converters for use in almost every major system developed by Bell Laboratories after 1965. In 1976, custom linear ICs represented about 70 percent of linear ICs manufactured for Bell System applications.

## V. SOLID-STATE CROSSPOINTS

As mentioned in Chapter 1, already in the mid-1950s, solid-state devices were seen as a tantalizing alternative to metallic switches for crosspoint applications.[38] In particular, the p-n-p-n switches developed at that time exhibited acceptable ON-OFF impedance ratios but did not gain a place in the newly emerging electronic switching systems. The reason was that newly developed ferreed switches could withstand higher voltages and could pass ringing signals, battery voltages, test signals, etc., whereas these functions had to be performed with additional circuits if the p-n-p-n switches were used. Solid-state switches were, however, used in time division offices in which the electrical environment was more favorable. In the 101ESS switch and, later, in the DIMENSION PBX series, these devices transmitted pulse amplitude-modulated (PAM) signals through the switching network.

Solid-state switching also received a big boost in the 4ESS switching equipment developed in the late 1960s and early 1970s for large offices providing long-distance calling. The network signals are coded in the companded pulse code modulation (PCM) format used for digital transmission systems, and the network fabric is constructed entirely from standard memory and logic ICs. (For additional information on electronic switching, see *Switching Technology (1925-1975)*, a companion volume in this series.)

P-n-p-n IC technology evolved as well during the 1960s and early 1970s. Air-isolated and dielectrically isolated p-n-p-n arrays were developed for low-voltage applications, and the latter technology was extended to give devices with a breakdown voltage of about 300 V. A junction-isolated, and therefore lower-cost, structure was developed that used features of both the CDI and CBIC structures described earlier; it was intended for use in advanced electronic key systems.

## VI. METAL-OXIDE-SEMICONDUCTOR (MOS) INTEGRATED CIRCUITS

### 6.1 MOS Device Structure

Two papers presented at the Solid State Device Research Conference held at the Carnegie Institute of Technology (later Carnegie-Mellon University) in 1960 provided the departure points for two key developments in IC technology. One, as mentioned earlier,[11] described and demonstrated the importance of epitaxial growth in device technology, and led to the development of the SBC technology. Developments in bipolar technology were then quite rapid, and the IC industry, based on this technology, was already flourishing by the mid-1960s. The other paper, by Kahng and Atalla, described the first MOS device, an insulated gate field-effect transistor (IGFET).[39] These devices are self-isolating and therefore need no separate isolation diffusion, thus making possible ICs of high functional density. The basic MOS transistor structure is shown in Chapter 1, Fig. 1-31. Because the MOS transistor is self-isolating, it is considerably smaller than a bipolar device laid out to the same lithographic design rules. An early MOS IC was reported by S. R. Hofstein and F. P. Heiman at RCA,[40] but exploitation was slower than it was for bipolar ICs, partly because the devices themselves were newer and presented new design challenges, but primarily because the available technology could not provide high yields of electrically stable devices.

M. Yamin and F. L. Worthing found ionic conduction in $SiO_2$ layers at temperatures as low as 150 degrees C.[41] Their finding demonstrated that some ionized impurity was moving in $SiO_2$ under the influence of electric fields to form a layer of fixed charge at the $Si$-$SiO_2$ interface that modifies the electrical properties of field-effect devices uncontrollably. D. R. Kerr, J. S. Logan, P. J. Burkhardt, and W. A. Pliskin[42] of IBM demonstrated that this unidentified impurity could be gettered by phosphorus glass, building on the earlier work of A. Goetzberger and Shockley, who used phosphorus glass for gettering metal precipitates from silicon to produce junctions of high breakdown voltage.[43] Finally in 1965, E. Snow, A. S. Grove, B. E. Deal, and C. T. Sah of Fairchild identified the ions as alkali ions, mainly sodium ions.[44]

Thus, the introduction of a phosphorus glass getter into the MOS process in the form of a thin layer over the regular oxide offered an approach that provided the needed MOS stability.

### 6.2 Dual Dielectric MOS Integrated Circuits

At Bell Laboratories, stable MOS transistors were produced by extending the BLSJ technology to these structures. Although the direct BLSJ structure using SiN as the sealing insulator over $SiO_2$ did not produce stable MOS devices, stable unencapsulated devices were created by H. E. Nigh, J. Stach, and R. M. Jacobs, using a dual-dielectric structure with $Al_2O_3$ in place of SiN under the gate electrode, with the entire device structure coated with silicone.[45] The dual-dielectric structure gave these devices a low threshold voltage—approximately 1 V—which allowed MOS devices to interface readily with TTL logic. This is not the case for the higher threshold voltage of single-dielectric metal gate MOS devices.

In 1967, development effort began on dual-dielectric MOS ICs for specific applications. The high functional density made possible with this technology was applied to the design of ICs for use in the 2ESS electronic switch peripheral decoder, a function that was being handled by bipolar SSI chips. This design, completed in 1969, replaced 11 SSI chips with a single chip containing 120 logic gates. It showed the potential for savings in system cost by integrating more functions on silicon with custom IC designs. It also demonstrated the compatibility of the low threshold voltage MOS technology with a bipolar device environment.

Shortly thereafter, development was completed for a family of MOS devices for the subscriber loop multiplexer system. This application represented the first use of LSI circuits in remote plant equipment. And soon thereafter, in 1972, designs for custom logic, rotary dial conversion, and memory chips for use in the TOUCH-A-MATIC* telephone were completed, making possible the introduction of LSI into subscriber equipment, and providing for the first time the required functions at a marketable cost.

In 1970 and 1971, the advent of a bipolar output stage on an MOS circuit and the depletion load, a current source-like load element, provided another major increase in the potential applications for MOS ICs. The combination provided higher speed, simplification of power requirements, and improved compatibility with bipolar circuits. These circuits were first used in the TSPS to provide decoding storage and drive functions for operator's console light-emitting diode (LED) displays, and other functions requiring compatibility with diode-transistor logic and TTL circuits. Depletion load designs were included in a 1974 LED driver used in the DIMENSION PBX and many other dual dielectric codes.

### 6.3 Silicon Gate MOS ICs and Semiconductor Memories

While exploring alternative MOS technologies, R. E. Kerwin, D. L. Klein, and J. C. Sarace in 1966 invented the self-aligned silicon gate MOS structure, in which the gate electrode is formed by deposited polycrystalline silicon.[46] [Fig. 2-11] In this system, a low threshold is possible even with a single $SiO_2$ dielectric. Since the silicon gate withstands high-temperature processing, it can be formed at an intermediate stage of device fabrication and can be used as a masking material to define source and drain regions, giving self-alignment of these regions and resulting in more compact devices. Futhermore, the deposited polysilicon can readily be oxidized, so that the normal metalization can be routed over the polysilicon. Since polysilicon has useful conductive properties, it can be used for device interconnections under the metal to provide additional interconnection capabilities. Finally, because of its high-temperature tolerance, a gettering phosphorous glass layer can be applied over the gate structure, assuring stability, while resulting in a low threshold voltage with a single dielectric gate insulator.

The full capability of this innovation was recognized only as the silicon gate LSI story unfolded; all active Bell System designs in the second half of the 1960s continued to rely on the established dual dielectric MOS technology, which offered all necessary characteristics.

However, F. Faggin and T. Klein at Fairchild developed silicon gate technology in the latter half of the 1960s,[47] and soon thereafter, the founders

---

* Registered trademark of AT&T Co.

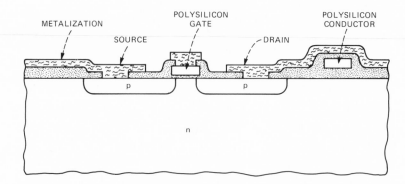

Fig. 2-11. The self-aligned silicon gate structure, which offers size reductions (compared to structures that are not self-aligned), since source and drain are automatically aligned to the polysilicon gate. Furthermore, polysilicon can be used as conductors, resulting in higher interconnection density.

of the Intel Corporation designed the first commercially successful 1024-bit (1K) memory in that technology. In 1971, this memory, the Intel 1103, established semiconductor memory as a serious competitor for the then prevailing magnetic core memory technology (see Chapter 5) and led to an industry-wide development of semiconductor memory, with the storage capacity per chip approximately doubling each year.

Before these developments, for Bell System applications, it was recognized in the mid-1960s that semiconductor ICs in the form of shift registers and small random access arrays would be faster and less costly than magnetic memory devices of similar performance.

The first Bell System application of a regular semiconductor memory was in the time- and space-division network of the 4ESS switch. This memory initially exploited the dual dielectric MOS technology in combination with thin-film interconnection technology to provide high speed not readily available in magnetic core technology. The first installation of this memory was in the 4ESS switching equipment in the Chicago 7 office.

By 1975, silicon gate technology was implemented in manufacture by Western Electric. The 4ESS electronic switch memory was redesigned into that technology, and all new system designs were then based on semiconductor memory. [Fig. 2-12] (Since then the level of integration has continued to evolve and Western Electric—now AT&T Technologies, Inc.—became in 1983 the first U.S. company to manufacture 256K memories.)

Semiconductor memories are generally volatile, in the sense that if the power supply is turned off or fails, the information is lost. Kahng and S. M. Sze introduced the concept of information storage in a floating gate, buried in the insulator near the insulator-semiconductor interface.[48] Charge placed on the floating gate in the oxide will stay there for many years unless deliberately erased. This can be used to control the threshold of

Fig. 2-12. Photos of a 64-bit (left, 1969) and a 4096-bit (right, 1975) memory circuit at the same scale. In the six years between the two designs, functional density increased by more than an order of magnitude, because of improvements in photolithography, device fabrication technology, and circuit design.

an MOS device, so that the presence or absence of charge can be detected by interrogating the MOS device. The basic concept was used in a number of different ways both with and without the floating gate by workers from various companies.

### 6.4 Complementary MOS ICs

While dynamic memories are typically implemented in a single-channel MOS technology, for static memories and for logic, complementary MOS (CMOS) technology further increased the scope of MOS applications by providing reduced power consumption over single-channel depletion load devices. In this structure, n- and p-channel devices, coupled together in a complementary configuration on the same chip, give a virtually ideal switch in which the individual transistors conduct current only during transitions.[49] This technology keeps power dissipation to a minimum, and submicrowatt standby power becomes possible. Forming both types of devices on the same chip called for more complex processing. Isolating n- from p-channel devices requires a p-tub to be formed in the n-type substrate with careful control of the impurity concentrations, another natural application for the ion implantation technology. As the level of integration increased, the low-power capability of CMOS gained in importance. The technology is particularly valuable for customer equipment where only line power is available.

A major application of CMOS technology in the mid-1970s was in the development of an 8-bit microprocessor designed to meet Bell System

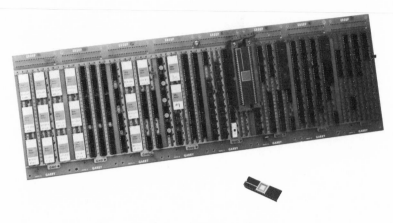

|— 5 INCHES —|

Fig. 2-13.   Photo of a WE 8000 microprocessor in 40-pin package. The accompanying breadboard of SSI and MSI simulated one-half the functional complexity of the WE 8000 processor.

needs. This device, the WE* 8000 microprocessor, has about 7000 transistors and is comparable in sophistication to a typical 1960-era central processor. It can execute a repertoire of over 400 instructions at a rate of 200,000 per second, and depending on the complexity of a given system, can work with a memory of up to one-half million bits. The complete microprocessor fits in a 40-pin package, shown in Fig. 2-13 alongside a breadboard of SSI/MSI packages set up to simulate just half of the function of the WE 8000 microprocessor.

### 6.5 Charge-Coupled Devices

As noted above in section III, a completely new concept in information storage, the charge-coupled device or CCD, was reported by W. S. Boyle and G. E. Smith in 1970.[50] While the device is implemented in single-channel MOS technology, no transistor action is involved. Instead, information is stored at the $Si$-$SiO_2$ interface under surface electrodes. [Fig. 2-14] When the surface electrode has a potential applied to it that can cause inversion, inversion can occur only if minority carriers are available. These can be supplied either by normal generation processes, which typically take hundreds of milliseconds or more at room temperature, or by injection from some external source, which can be made to happen in well under a microsecond. Charge injected into one such layer can be transferred to adjacent devices by overlapping their fringing fields. Logical

---

* Trademark of AT&T Technologies, Inc.

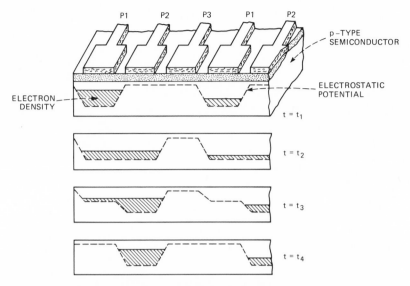

Fig. 2-14. CCD structure and potential diagram. The dashed lines indicate the variation in potential at the semiconductor surface underlying the three series of electrodes (P1, P2, P3). The hatched regions represent minority electron carrier densities. Minority carriers are transferred from beneath P1 to beneath P2 by first increasing the voltage on P2 and then reducing it on P1.

Fig. 2-15. W. S. Boyle (left) and G. E. Smith (right) demonstrating a TV camera using a CCD as a solid-state replacement for a conventional camera's vacuum tube.

1s and 0s, represented by the presence or absence of minority carriers, can therefore be transferred from device to device, daisy chain fashion, at megabit rates. The charge can be injected either electrically or by illuminating the device to accelerate the generation of minority carriers. The former case gives a high-density serial memory device, the latter an imaging array.

CCDs have many important applications. The first of these to be actively developed was in image sensing. A PICTUREPHONE visual telephone service imaging device was realized early in 1973, and in mid-1975 a charge-coupled image sensor was developed that provided black and white pictures in the standard 525-line television format. Figure 2-15 shows the inventors demonstrating a TV camera using a CCD sensor.

## VII. PHOTOLITHOGRAPHY, MASK MAKING, AND COMPUTER-AIDED DESIGN

One major aspect of technology fundamental to all forms of ICs, but touched on only briefly so far, is that of photolithography and mask making. As pointed out above, the pioneering work in photolithography was by Andrus in the mid-1950s,[7] and his basic processes are still in use today in all IC technologies. Major advances had come about, however, in all aspects of device photolithography, including photoresists, wet chemistry, etching processes, materials, exposure systems, and the generation of patterns (artwork) for masks, as well as in the fabrication of the masks themselves. Initially, a draftsperson prepared the pattern for a complete slice of silicon by hand, one drawing for each mask level needed. Since many devices were made on a single slice, and since they all used the same pattern, the original process was soon replaced by one in which the artwork for a single device or cell was prepared by the draftsperson, photographed, and stepped and repeated until a full array was completed. The step-and-repeat process was first used in Bell Laboratories about 1960. As performance requirements increased and areas got smaller and tolerances tighter, the hand-drawn pattern was replaced by the cutting of a large, dimensionally stable sheet of colored plastic called rubylith on a precise machine called a coordinatograph. This sheet was then photographed, reduced, and stepped and repeated to form the mask.

Mask shops were set up at several Bell Laboratories locations, a major one being established at Allentown in 1965. With the establishment of these shops, the process of hand cutting rubylith was soon replaced by one using an automatic coordinatograph, controlled by computer, as the need for the complex masks of ICs increased. This machine reduced artwork generation from about one week to about three hours for a typical IC mask level of the day and reduced errors by about a factor of 100. By

1967, however, it was clear that the automatic coordinatograph would not be able to handle the complex IC masks of the future and, as a consequence, development of a new system was undertaken. The heart of the new system was the primary pattern generator (PPG), which uses a computer-controlled, raster-scanned, modulated laser beam to expose an 8- by 10-inch photographic glass plate. The pattern, consisting of 35-$\mu$m dots in a 26,000-by-32,000 array, is formed in about 12 minutes, independent of the complexity of the pattern. This development removed major bottlenecks in the mask-making operation. PPG development was carried out in departments under J. W. West and K. M. Poole at the Murray Hill location of Bell Laboratories.[51]

A companion development, a step-and-repeat camera using air bearings to float the photographic plate and to maintain precise focus, was developed at the Allentown location of Bell Laboratories in a department under F. L. Howland.[51] Plates produced by the PPG were photoreduced twice: first to form a 10X reticle and then in the step-and-repeat camera to make the actual masks. Limitations in the optical quality of available photographic emulsions led to the development by M. Feldman and J. W. West of the laser reticle generator (LRG).[52] The LRG produced reticles in a one-step process by laser machining iron oxide films on appropriate substrates.

As good as these systems were, the ever-decreasing line width required for ICs soon approached their fine-line capability because of the use of visible light to generate the masks. To overcome this limitation, an electron beam exposure system (EBES), initially writing at 0.5-$\mu$m addresses, was developed at Bell Laboratories by D. R. Herriott, R. J. Collier, D. S. Alles, J. W. Stafford, and others.[53] By 1975, EBES was phased into the fabrication of mask masters at Bell Laboratories.

The increasing quantities of masks, as well as the decreasing permissible numbers and sizes of defects, led to the development of an automatic mask inspection system (AMIS) by J. H. Bruning, M. Feldman, T. S. Kinsel, E. K. Sittig, and R. L. Townsend.[54] AMIS used a laser flying spot scanner to compare adjacent chips on a mask, and detected differences between these chips corresponding to defect sizes down to 2 $\mu$m.

A second aspect of technology common to the design of all ICs is the use of computers as a design aid tool. Early SICs, consisting of a few resistors and transistors on a single chip, were relatively easy to design and fabricate. Circuit design of minimal complexity was based on simple models for single transistors derived at Bell Laboratories during the early days of the transistor era (such as the Ebers-Moll representation[55]), while layout of the ICs was largely manual, as indicated above. Since LSI circuit chips of the mid-1970s already consisted of 10,000 or more interconnected devices, the design of such circuits would have been impractical without an extensive system of machine aids. Such systems evolved at Bell Laboratories in a timely manner so that, from the early days of IC design, computers were used extensively in circuit analysis (both linear and digital), logic simulation, artwork and electrical test program generation, electrical testing, and test data analysis.

Although large mainframe computers were used initially to perform these functions, stand-alone systems, which were based on minicomputers and possessed interactive graphical capability, were soon developed under the leadership of D. Katz in the early 1970s.[56,57] These systems were augmented by software that enabled most of the required design functions to be performed.[58]

As the complexity and the performance of the circuits increased, significant improvements in device models were required. A major breakthrough was made by H. K. Gummel with the development of an efficient approach to one-dimensional computer modeling of transistors[59] and, jointly with H. C. Poon, with the development of compact transistor models usable in IC analysis.[60,61]

Other programs relating to complex circuit analysis and design were developed in various Bell Laboratories organizations, too many to be noted here. Outside commercial and educational organizations also contributed in this area.

For the transmittal of graphic design information between different organizations, a universal graphic computer language, XYMASK, was developed by S. Pardee and B. R. Fowler for describing IC or printed wiring board artwork and mask information.[62]

Computer-aided design plays an increasingly important role in the development of SICs. Appropriately enough, computers are becoming smaller and more powerful because of improved SICs. Therefore they will be still more useful, indeed essential, in coping with the growing complexity of future SIC designs.

## VIII. CONCLUSION

In this chapter, we have attempted to tell the story of technology development, device structure innovation, and applications, intertwining them frequently, because the way that developments actually came about was through continuous interaction between technology developer, device designer, and systems user. The story is, of course, a continuing one. Even at the time of publication of this book, it was still nowhere near its conclusion. It is clear that SICs have made and are continuing to make a major impact on telecommunications. By 1975, two Western Electric locations were already producing SICs of over 400 designs used across the board in switching, transmission, and customer equipment, in applications ranging from telephone sets to 4ESS switching equipment. Subsequently the total number of applications continued to grow at a rapid rate. Many new communication services became possible for the customer, and the impact on telecommunications and society in general was profound indeed. As this narrative makes clear, advances in SICs do not occur through radical departures from current technology but rather through building on the existing base, much of which, as this chapter has described, was developed at Bell Laboratories.

## REFERENCES

1. J. S. Kilby, "Invention of the Integrated Circuit," *IEEE Trans. Electron Dev.* **ED-23** (July 1976), pp. 648-654.
2. R. N. Noyce, U.S. Patent No. 2,981,877; filed July 30, 1959; issued April 25, 1961.
3. M. F. Wolff, "The Genesis of the Integrated Circuit," *IEEE Spectrum* **13** (August 1976), pp. 44-53.
4. C. S. Fuller and J. A. Ditzenberger, "Diffusion of Boron and Phosphorus into Silicon," *J. Appl. Phys.* **25** (November 1954), pp. 1439-1440.
5. C. S. Fuller and J. A. Ditzenberger, "Diffusion of Donor and Acceptor Elements in Silicon," *J. Appl. Phys.* **27** (May 1956), pp. 544-553.
6. C. J. Frosch and L. Derick, "Surface Protection and Selective Masking during Diffusion in Silicon," *J. Electrochem. Soc.* **104** (September 1957), pp. 547-552.
7. J. Andrus, U.S. Patent No. 3,122,817; filed August 15, 1957; issued March 3, 1964.
8. I. M. Ross, L. A. D'Asaro, and H. H. Loar, paper presented at Solid State Device Research Conference, Lafayette, Indiana (June 1956).
9. S. L. Matlow and E. L. Ralph, U.S. Patent No. 2,984,775; filed July 9, 1958; issued May 16, 1961.
10. G. E. Moore and R. N. Noyce, U.S. Patent No. 3,108,359; filed June 30, 1959; issued October 29, 1963.
11. H. C. Theuerer, J. J. Kleimack, H. H. Loar, and H. Christensen, "Epitaxial Diffused Transistors," *Proc. IRE* **48** (September 1960), pp. 1462-1463.
12. J. A. Hoerni, "Application of the Free-Electron Theory to Three Dimensional Networks," paper presented at Professional Group on Electron Devices Meeting, Washington, D.C. (October 1960).
13. O. L. Anderson, H. Christensen, and P. Andreatch, "Technique for Connecting Electrical Leads to Semiconductors," *J. Appl. Phys.* **28** (August 1957), p. 923.
14. B. T. Murphy, U.S. Patent No. 3,321,340; filed October 20, 1961; issued May 23, 1967.
15. J. R. Biard, U.S. Patent No. 3,463,975; filed December 31, 1964; issued August 26, 1969.
16. D. Kahng, "Conduction Properties of the Au-*n*-TYPE-Si Schottky Barrier," *Solid-State Electron.* **6** (May/June 1963), pp. 281-295.
17. D. Kahng and L. A. D'Asaro, "Gold-Epitaxial Silicon High-Frequency Diodes," *Bell Syst. Tech. J.* **43** (January 1964), pp. 225-232.
18. D. Kahng and M. P. Lepselter, "Planar Epitaxial Silicon Schottky Barrier Diodes," *Bell Syst. Tech. J.* **44** (September 1965), pp. 1525-1528.
19. E. R. Chenette, J. J. Kleimack, R. Edwards, and R. A. Pedersen, "Integrated Schottky-Diode Clamp for Transistor Storage Time Control," *Proc. IEEE* **56** (February 1968), pp. 232-233.
20. M. P. Lepselter, "Beam-Lead Technology," *Bell Syst. Tech. J.* **45** (February 1966), pp. 233-253.
21. E. Snow, A. S. Grove, B. E. Deal, and C. T. Sah, "Ion Transport Phenomena in Insulating Films," *J. Appl. Phys.* **36** (May 1965), pp. 1664-1673.
22. J. V. Dalton and J. Drobek, "Structure and Sodium Migration in Silicon Nitride Films," *J. Electrochem. Soc.* **115** (August 1968), pp. 865-868.
23. G. H. Schneer, W. van Gelder, V. E. Hauser, and P. F. Schmidt, "A Metal-Insulator-Silicon Junction Seal," *IEEE Trans. Electron Dev.* **ED-15** (May 1968), pp. 290-293.

24. J. S. Mayo and M. G. Stickler, "Integrated Circuits for a Digital Transmission System," *Bell Lab. Rec.* **44** (October/November 1966), pp. 324-327.

25. B. T. Murphy, V. J. Glinski, P. A. Gary, and R. A. Pedersen, "Collector Diffusion Isolated Integrated Circuits," *Proc. IEEE* **57** (September 1969), pp. 1523-1527.

26. K. Hart and A. Slob, "Integrated Injection Logic: A New Approach to LSI," *IEEE J. Solid-State Circuits* **SC-7** (October 1972), pp. 346-351; see also C. M. Hart and A. Slob, "Integrated Injection Logic—A New Approach to LSI," *IEEE Int. Solid-State Circuits Conf. Digest of Tech. Papers*, Philadelphia, Pennsylvania (February 16-18, 1972), pp. 92-93.

27. H. H. Berger and S. K. Wiedmann, "Merged-Transistor Logic (MTL)—A Low-Cost Bipolar Logic Concept," *IEEE J. Solid-State Circuits* **SC-7** (October 1972), pp. 340-346; see also H. H. Berger and S. K. Wiedman, "Merged Transistor Logic—A Low-Cost Bipolar Logic Concept," *IEEE Int. Solid-State Circuits Conf. Digest of Tech. Papers*, Philadelphia, Pennsylvania (February 16-18, 1972), pp. 90-91.

28. R. S. Ohl, U.S. Patent No. 2,750,541; filed January 31, 1950; issued June 12, 1956.

29. W. Shockley, U.S. Patent No. 2,787,564; filed October 28, 1954; issued April 2, 1957.

30. A. U. MacRae, "Recent Advances in Ion Implanted Junction-Device Technology," in *Ion Implantation in Semiconductors*, ed. I. Ruge and J. Graul (Berlin: Springer-Verlag, 1971), pp. 329-334.

31. W. M. Gibson, C. Erginsoy, H. E. Wegner, and B. R. Appleton, "Direction and Energy Distribution of Charged Particles Transmitted through Single Crystals," *Phys. Rev. Lett.* **15** (August 1965), pp. 357-360.

32. H. M. B. Bird and B. Weissman, "The Development of Ion Implantation Systems," *W. Elec. Eng.* **25** (Fall 1981), pp. 32-45.

33. P. T. Panousis and R. L. Pritchett, "GIMIC-O—A Low Cost Non-Epitaxial Bipolar LSI Technology Suitable for Application to TTL Circuits," *IEEE Int. Electron Dev. Meeting Tech. Digest*, Washington, D. C. (December 9-11, 1974), pp. 515-518.

34. K. H. Olson and F. D. Waldhauer, "Development of a Low Level Amplifier with a 700 MHz Bandwidth," paper presented at IEEE International Electron Devices Meeting, Washington, D. C. (October 11-13, 1971), paper 13.1.

35. W. H. Eckton, Jr. and T. E. O'Shea, "Development of an ECL Gate with a 300 ps Propagation Delay," paper presented at IEEE International Electron Devices Meeting, Washington, D. C. (October 11-13, 1971), paper 13.2.

36. M. P. Lepselter, H. A. Waggener, and R. E. Davis, "Beam-Leaded and Intraconnected Integrated Circuits," paper presented at Electron Devices Conference, Washington, D. C. (October 29-31, 1964).

37. P. C. Davis, V. R. Sarri, and S. F. Moyer, "High Slew Rate Monolithic Operational Amplifier Using Compatible Complementary PNP's," *IEEE J. Solid-State Circuits* **SC-9** (December 1974), pp. 340-347.

38. J. L. Moll, M. Tanenbaum, J. M. Goldey, and N. Holonyak, "P-N-P-N Transistor Switches," *Proc. IRE* **44** (September 1956), pp. 1174-1182.

39. D. Kahng and M. M. Atalla, "Silicon-Silicon Dioxide Field Induced Surface Devices," paper presented at the Solid State Device Research Conference, Pittsburgh, Pennsylvania (June 1960).

40. S. R. Hofstein and F. P. Heiman, "The Silicon Insulated-Gate Field-Effect Transistor," *Proc. IEEE* **51** (September 1963), pp. 1190-1202.

41. M. Yamin and F. L. Worthing, "Charge Storage and Dielectric Properties of $SiO_2$ Films," *Extended Abstracts, Electron. Div., Electrochem. Soc. Meeting* **13**, Toronto, Canada (May 3-7, 1964), pp. 182-184.

42. D. R. Kerr, J. S. Logan, P. J. Burkhardt, and W. A. Pliskin, "Stablization of $SiO_2$ Passivation Layers with $P_2O_5$," *IBM J. Res. Develop.* **8** (September 1964), pp. 376-384.

43. A. Goetzberger and W. Shockley, "Metal Precipitates in Silicon *p-n* Junctions," *J. Appl. Phys.* **31** (October 1960), pp. 1821-1824.

44. E. Snow, A. S. Grove, B. E. Deal, and C. T. Sah, "Ion Transport Phenomena in Insulating Films," *J. Appl. Phys.* **36** (May 1965), pp. 1664-1673.

45. H. E. Nigh, J. Stach, and R. M. Jacobs, "A Sealed Gate IGFET," *IEEE Trans. Electron Dev.* **ED-14** (September 1967), p. 631.

46. R. E. Kerwin, D. L. Klein, and J. C. Sarace, U.S. Patent No. 3,475,234; filed March 27, 1967; issued October 28, 1969.

47. F. Faggin and T. Klein, "Silicon Gate Technology," *Solid State Electron.* **13** (August 1970), pp. 1125-1144.

48. D. Kahng and S. M. Sze, "A Floating Gate and Its Application to Memory Devices," *Bell Syst. Tech. J.* **46** (July/August 1967), pp. 1288-1300.

49. F. M. Wanlass and C. T. Sah, "Nanowatt Logic Using Field-Effect Metal-Oxide Semiconductor Triodes," *Int. Solid-State Circuits Conf. Digest of Tech. Papers* (February 1963), pp. 32-33.

50. W. S. Boyle and G. E. Smith, "Charge Coupled Semiconductor Devices," *Bell Syst. Tech. J.* **49** (April 1970), pp. 587-593.

51. See special issue on Device Photolithography, *Bell Syst. Tech. J.* **49** (November 1970), pp. 1995-2220.

52. J. M. Moran and R. L. Ruth, "Meeting New Mask-Making Needs," *Bell Lab. Rec.* **57** (July/August 1979), pp. 192-198.

53. D. R. Herriott, R. J. Collier, D. S. Alles, and J. W. Stafford, "EBES, A Practical Electron Lithographic System," *IEEE Trans. Electron Dev.* **ED-22** (July 1975), pp. 385-392.

54. J. H. Bruning, M. Feldman, T. S. Kinsel, E. K. Sittig, and R. L. Townsend, "An Automated Mask Inspection System—AMIS," *IEEE Trans. Electron Dev.* **ED-22** (July 1975), pp. 487-495.

55. J. J. Ebers and J. L. Moll, "Large-Signal Behavior of Junction Transistors," *Proc. IRE* **42** (December 1954), pp. 1761-1772.

56. D. Katz and S. Pardee, "Programs and Machines in Design and Layout of Integrated Circuits," *IEEE Int. Convention Digest*, New York (March 22-25, 1971), pp. 634-635.

57. D. Katz, "Logic Gate Arrays: 'Catalog' ICs with Custom Advantages," *Bell Lab. Rec.* **55** (July/August 1977), pp. 187-191.

58. B. R. Chawla, H. K. Gummel, and P. Kozak, "MOTIS—An MOS Timing Simulator," *IEEE Trans. Circuits Syst.* **CAS-22** (December 1975), pp. 901-910.

    G. Persky, D. N. Deutsch, and D. G. Schweikert, "LTX—A Minicomputer-Based System for Automated LSI Layout," *J. Design Autom. Fault-Tolerant Comput.* **1** (May 1977), pp. 217-255.

    D. N. Deutsch, "A Dogleg Channel Router," *Proc. 13th Design Automation Conf.*, San Francisco, California (June 1976), pp. 425-533.

59. H. K. Gummel, "A Self-Consistent Iterative Scheme for One-Dimensional Steady State Transistor Calculations," *IEEE Trans. Electron Dev.* **ED-11** (October 1964), pp. 455-465.

60. H. K. Gummel and H. C. Poon, "A Compact Bipolar Transistor Model," *IEEE Int. Solid-State Circuits Conf. Digest of Tech. Papers*, Philadelphia, Pennsylvania (February 18-20, 1970), pp. 78-79.

61. H. K. Gummel and H. C. Poon, "An Integral Charge Control Model of Bipolar Transistors," *Bell Syst. Tech. J.* **49** (May/June 1970), pp. 827-852.

62. B. R. Fowler, "XYMASK," *Bell Lab. Rec.* **47** (July 1969), pp. 204-209.

# Chapter 3

# Electron Tubes

*Early fundamental work on high-vacuum tubes, which transformed the electron tube into a practical device, soon led to increased dependence on tubes in telecommunications systems and to great emphasis on higher frequencies, greater bandwidths, and enhanced reliability. In addition to extensive use in carrier transmission systems, another major use of tubes was in controlling the ringing voltage of multiparty telephones. During World War II, Bell Laboratories and Western Electric tube development and manufacture provided major support to the military effort. Early tubes for radar preamplifiers and amplifiers were designed by Bell Laboratories engineers. Reflex and magnetron oscillators also provided the key technology for radar systems. Significant magnetron advances included tunable frequencies and size and weight reductions. In the postwar period, much attention was paid to a better understanding of the physics of cathode structures, culminating in the coated powder cathode. Klystron, microwave triode, and traveling wave tube (TWT) performance was enhanced in this period, and the Telstar project capitalized on TWT advantages in one of the most dramatic communications experiments of the 1960s. Vastly increased reliability was achieved in electron tubes for submarine cable systems; the first long undersea cable system operated for 22 years under the Atlantic Ocean without a tube-related failure. Among the last electron tubes developed by Bell Laboratories were a camera tube used in the experimental PICTURE-PHONE\* visual telephone service and memory tubes used in major experiments in electronic telephone switching.*

## I. EARLY WORK ON ELECTRON TUBES

As was briefly discussed in Chapter 4, section 4.2.2 of another volume of this series subtitled *The Early Years (1875-1925),* the application of the vacuum tube to telephony began with H. D. Arnold's recognition of the enormous potential of the crude three-electrode tube (triode) demonstrated

---

Principal authors: V. Rutter, L. Von Ohlsen, and W. Van Haste

133

to the Bell System by L. De Forest on November 1, 1912, and with the subsequent decision to purchase patent rights for the Bell System. Less than three years later, a greatly improved triode was the active device in amplifiers strung at intervals across the country, which inaugurated transcontinental telephony in January 1915. Arnold contributed greatly to this remarkable progress in vacuum tube performance by his leadership in three major types of studies: vacuum technology, cathode emission, and circuit properties.

*Vacuum technology*. Arnold showed that the erratic behavior of De Forest's 1912 triode was caused by poor vacuum; using better exhaust techniques, he improved conditions inside the tube envelope, with the result that triodes of a given design behaved much more reproducibly and reliably. With these tubes, practical repeaters could be designed, good enough to make possible transcontinental telephony.

*(Oxide) cathode emission*. Arnold studied the effect, observed by A. Wehnelt in Germany, that coating the cathode metal with alkaline earth oxides greatly increased electron emission and, with his associates, developed the combined filament using Wehnelt's discovery. This filament employed a core composed of 90- to 95-percent platinum into which 5 or 10 percent of another metal, usually iridium, nickel, and/or cobalt, was alloyed. The alkaline earth oxides were chemically combined with this core by heating to incandescence in air during the coating process. The resulting filament was, for about 12 to 15 years, the most efficient, reproducible, and rugged electron emitter known. From about 1926 to 1937, the combined filament was largely replaced in the Bell System by the uncombined type, in which the alkaline earth carbonates were applied to a base of nearly pure nickel, the carbonates being converted to oxides at an appropriate portion of the exhaust process. This change provided an uncombined coating on unipotential cathodes with nickel bases, which was less expensive than a combined coating on platinum. With improvements made in obtaining good vacuum, the more sensitive uncombined coating could be used in a wide variety of devices. (For a further discussion of cathode improvement, see section 4.1.)

*Circuit properties*. The behavior of the triode as a circuit element was advanced greatly by Arnold and his Bell System colleagues. This advance was based on a better understanding of the internal electrode geometry of the triode as well as of the interaction of the triode with the remainder of the circuit in which it was used.

By 1926, 15 codes of pre-1925 tubes were in manufacture at the Bell Laboratories tube shop at 395 Hudson Street in New York City. These included water-cooled power tubes for broadcasting, a small "peanut" tube for receiving (the 215A), small power tubes for public address amplifiers, and tubes for telephone repeater use. Of these last types, which were used in large numbers, the 101D was most outstanding. Figure 3-1

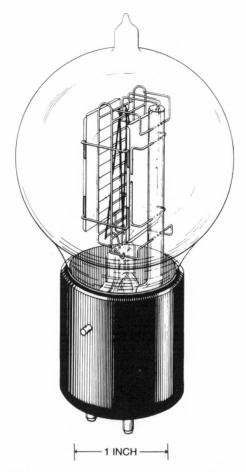

Fig. 3-1. The 101D high-vacuum triode. By 1926, when the tube had been in use for 12 years, it was the most reliable and long-lived tube ever produced. It was used in telephone repeaters.

shows its structure. It was a high-vacuum triode using a combined filament consuming 4.5 watts (W). By 1926, it had been in use for 12 years and was at that time the most reliable and long-lived tube produced anywhere. It was singled out for special comment in a 1926 *Bell Laboratories Record* article by M. J. Kelly, who was later to become president of Bell Laboratories.[1] After 1925, the 101D received many improvements in both its mechanical and filament designs. One design lowered the filament current to 1 ampere (A)—a one-third reduction from the earlier 101D. The tube was characterized by an amplification factor of 6, a plate resistance of 5700 ohms, and a plate current of 7 milliamperes (mA).

The period from 1925 to 1930 saw many accomplishments in the vacuum tube laboratory. Simultaneously, radio ("wireless") systems were being studied both at long and short wavelengths. Commercial broadcasting was expanding. All these required transmitting oscillators and amplifiers, and the Bell Laboratories group was a leader in developing these devices, in addition to developing repeater tubes for the telephone plant. It was in this period that many advanced techniques were established. Included were design and construction of high-vacuum exhaust systems, the double-ended construction techniques used for transmitting devices to limit surface leakage and to provide low-impedance electrode leads, and vastly increased understanding of the factors that affect operation lifetimes. By the end of 1930, tube lives up to 20,000 hours were being obtained with some types, and over a quarter-million tubes of all types were in operation in Bell System circuits.[2,3]

## II. TELEPHONY AND BROADCASTING IN THE 1930'S

In May 1931, the Supreme Court decided that the idea of a *high*-vacuum tube, as compared with a vacuum tube, was not patentable.[4] Thus ended a 16-year controversy between General Electric and Bell Laboratories on the priority of conception of the high-vacuum tube, leaving Bell Laboratories designers with an open field for application of the tube technology. Work on a large variety of tubes was in progress, and the variety continued to expand during the next decade.

### 2.1 High-Power Broadcast Tubes

During this period, much attention was paid to high-frequency applications for radio-telephone circuits and to commercial AM broadcasting, and in both of these areas, power tubes were developed. Further, a considerable amount of work in aircraft, police, and marine radio was in progress, and these areas also would benefit from transmitting-tube development, especially the development of smaller, radiation-cooled devices. Water-cooled tubes continued to be needed for high-power broadcasting and transoceanic service. Figure 3-2 shows one of these devices.

The 265A, a 100-kilowatt (kW) tube developed in the early 1930s, is representative of the high-power devices of the broadcast class. For the increasingly needed low- and medium-power transmitter, which served the broadcast needs of smaller communities and was used for base station aircraft and police transmitters, a significant advance was the 1932 development of three radiation-cooled tubes, whose principal advantage was economy in circuit design. These tubes were the 270A (500 W), 251A (1500 W), and 279A (2000 W). Because the power dissipated in the tube must be lost by radiation, the anodes of these tubes used molybdenum, which ran continuously at "cherry red" temperatures. Also, fins and

|←— 5 INCHES —→|

Fig. 3-2. The 240A water-cooled power tube. Before the development of radiation-cooled tubes, this type was used in high-power broadcasting and transoceanic service.

roughened surfaces increased the heat radiation. Molybdenum grids (carbon coated to reduce grid electron emission), thoriated tungsten filaments, and hard or Nonex*-type glass all aided in providing dependability at high operating temperatures.[5] The technical maturity of these devices was confirmed by the fact that all of these important features were still found in medium-power transmitting tubes manufactured over 50 years later.

In the mid-1930s, Bell Laboratories undertook development of a new line of high-power transmitters at the Whippany, New Jersey facility. The design was based on the Doherty linear amplifier, designed by W. H. Doherty. For a description of the amplifier and a discussion of its effect on broadcast transmitter design, see *The Early Years (1875-1925)*, p. 448.

One of the new tubes, the 298A, a 100-kW double-ended tube, was made available in the mid-1930s. This water-cooled tube had a maximum plate dissipation of 100,000 W and a maximum plate voltage of 20,000 volts (V). The anode was made of copper with an integral copper water jacket. The upper frequency limit of the tube was 20 megahertz (MHz). However, when operating under maximum power ratings, the tube was limited to 4 MHz, since radio-frequency (RF) power dissipation at higher frequencies exceeded the capability of the tube.

A new 50-kW transmitter used two of the 298A tubes. Two transmitters were made by Northern Electric from the Western Electric design and placed in service in Canada in the winter of 1937-8. The first Western Electric transmitter was installed in early 1938 in Louisville, Kentucky with the call letters WHAS.[6] There was also an interest in "super-power" 500-kW transmitters, which continued for the balance of the 1930s with the design of a 500-kW Doherty-type amplifier at the Whippany Laboratory.

Also in the latter half of the 1930s, the Bell Laboratories tube laboratory in New York City was working on a new double-ended water-cooled tube. This tube, later coded 320A, had a peak power capacity of 250 kW, the most powerful sealed-off vacuum tube ever made to that time. [Fig. 3-3]

---

* Trademark of Corning Glass Works.

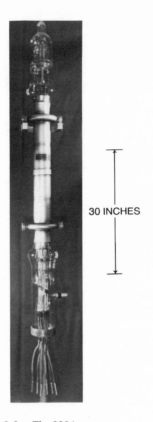

30 INCHES

Fig. 3-3.   The 320A vacuum
tube, a double-ended water-
cooled device. It was the first
tube to use jeweled bearings
or guides. Used in the 1930s
for high-power broadcasting,
the tube was no longer in demand after the FCC decided
to limit broadcast power in
the United States to 50 kW.

The size and power of the 320A presented many processing and engineering problems. When the tube was in operation, the filament strand
increased in length by half an inch. The tube design had to include a
tensioning device to permit expansion and at the same time keep the
filament from sagging or bowing. Attached to one end of the filament
strands were polished tungsten rods mounted in synthetic sapphire guides.
The design eliminated the problems of seizure that resulted when metal
such as molybdenum was used to guide the filament support rods during

the expansion cycle. The grid of the tube was fastened at one end and was allowed to expand in jeweled guides to accommodate changes in dimension produced by its 2000-W dissipation. This was the first use of jeweled bearings or guides in a vacuum tube.

The 320A was a highly successful tube. Eight were installed (for a peak power of 2000 kW) in a 500-kW transmitter built in Mexico by Continental Electronics, an independent firm licensed by Western Electric to use the Doherty circuit. This station, XERA, operated for several years with no tube failures. The market for the 320A disappeared, however, when the Federal Communications Commission decided to limit broadcast power in the United States to 50 kW.

Shortly after World War II, Western Electric stopped manufacture of broadcast transmitters. The Western Electric tube division negotiated an agreement to transfer manufacture of their transmitting tubes to Machlett Laboratories in Connecticut. Thus, the needs of both the Bell System radio-telephone systems and commercial broadcasters were satisfied. Technical assistance from both Bell Laboratories and Western Electric was supplied to Machlett during manufacturing start-up.

## 2.2 The Trend to Higher Frequencies

In the receiving and telephone repeater areas, the concept of a screen grid, inserted between the control grid and anode, was well known by 1927 as a means of reducing interelectrode capacitance and thus increasing amplification at higher carrier frequencies. This tetrode concept and the pentode, with a third grid between screen grid and plate to suppress plate secondary electrons, were emphasized in the early 1930s. The climax of this work was the development, by the end of 1937, of the 310A and 311A amplifier tubes for carrier transmission systems and two companion types, 328A and 329A, designed for a different heater voltage to be used with ballast lamp current regulation.[7] [Figs. 3-4(a) and 3-4(b)] These pentode devices incorporated concepts that were to be applied in small tubes well into the era of solid state: equipotential cathodes with which an ac heater could be used, slotted mica insulators to reduce dc leakages, close electrode spacings, and shielding to reduce magnetic pickup (hum) from the heater. The extent of progress may be appreciated by comparison of the 310A pentode with the 102F filamentary triode, which was at the time the primary voice frequency amplifier in use. (See *The Early Years (1875-1925)*, pp. 844-845.) The 310A delivered a voltage gain of 44 dB, whereas the 102F delivered 26-dB gain, using similar plate voltage but usable only at much lower frequencies. The required power drive on the 310A grid was less than one-third that of the 102F.

During the early 1930s, appreciation of the broadband advantages of communication at VHF (very high frequencies—above 30 MHz) provided

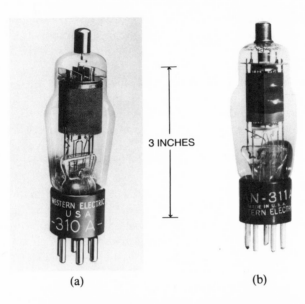

3 INCHES

(a)                                          (b)

Fig. 3-4.   Amplifier tubes for carrier transmission systems.
(a) The 310A pentode offered a gain of 44 dB with less
than one-third the required drive of its predecessor, which
gave only 26 dB of gain. (b) The 311A pentode amplifier
tube. Both designs offered equipotential cathodes, slotted
mica insulators, close electrode spacing, and shielding
from the heater.

a new stimulus toward the understanding and practice of the tube art.
Tubes produced prior to this time would not give usable results at VHF
because: (1) the cathode-to-plate transit time (the time for an electron to
travel across the cathode-plate distance) was an appreciable part of the
period for VHF, and (2) the shunting effects of tube interelectrode capacitance and distributed inductance could eliminate amplification at VHF.
The transit-time problem in conventional tube structures can be reduced
by decreasing the mechanical dimensions and using higher electrode voltages. The second problem required formulation of an electrical model
(equivalent circuit) of a tube. As shown in Fig. 3-5, for a triode tube,
resistance, capacitance, and inductances are present both for each electrode
and between electrodes. Additionally, capacitances exist from each electrode
to the outside world. As frequency of operation is increased, the deleterious
effect of all these is increased. At a sufficiently high frequency, the input
capacitive reactance of the grid decreases so that the input signal flows
through this reactance and gain disappears. Reduction of the series resistance and inductance in each lead calls for large leads of short length. An

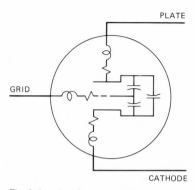

Fig. 3-5. An electrical model for a very high-frequency triode. The shunting effect of grid capacitance, the resistance, and the inductance had to be reduced.

important contribution to understanding these effects in high-frequency grid-controlled tubes was made by F. B. Llewellyn and L. C. Peterson.[8]

One of the first VHF tubes was the 304A, which could be used to amplify or oscillate at frequencies up to 350 MHz.[9] [Fig. 3-6] Note that the general construction of this tube did not depart radically from its contemporaries except that grid and plate leads were heavy and short, and both were brought out through the top of the envelope. The inter-electrode capacitances were not much lower than those of contemporary tubes, but the series lead inductances were; this was the key to the tube's performance at frequencies nearly double the capabilities of others.

The next step in raising frequency occurred with elimination of the tube base, a search for ways to obtain wider spacings among the lead wires, and rearrangement of the control grid into straight wires. Such a tube, shown in Fig. 3-7(a), an experimental device only, could be operated to 740 MHz. Another tube, with further refinement, would deliver power at 1200 MHz. With these same techniques, a double-pentode negative grid amplifier was devised to deliver gain up to 300 MHz.[10] [Fig. 3-7(b)] These and similar tubes were helpful in basic work with microwaves that proved vital during World War II.

The growth of sound motion pictures required high-gain, low-noise amplifiers. Bell Laboratories efforts in this area resulted in understanding the causes of audio frequency hum arising from the heater electric field, from the heater magnetic field, and from the heater-to-output circuit because of interelectrode capacitances or direct leakage. Several tubes, such as the 259B to 262A, made quiet amplifiers that greatly benefited the motion picture industry. These tubes also served to develop methods and techniques that were later useful in the 310A and 311A for telephone repeaters.

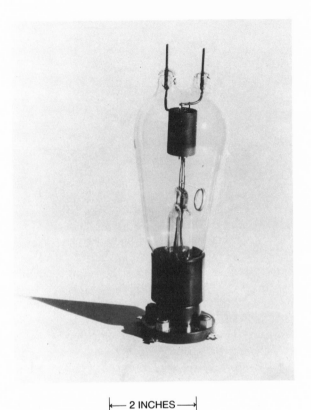

|←— 2 INCHES —→|

Fig. 3-6.   The 304A vacuum tube, designed for frequencies up to 350 MHz. Much lower series lead inductances were the key to doubling the frequency capabilities of this device over those of its predecessors.

## 2.3 Cathode Ray Tubes

Other classes of tubes during this period included cathode ray tubes, phototubes, and gas tubes. In 1922, J. B. Johnson introduced a gas-focused, 300-V cathode ray tube for laboratory use. This tube was later extensively redesigned for production, coded series 224, with three different fluorescent screens, but from a modern perspective, performance was limited and service life was short.[11]

The next developments were an electron gun with electrostatic focusing (tubes coded 325 and 326)[12] and a tube coded 330 with three electron guns,[13] which simultaneously displayed three signal traces on the screen. The 326 and 330 tubes, shown in Fig. 3-8, were superior in performance and reliability to others available at that time, and they found a ready market.

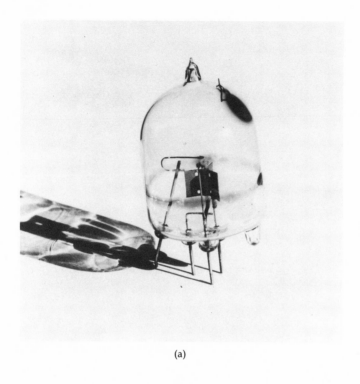

(a)

(b)

|←———— 1 INCH ————→|

Fig. 3-7.   Tubes representing further pre-World War II progress toward achieving high-performance VHF operation: (a) an experimental 740-MHz tube in which the base has been eliminated, and (b) a double-pentode negative grid amplifier that delivered gain at 300 MHz.

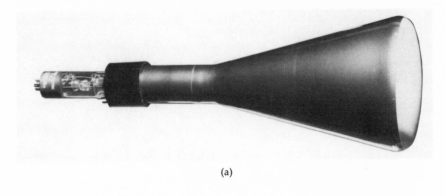

(a)

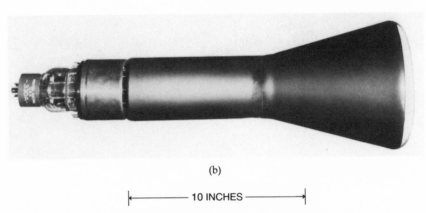

(b)

|←——————— 10 INCHES ———————→|

Fig. 3-8.  Two Western Electric cathode ray tubes of the 1930s: (a) the 326A, which used electrostatic focusing; (b) the 330A, which had three electron guns.

An interesting project of the time was a special cathode ray tube, needed for litigation reasons, that was designed to display transient voltages on telephone lines induced from large power surges on railroad power lines. For a photographic display, both a finely focused spot on the screen and roughly a tenfold increase in electron current was achieved with a condensing lens based on a frustum of a cone that allowed the use of electrons from the entire active area of the cathode.[14]

In the mid-1920s, C. J. Davisson analyzed the focusing of electrons in fields set up in and around the apertured electrodes then in use in electron guns. Later, when the television development people were preparing for TV transmission experiments between New York City and Washington, D. C. (see *The Early Years (1875-1925)*, Chapter 7, section 9.8), Davisson was asked to design and have built a high-quality picture-receiving tube. The resulting tube was extremely long to avoid distorting the raster, and it operated in a high-voltage, low-current mode. [Fig. 3-9] Several such tubes were built and performed excellently in a number of demonstrations.

Another special cathode ray tube was configured in such a way that the electron beam struck the side walls instead of a screen at the end of the tube. By analyzing speech according to energy at various frequencies compared to time, by tracing these quantities on the phosphorescent side walls, and by rotating the tube around its long axis, researchers were able to display "visible speech" spectrogram information continuously instead of as a series of still pictures.[15]

### 2.4 Phototubes

Efforts in the phototube area began in the early to mid-1920s with the need to detect light transmitted through the sound tracks of motion picture film. Initial tubes, the 1A and 2A, using potassium hydride as the photosensitive material, met with indifferent success. Developments based on

Fig. 3-9. C. J. Calbick with the picture tube designed for television demonstrations by C. J. Davisson. The tube was extremely long to avoid raster distortion.

cesium oxide/silver soon led to the superior 3A phototube, produced at the Bell Laboratories Hudson Street tube shop in New York City.[16,17] Other versions, the 5A and 6A, were developed for burglar alarms and for monitoring the recording of sound on film.

This background in phototube work led to a forerunner of the TV camera tube. In the early Bell Laboratories experiments, the subject was illuminated by a rapidly moving, scanning spot of light. In the camera, this illuminated raster was recorded on a large flat rectangular cathode. Photomultiplier tubes, some with as many as nine or ten stages, were also developed, and a special tube for extremely low light levels was built for astronomical observations.

Early in World War II, the British asked for a special, highly secret phototube, which was designed by Bell Laboratories and manufactured by Western Electric, RCA, and others.[18] Later, it was learned that this tube was put in the nose of a surface-to-air projectile. When such a projectile passed into the shadow of an aircraft's wing, the change in light level triggered a circuit that detonated an explosive charge. Postwar reports indicated that this projectile was so effective that it sharply reduced the number of German daylight bombing raids.

## 2.5 Gas Tubes

Hot-cathode gas tubes were used throughout industrial electronics as diode rectifiers in unregulated rectifiers, or with a control electrode, as thyratrons in regulated rectifiers. For various applications in the Bell System, several thyratrons were developed, including a unique all-metal design using mercury vapor. For high-power uses, this tube used a radiator for control of vapor pressure that provided a long life.

The cold-cathode gas tube was another type of electron tube that was given attention at Bell Laboratories and that found many uses in telephony, some leading the way to later device uses. But the cold-cathode tube used a glow discharge for conduction of current rather than the low-voltage discharge supplied by a thermionic cathode, so it could carry tens of milliamperes rather than the amperes of a thyratron. The forms familiar to industry in the 1930s were: (1) the common voltage regulator, a two-electrode device that used the spread of the glow discharge to provide a regulated voltage over a current range, and (2) another two-electrode tube designed for the discharge to provide a visual signal as an indicator lamp.

In the early 1930s, Bell Laboratories added a control electrode. A fairly well-controlled voltage on this electrode of about 70 V triggered the discharge between the main anode and cathode. The tube then became a two-state switch triggered by an isolated signal; it was used widely by Bell Laboratories in Bell System switching and relay circuits. Holding-off voltages were on the order of 150 to 300 V in the main gap while the tube was not operating. When a tube was conducting, the voltage drop across the discharge was about 75 V, providing enough voltage and current to operate a relay, another gas tube, or other devices. Some of the features that made it attractive were (1) the lack of continual power requirements for heating a cathode, particularly in applications where the tube is not operating most of the time; (2) the isolation between the control voltage and the main gap operating voltage; (3) the stability of the firing voltage in the control gap; (4) the small size—similar to a miniature or slightly larger radio tube; and (5) the visible discharge, useful as an indicator. One disadvantage quickly recognized was that the oscillations present in the normal glow discharge interfered with the transmission of voice signals through such a tube.

Many applications were developed in the 1930s. Probably the largest was that for controlling the ringing voltage to one of four parties in multiparty telephones. By 1939, this usage was quoted as using "hundreds of thousands" of tubes,[19] and the application continued into the 1950s, when new designs were developed for mounting in telephone sets, providing extra features for ringing control in eight-party applications. The low power requirements were particularly important where the operating power for ringing a customer's telephone came from the central office. Other applications for general-purpose tubes included message registers, line identifiers, and indicators in various units of office equipment.

Early studies of the cold-cathode discharge provided an understanding of the relationship between current density and useful cathode life, so that highly reliable circuit functions could be obtained.[20]

## 2.6 The Design of Electron Beams

A final development is most significant. In 1940, J. R. Pierce made an outstanding contribution to the design of electron beams for cathode ray, television, klystron, and traveling wave tubes.[21,22]

Pierce's methods covered the design and shaping of auxiliary electrodes, which were connected to the cathode and grids of electron guns, and also gave the proper boundary potentials around the electron stream so that space charge repulsion would, to an excellent approximation, not expand the beam and lose electrons. The resulting beams were of uniform cross-sectional density and were far more efficient than those designed by previous methods. Pierce's methods were universally adopted, and they improved the performance of electron beams so that many additional but less revolutionary improvements became practical. (For more information on this subject, see another volume in this series subtitled *Communications Sciences (1925-1980)*, Chapter 4, section 1.1.)

## III. MILITARY DEVELOPMENT DURING WORLD WAR II

Having established their capabilities in the areas of research, development, and manufacture, Bell Laboratories and Western Electric became a vital part of the tremendous World War II effort. The dimensions of military action created a need for an immense amount of communications equipment, much of it of new and complex design. The experience gained from the study of telephone tubes proved adaptable to military equipment.

One example of Bell Laboratories development capabilities and Western Electric's production experience and capacity to meet special military needs was the "three-day" design of an amplifier tube to be used in a captured German telephone system left behind as General Omar Bradley's forces advanced in Western Europe. In the retreat, the Germans had removed the tubes from the communications equipment to render it useless. However, one tube was found and sent to Bell Laboratories for urgent design effort on a replacement that would be physically and electrically equivalent. The design was completed in three days, information was given to Western Electric engineers, and in three weeks, 1000 tubes operable in the captured equipment were sent overseas to reactivate the telephone system.

Particularly applicable was the experience accumulated over 10 years on high-frequency techniques and on radio equipment for aircraft use. The 1930s had been a decade of steady progress, with ever-lighter-weight multichannel transmitters and receivers. In the area of radar, however, Bell Laboratories had no backlog of experience, especially with respect to tubes that would meet peak or pulse transmitter power requirements of hundreds or thousands of kilowatts. Effort and ingenuity had to substitute for experience as a newly devised tube, the magnetron, was developed. (See section 3.3)

### 3.1 Receiving-Type Tubes

Military requirements for general-purpose tubes are closer to requirements for Bell System use than to those of the entertainment or industrial fields. These are a long life (although the military always had to be ready to sacrifice long life for high performance), ruggedness, high transconductance, low noise, and high input impedance. By 1941, the 386A and 717A represented the latest thinking on high-gain and high-figure-of-merit tubes for use in the telephone plant. The 717A pentode and the 6AC7 commercial pentode became the first tubes used in preamplifiers and intermediate amplifiers used in radar. Intermediate-frequency (IF) amplifiers using the 717A had a gain of 85 dB at center frequencies of 30 and 60 MHz and a bandwidth of 4 MHz, with a weight of just under two pounds. A large part of the Western Electric-manufactured airborne bombing radar equipment used this amplifier, with few additional changes as time progressed.

The search for more compact and lightweight amplifiers resulted in the famous 6AK5 pentode,[23] which delivers gain of approximately the product of grid plate transconductance ($g_m$) and the load resistance. In the case of the triode, the gain is not this high, because the effective load resistance is shunted by the tube plate resistance, which in a triode is usually lower than, or comparable to, the load resistance. The 6AK5, with plate voltage of 250 V, yields $g_m$ of 5000 microsiemens (mS) and may be compared with the World War I "peanut" tube of much the same size, which yielded a $g_m$ of 420 mS. The 6AK5 thus represented a major improvement over earlier small tubes. Figure 3-10 shows the 6AK5 with its immediate predecessors, the 717A and the 386A. A companion of this tube, the 6AJ5, was developed for use in aircraft equipment, at plate and screen voltages of 25 to 30 V. The lower voltages were necessary because of the limitations of aircraft power supplies. The 6AJ5 was a lower-$g_m$ tube, of about 2000 mS, but it served well as an airborne IF amplifier. The major developments in such tubes included the oxide-coated cathode, a flat glass plate through which connecting pins were sealed, direct mounting of the tube structure on these pins with short leads (low inductance), and small elements mounted on low-leakage coated mica supports. As a result, high gain to at least 100 MHz could be achieved.

├─ 1 INCH ─┤

Fig. 3-10. The famous 6AK5 pentode (center), a major improvement over its immediate predecessors, the 717A (left) and the 386A (right). It remained for decades the standard for moderate- to high-performance, low- and medium-frequency tubes.

The basic structural features of the 6AK5 increasingly became the standard for moderate- to high-performance, low- and medium-frequency tubes, and remained so for decades; the major change after World War II was the inclusion of more than one tube within a single envelope. One of the first of these modern multifunction devices was the Western Electric 396A, with two entirely separate triodes in the same envelope.[24] However, one even earlier Western Electric development, the 292A/303A, performed multiple functions with a common cathode.

### 3.2 Reflex Oscillators

A major need early in the war was for a source of local oscillator signals at centimeter wavelengths to be used in radar systems. The need arose because no sufficiently low-noise detectors were available to allow direct rectification at ultrahigh frequency (UHF) or microwave frequencies. Further, detection at the incoming frequency would result in loss of selectivity, so it was necessary to heterodyne down (or "beat" down) the reflected radar pulse and amplify at a lower frequency of 30 to 70 MHz. Because available mixers would yield an optimum signal-to-noise ratio over a very narrow range of beating signal power, the general need was for beat oscillator power on the order of 20 milliwatts (mW).

Soon after the invention of the klystron by the Varian brothers[25] at Stanford University became generally known, Pierce and W. G. Shepherd of Bell Laboratories invented the reflex klystron oscillator[26] (see *Communications Sciences (1925-1980)*, Chapter 4, section 1.2). This microwave oscillator played a particularly important role in radar. Because of the need to stay near to, but separated by the IF from the incoming pulse frequency, the reflex oscillator needed to be stable and, ideally, follow variations in transmitter frequency automatically. Much of the rapid improvement of the performance of this oscillator to meet specific requirements came from Bell Laboratories efforts.

The reflex oscillator extends the frequency range far beyond that of a gridded triode or pentode, where performance deteriorates at higher frequencies and where electron transit time from cathode to plate is a significant fraction of the period of oscillation. In the reflex klystron, the electron beam is bunched—i.e., velocity modulated—by the RF electric field of the resonant cavity of the tube structure. A schematic cross section of a reflex oscillator is shown in Fig. 3-11. Dimensions are such that C, a round, pillbox-shaped cavity, and the grids that cross two holes within the cavity, are self-resonant at a desired microwave frequency. Electrons emitted from a hot cathode K are accelerated toward and through the grids by a positive potential on the cavity. The repeller R is operated at a negative voltage with respect to the cathode. Upon proper adjustment of voltages of the retarding field, the bunched electrons can be made to

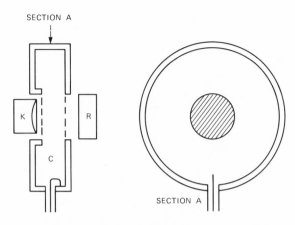

Fig. 3-11.   Schematic of a reflex oscillator. Hot cathode K emits electrons through grids toward repeller R. Dimensions are such that cavity C and the grids are self-resonant at the desired microwave frequency.

traverse the grid structure in the correct phase to give up energy to the circuit. The tube oscillates, and RF power is delivered to an external microwave circuit coupled to the cavity. The advantages of the klystron are that it is basically a high-frequency device; it is compact and requires only moderate voltages; it does not require an external beam focus arrangement; and, importantly, it can be tuned over an appreciable frequency range by merely changing repeller voltage. Thus, it can be a tunable microwave source or, as later used, it can be frequency modulated by coherent information impressed on the repeller voltage. A primary problem with the device is its low overall efficiency of only a few percent. This limitation, however, is not very important in local oscillator use, where only a few milliwatts of power is required.

The cavity of the reflex klystron can be either external or internal to the vacuum envelope. The Western Electric 707A was the first reflex klystron to be extensively used in a radar application; it had an external cavity. The tube, which can be seen in Fig. 3-12, used two Houskeeper seals of copper to glass (see *The Early Years (1875-1925)*, p. 849). These early seals were also the cavity contacts.

The 707A operated at 3 gigahertz (GHz) and was the first reflex oscillator to be designed for voltages as low as 300 V, an advantage in the design of radar equipment, because the oscillator can be driven from the low-voltage supply that powers IF amplifiers, preamplifiers, etc. The use of fine mesh wire grids instead of coarse wire or no grids increased the efficiency of beam bunching and increased the range of electronic tuning.

For extension to higher frequencies, and for a compact manufacturable structure, the tuning cavity needs to be integrated with the electronic array.

|←——— 2 INCHES ———→|

Fig. 3-12. The 707A, the first reflex klystron designed for voltages as low as 300 V, an advantage for radar equipment.

The basic design that resulted, after the usual development problems, was typified by the 2K25, a 10-GHz oscillator, shown in Fig. 3-13. While the 707A set the precedent in voltage range, the 2K25 became the prototype for the mechanical configuration of later oscillators. From the figure, it can be seen that the outer shell, plus added grid supports, form the resonant cavity. The base is standard octal, and the coupling loop is part of a coaxial line that, at the base end, can be inserted parallel to the electric field lines in a waveguide operating in the $TE_{10}$ mode.

Mechanical tuning was needed so that the center frequency for any electronic tuning could be varied. This was achieved, but with considerable difficulty. For adequate control, it is necessary to set grid spacing reproducibly to a few millionths of an inch, without backlash. This was finally accomplished by the mechanism shown in Fig. 3-13 and described in principle by Fig. 3-14. Turning the threaded screw varies the overall length of the bow arrangement by extremely small increments. The screw threads into a nut in each bow, one of which has a right-handed thread, the other, a left-handed one. By coupling to one grid, the mechanical spacing between grids is varied.

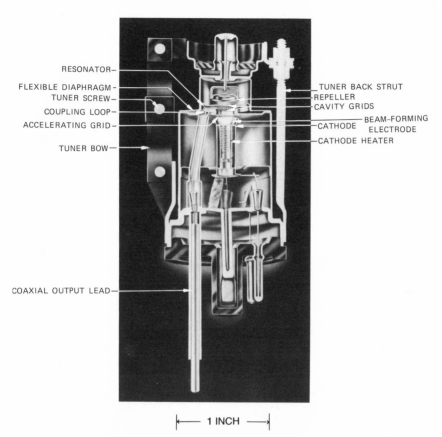

RESONATOR—

FLEXIBLE DIAPHRAGM—
TUNER SCREW—
COUPLING LOOP—
ACCELERATING GRID—

TUNER BOW—

COAXIAL OUTPUT LEAD—

TUNER BACK STRUT
REPELLER
CAVITY GRIDS
BEAM-FORMING
CATHODE    ELECTRODE
CATHODE HEATER

|← 1 INCH →|

Fig. 3-13.  Internal structure of the 2K25 10-GHz oscillator tube, in which the tuning cavity was integrated with the electronic elements to form a compact structure.

Additional changes in the 2K25 type of device were made to satisfy military requests for plug-in replacement—i.e., interchangeability. These included changes to the electron gun and repeller shapes to eliminate discontinuities in output, and improvement to the tube-waveguide match.

Perhaps the last major contribution by Bell Laboratories to the reflex klystron during World War II was the development of thermally tuned tubes. The need for such tuning arose from the desirability of reducing adjustments in the radar systems to a minimum, and the need to cope with the possibility of enemy jamming. The latter problem, in particular, dictated the need for fast frequency change; both transmitter and receiver would have to be changed by the same interval. The reflex oscillator requirement was for a tuning range greater than 1000 MHz.

The first tube of this type made in quantity, the 2K45, employed a triode built into the reflex oscillator envelope in a way such that the anode,

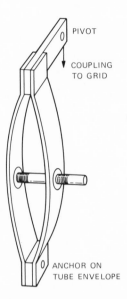

Fig. 3-14. Mechanical tuner for the
reflex oscillator. This component allowed reproducible grid spacings set
to a few millionths of an inch for
variable electronic tuning.

a channel-shaped member, is an element with a high coefficient of thermal
expansion. Welded to this at the ends is a bow of low thermal expansivity
material. Incoming electron current heats the anode, which then expands,
causing the bow center to move toward the anode. The bow is coupled
mechanically to a cavity grid, thus effecting spacing change. The construction is illustrated by Fig. 3-15.

The 2K45 also was the first major design to use an electron gun of
spherical rather than cylindrical geometry. The spherical gun eliminated
output discontinuities and a gun grid, and lowered grid losses in the interaction space.

During World War II, device people in industrial, university, and government laboratories cooperated effectively. V. Neher of the Massachusetts
Institute of Technology Radiation Laboratory had been working on a model
of a UHF klystron that could be manufactured in quantity. Neher approached Bell Laboratories to develop a design for possible manufacture
by Western Electric and/or other manufacturers.

As a result of this interaction, the 2K50 klystron oscillator was developed
at Bell Laboratories by the J. O. McNally-V. L. Ronci klystron groups. The
tube, with an upper frequency of 25 GHz, had a resonant circuit and a
means of tuning in an integral package. Tuning was accomplished thermally

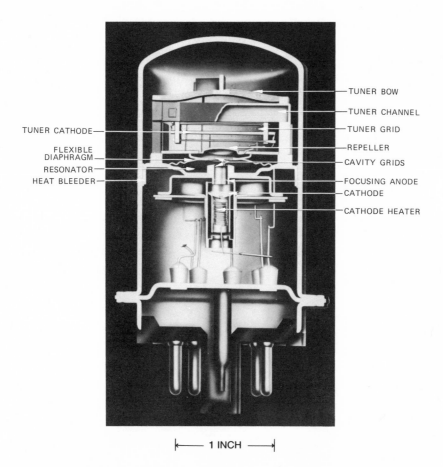

TUNER CATHODE—

FLEXIBLE
DIAPHRAGM—
RESONATOR—
HEAT BLEEDER—

—TUNER BOW
—TUNER CHANNEL
—TUNER GRID
—REPELLER
—CAVITY GRIDS
—FOCUSING ANODE
—CATHODE
—CATHODE HEATER

|←——— 1 INCH ———→|

Fig. 3-15.   Construction details of the 2K45 klystron, a thermally tuned tube for radar applications. Temperature changes in elements at the top changed cavity grid spacing and thus the operating frequency.

and controlled by a triode section in the tube. The tube output was delivered through a window rather than a coaxial lead as in older reflex oscillators. It was designed for direct coupling to a waveguide by a plastic connector. A slot in the connector was used to orient the waveguide with the waveguide section of the tube. Because of the end of World War II, the 2K50 was not manufactured by Western Electric.

In all, a great variety of oscillators were developed for military use. One design with the 2K45 was used in a pulse position modulation system for communication. A measure of the merit of Western Electric tubes is that of eleven Army-Navy preferred types, nine were of Bell Laboratories design. Figure 3-16 shows a list of the devices plotted against frequency.

Further expansion of the uses of the klystron would come about with its application as a power output tube in microwave radio relay systems as the decade closed.

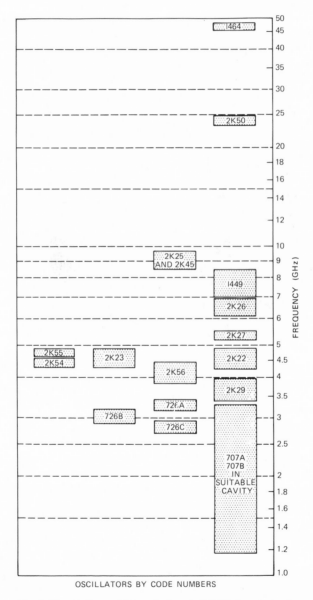

OSCILLATORS BY CODE NUMBERS

Fig. 3-16. Reflex oscillators designed at Bell Laboratories for military use, plotted against frequency.

### 3.3 Magnetrons

Probably no single development of the electron tube art of the period was as crucial to the outcome of the war as the resonant cavity magnetron oscillator. Without this oscillator, there was no way of reliably generating centimeter wavelength pulse power in the high-kilowatt and megawatt range, a range needed for effective application of radar over usable distances under all conditions of atmospheric water vapor content. Previous to 1940, radar development was being conducted by use of triode tubes that would, at best, deliver only a few kilowatts of microwave pulse power.[27] In 1940, however, an early British magnetron was brought to Bell Laboratories for study, and it was immediately apparent that a breakthrough in the power-frequency limitation had been found. The British device was reproduced in the Bell Laboratories tube shop at West Street in New York City, and from that point, a continuously accelerating research and development program began. This resulted in major contributions by J. B. Fisk, P. L. Hartman, H. D. Hagstrum, and others at Bell Laboratories (see *Communications Sciences (1925-1980)*, Chapter 4, section 2.1). At first, Bell Laboratories carried nearly the entire development load for the United States; later, the Massachusetts Institute of Technology Radiation Laboratory shared the responsibility.

The operation of a multicavity magnetron oscillator is more difficult to explain than the operation of most tube types. The basic requirement is that electrons emitted from a cathode, and drawn toward an anode, be bunched in some way to give up energy to an RF field existing in a resonator system. This requirement is identical to that for a reflex oscillator, as previously described. However, in the magnetron oscillator, a magnetic field is introduced at right angles to the dc electric field. In this crossed electric-magnetic field, individual electron motion becomes cycloidal, and the average path length before interception by the anode is greatly extended. The corresponding interaction path length is also extended, leading to the possibility of efficient energy transfer from electrons to the resonator RF field.

Figure 3-17 shows the basic parts of a magnetron oscillator for centimeter wavelength RF generation. The multicavity resonator anode consists of six or more individual resonators, each consisting of the hole and an entrance slot to the hole. Each individual resonator is coupled to its neighbors. Coaxially mounted inside the resonator is a cylindrical cathode. The space between cathode and resonator is the electron interaction space, where electrons become bunched in proper phase to yield net energy to the resonators as they pass each slot. Not shown is the magnetic field, applied vertically through the structure. In typical radar operation, a control circuit applies a high positive voltage (with respect to the cathode) to the anode in short pulses; peak pulse currents may reach many tens of amperes.

|←———— 5 INCHES ————→|

Fig. 3-17.   Basic structure of magnetron oscillator for centimeter wavelength RF generation. A series of resonators with radial slots surrounds a central cylindrical cathode.

The multicavity arrangement is closed upon itself, and therefore self-exciting, so that oscillation occurs during the pulse, the frequency being determined by the dimensions and configuration of the resonator structure.

The multicavity structure of the device allows different modes of oscillation, and therefore different frequencies. A primary problem with magnetron design, and this was the case in 1940, is how to provide a clean single-frequency output. An important objective was to widen separations between operating frequency and undesired modes. This objective was accomplished by alterations to the resonator structure. Two methods were used. In the first, which was applied to 10-cm and 3-cm magnetrons, coupling between resonators was increased by strapping, i.e., the tying of alternate anode segments, at the capacitance part of the resonators, with wire straps at each end of the anode block. Since the efficient mode operation was the $\pi$ mode, in which alternate anode segments are 180 degrees out of phase, the straps would not adversely affect the operation in this mode but would tend to disturb the other competing modes. In the second method of anode design, alternate cavities, made of simple wedge-shaped resonators, were of different lengths. The $\pi$ mode frequency was situated in a large gap in the mode spectrum, approximately midway between the frequency associated with the small resonators and that corresponding to the large resonators. This design was particularly good for 1-cm and shorter-wavelength magnetrons, where strapping would be extremely difficult.

(See another volume in this series subtitled *National Service in War and Peace (1925-1975)*, Chapter 2, section IV.)

The first magnetron designs produced, the 700A to 700D, employed techniques similar to those used to produce the original British device. These units were for a 40-cm wavelength and necessarily departed considerably from the dimensions of the British model, which operated at 10 cm. Initial models were not strapped, but when strapping was introduced, the efficiency was increased to the level of later devices—as high as 60 percent. The 700 series was applied to advantage in early radar for controlling guns on naval ships.

Resonator strapping was used in immediately succeeding designs, the 728 series and the 5J23. In this period, the noncontacting coaxial load-coupling scheme was devised; this is a means of reflecting, by a slot one-half-wavelength long, a metallic short circuit to the point where the coaxial line and the magnetron loop pickup line meet. Such a joint eased the requirements on the glass seal across the coaxial output of the magnetron.

When a new frequency allocation required redesign of several magnetrons, it was possible to predict performance by previously accumulated knowledge of the effect of design parameters, and the results of mechanical scaling had reached the point where effects on performance could be predicted. The 4J21 to 4J30, for the 20- to 30-cm band, were designed and built in a reasonably straightforward manner. They used a double-ring-strapping arrangement with fine frequency control being attained by strap size variation. The contact-free, or choke, coupling was used. A pulse-power output of 750 kW at 50-percent efficiency was obtained.

As frequency of devices was increased, it became preferable to have power coupled directly to a waveguide. This required development of a coaxial-to-waveguide transition, adaptable to a given design. A more direct way of coupling the magnetron power to the external waveguide was devised and first used in K-band magnetrons by S. Millman at the Columbia Radiation Laboratory.[28] In this design, a low-impedance, quarter-wave slot transformer connects the back of one of the magnetron resonator sectors to a short section of K-band waveguide inside the vacuum envelope. A properly designed circular glass window in a Kovar* cup, isolated by choke joints, provides the vacuum seal. This design was adapted to 3-cm magnetrons by using a quarter-wave transformer of H-shaped cross section. (See *National Service in War and Peace (1925-1975)*, p. 121.)

For the same reasons noted in the discussion of reflex oscillators, tunable frequency became an early goal. Work on tuning followed the course of operating on a single resonator and of tuning all resonators simultaneously.

---

* Trademark of Carpenter Technology Corp.

The former can be achieved by coupling to the outside and then varying the impedance outside the cavity. This asymmetrical tuning scheme does not result in a wide tuning range, and moding (i.e., shifting of operation to an undesired frequency) is difficult to avoid. It was found that the most acceptable arrangement is to provide a means of varying capacitance to resonator straps. A vacuum diaphragm is then needed to attain smoothly controllable movement of pins inserted into each resonator within the envelope. Figure 3-18 shows this arrangement as it was incorporated into the 4J51 tube.

With the introduction of strapping and tuning methods, the basic magnetron techniques were available, and a great variety of devices in various power and frequency ranges were designed and manufactured. A further mechanical simplification was made, however, by producing a packaged tube, i.e., one in which the external magnet was an integral part of the device. Magnet weight and size could thereby be reduced significantly. A photograph of the 2J55, the packaged version of one of the most extensively manufactured designs (725A), is shown in Fig. 3-19.

|←——— 5 INCHES ———→|

Fig. 3-18. The 4J51 magnetron, which had a vacuum diaphragm to attain smoothly controllable movement inside an envelope of resonator pins, allowing tunable frequency.

|← ——— 3 INCHES ——— →|

Fig. 3-19. The 2J55 packaged tube magnetron, which featured an external magnet as an integral part of the device. Magnet weight and size were reduced considerably in this design.

A special problem of the magnetron is the cathode, because of the enormous peak currents it must supply in short pulses. Also, it is subjected to intense back bombardment from electrons that gain energy and return to the cathode. Generally, the use of a sprayed oxide cathode is satisfactory in those tubes operating at longer wavelengths, above about 10 cm, where peak emission of 10 to 30 $A/cm^2$ is required. However, when short wavelengths and high power are to be attained, the small size requires both close anode-cathode spacing and high currents. Arcing becomes a primary problem, and the cathode surface is rapidly destroyed. Studies of these difficulties eventually led to the matrix cathode, in which, at first, finely divided nickel and barium plus strontium carbonates were mixed and applied, held in place by a nickel mesh. The final version consisted of a coarse nickel powder and the carbonates, the mixture being sintered onto the surface of a machined nickel cathode base. Whereas lives of a few tens of hours were experienced in use of simple oxide-coated cathodes in the higher-frequency tubes, the matrix cathode extended the life to a few thousand hours.

Adapting a device as complicated as the magnetron to economical volume manufacture required close cooperative efforts between Bell Laboratories and Western Electric. The magnetron requires massive machine operations to close tolerances in copper. It also requires glass-to-metal seals of highest integrity, reproducible assembly operations, air exhaust, and cathode activation. Western Electric met these challenges. Many of the 75 different codes developed by Bell Laboratories were manufactured in quantity for the armed services. These ranged in frequency from 600 MHz to 30,000 MHz and in power up to 3 megawatts (MW).

The major Bell Laboratories contributions to the war effort may be summarized as follows: (1) scaling of desirable strapping features of the magnetron anode for obtaining adequate mode separation and clean single-frequency oscillation from 3 GHz to 10 GHz; (2) obtaining manufacturable and reproducible output circuits; (3) making the magnetron tunable; (4) developing a rugged cathode; (5) developing packaged tube-magnet structures; (6) achieving completed device designs that could be reproducibly manufactured using reasonable parts tolerances.

In the postwar period, two magnetrons—one X-band at 250 kW and the other S-band at 1000 kW—were produced in large numbers by Western Electric until 1968. (See *National Service in War and Peace (1925-1975)*, p. 372.) The most innovative postwar improvement in magnetron design was the coaxial cavity magnetron conceived by J. Feinstein at Bell Laboratories in 1954.[29,30] It achieved mode separation, high efficiency, stability, and ease of mechanical tuning by surrounding the conventional magnetron anode configuration with a high-Q coaxial cavity. The $TE_{011}$ mode of the cavity was coupled to the anode through slots in the back wall of every other quarter-wave cavity. As a result, the magnetron output frequency could be tuned simply by moving a single plunger in the outer coaxial cavity. The 7208 Ku-band (15.5–17.5 GHz) magnetron was produced by Western Electric in Laureldale, Pennsylvania. [Fig. 3-20] It was employed in aircraft radar, and, 30 years after its invention, a wide variety of modified versions of the coaxial cavity magnetron were still being manufactured by several companies for the military.

### 3.4 Gas-Filled Tubes for Radar

A formidable problem in radar operation existed in designing a transmit-receive switch (commonly referred to as a TR box). It had to isolate the radar receiver so that it was not damaged or even overloaded by transmitted pulses of high-kilowatt or megawatt peak power; then in nanoseconds to microseconds, it had to switch so that a microwatt radar return signal was applied to a receiver operating at full sensitivity. Without such a device, it would be necessary to provide separate antennas for transmitter and receiver, which would be costly, too large, and too complex. Even then some kind of protection would be required to prevent receiver overload.

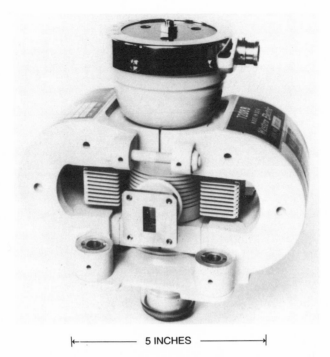

|←——————— 5 INCHES ———————→|

Fig. 3-20. The 7208 coaxial cavity magnetron, employed in military applications for over 30 years. Motion of a single plunger tunes the device.

Two primary functions in a radar transmit-receive array that can be fulfilled by gas tubes are: (1) the switching of stored charge in a pulse-forming network to a transmitting tube such as the magnetron, and (2) the switching of the antenna from transmitter to receiver. Radar advances at Bell Laboratories required rapid progress in the design of tubes for both applications.[31]

Initially, an open-air rotary, motor-driven spark gap was used to pulse the transmitting magnetron from the network. At high voltage, this arrangement operates well. At lower voltages of less than 5 kilovolts, initiation of the discharge is erratic; it was made successful by introducing auxiliary sharp points on each cathode, which started the discharge. However, the device was too bulky for airborne use and became erratic at high altitude. Fixed-gap tubes were then developed, generally using aluminum cathodes and a hydrogen and argon gas fill. A longer-life type used a sintered iron matrix saturated with mercury. The 1B42 is typical of this class, with over 1000 hours of life. Four basic switching tubes in this class were developed.

The earliest radar application was of the 709A vacuum oscillator tube with the addition of internal gas to provide an arc under transmitted power

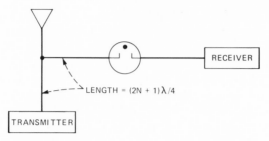

Fig. 3-21.  Schematic of the shunt-branching circuit incorporating the 709A tube used as a radar transmit-receive switch, thus allowing the use of a single radar antenna.

level. A tuned cavity around the tube further increases voltage at a given power, protects the tube from high ambient RF fields, and lowers arc extinguish time. The tube was used in a shunt-branching circuit as shown in Fig. 3-21. Three tubes were designed specially for use in a TR box. These were the 721A, 724B, and 1B23, as shown in Fig. 3-22. It was found that the best gas is a combination of water vapor plus hydrogen. The tubes were generally applied in series with the receiver line. For transmitting, the TR tube prevents much power from reaching the receiver; for receiving, the TR position is adjusted so that it is an even number of quarter-wavelengths from the effective short circuit represented by the transmitter, so that maximum voltage is induced across the receiver input. These tubes were made in quantity: in 1944, some 400,000 of the three types for TR use were manufactured.

Fig. 3-22.   Three tubes designed for transmit-receive use: left to right, 721A, 724B, 1B23.

## IV. POST-WORLD WAR II PERIOD

As the demand for telephone service rapidly increased after the close of World War II, the need for outside telephone plant increased as well. Further, it was expected that television would also grow rapidly. The areas of major tube application were radio relay and carrier transmission systems.

Postwar developments can be appreciated only with the perspective of a few earlier events. During the late 1930s, there was heightened activity in the development of new types of transmission systems requiring tubes capable of performing at higher frequencies. The 380-series tubes—primarily designed for low-power applications through audio and ultrahigh frequencies—were useful in both radio and carrier transmission systems. The importance of these tubes in World War II and in the postwar era is covered in section III of this chapter.

The 384A and 386A were used in repeaters of the L1 coaxial system, first made available in 1939 for a field trial between New York City and Philadelphia, Pennsylvania. Commercial service of an L1 system between Stevens Point, Wisconsin and Minneapolis, Minnesota was established in June 1941.[32] The L1 system could provide 480 telephone channels or one television channel on a pair of coaxial cables.

The UHF tubes were used in radio telephony as well as in wire transmission. As early as 1941, a 12-channel, 150-MHz system was in operation across Chesapeake Bay. The successful operation of the multiplex radio link demonstrated that radio telephony had an important place in the new era of communications systems. One of the UHF tubes developed for the Chesapeake Bay crossing was the 363A. The tube is shown in Fig. 3-23. It was a lined-up grid pentode. The plate was solely supported by its lead from the glass envelope, and no insulating material within the tube was subjected to high RF potentials. The tube operated most efficiently between 30 and 85 MHz, but could be operated above 85 MHz by reducing the plate dissipation.

The Chesapeake crossing covered 25 miles from Cape Henry, Virginia to Cape Charles, Maryland. Inserting this radio link in a K carrier system saved about 400 miles of wire carrier.

During World War II, as much nonmilitary work as possible was continued, and in early 1944, a permit was obtained for a repeatered radio relay system between New York City and Boston, Massachusetts. (See *Communications Sciences (1925-1980)*, Chapter 5, section 2.2.) This system, designated TDX, used two-cavity klystrons, coded 402A, as output tubes.[33] The system became operational by late 1947. Two years later, long-haul transmission was improved by introduction of the TD-2 system, using a microwave triode as an output amplifier.[34] Later, the TH and TD-3 systems were developed, using traveling wave tubes as output amplifiers.

The history of carrier systems and the application of electron tubes in them is longer than that of radio systems. The significant contributions

|←——— 2 INCHES ———→|

Fig. 3-23.   The 363A pentode, used in ultrahigh-frequency radio telephony transmission across the Chesapeake Bay in 1941. It operated most efficiently between 30 and 85 MHz.

following World War II were in the area commonly referred to as frame grid tubes; these were of much higher performance, and led to expanded bandwidth capability. The same devices also found application in radio systems as IF amplifiers of high performance.

Of great importance to the growing body of tube knowledge, although not produced in quantity, were tubes for continued military work and for submarine cable use. All-important to each type of tube is the understanding of cathode operation, which accumulated in depth in the two decades following the end of World War II.

### 4.1  Cathode Improvement

As noted earlier, estimates of average life of oxide cathodes in repeater-type tubes had been 20,000 hours as early as 1930. Most progress had been made by empirical methods, such as optimizing the processing cycles and obtaining maximum cleanliness, and by conservative operation at low current densities. The result of such progress, typified by the 310A, was a projected average life of 50 years. Historically, the expected maximum allowable current density for an oxide cathode was 0.5 A/cm$^2$; even at that level, life tended to be only a few thousand hours. Therefore, typical telephone tubes were run at densities of less than 0.1 A/cm$^2$. With the advances being made in tubes for radio relay, operation at higher densities or at moderate densities for very long lives became economically important. It was expected that the only way predictable long life could be attained would be by a more complete understanding of the cathode than existed at the close of World War II.

It was long known that chemical reactions occur in the oxide cathode between the supporting metal base and the coating of barium, strontium, and sometimes calcium oxides. The base metal was invariably nickel. This system was found by experiment to yield the highest electron emission per watt of heating power attainable in a readily manufacturable form. However, it was always true that the emission and life depended critically on the composition of the nickel base alloy; tube manufacture required a life test run of each new nickel alloy melt, and acceptance or rejection was based on this test. It was also found quite early that the nickel itself does not react perceptibly with the oxides, and very pure nickel resulted in lower emission than suitable nickel alloys. Thus, the materials in solution in the nickel-base metal alloy were of critical importance. Some elements, like silicon, readily react with the oxides to produce free barium, but result in a compound at the nickel-oxide interface that presents an impedance in series with the cathode. Others, like magnesium, tungsten, and zirconium, readily react without forming a harmful interface layer. Progress in cathode understanding would require understanding of the characteristics of these impurities in the base metal and their control.

Research in the metallurgical department resulted, by the early 1950s, in the ability to produce nickel with no impurity above the trace level (less than 50 parts per million) and, importantly, the ability to reintroduce controlled amounts of desired elements.[35] In the same time period, much had been learned about the diffusion rates of the elements in nickel and their reaction with alkaline earth oxides.[36] With this data, it became possible to predict the most favorable elements to add to nickel cathode bases, the operation temperatures to obtain the optimum element arrival rate, and the length of life of devices. Specifically, the following main conclusions were drawn:

(1) There is an optimum rate of atomic barium (and/or strontium) liberation, by reaction between a reducing element diffusing from the nickel base and the oxides, for a given current density of emission. This rate is $10^{-8}$ micromole/cm$^2$/sec for an emission density of 200 mA/cm$^2$, for example.

(2) At a given temperature, thickness of the nickel base, and concentration of the reducing element, the barium liberation rate can be calculated.

By 1960, the theory had been tested by extended life studies on both laboratory and production tubes, and the modifying effects of internal tube environment were recognized.[37] [Fig. 3-24]

During about the last two decades in which the vacuum tube was the predominant amplifying device, Bell System needs were mostly in two categories: (1) the microwave types—traveling wave tubes, triodes, and klystrons; and (2) grid-controlled carrier or receiving types. The former can possibly use cathodes in a wide range of nickel thicknesses, although the required current density is high in order to obtain good electron-circuit coupling and usable gain with reasonable tube size. The latter are limited to thin nickel cathodes because of requirements for low heater power and the need for a large area of electron emission. In this category, the density of emission requirements was initially moderate, but increased with progress in design.

Laboratory life studies indicated zirconium to be the preferred additive element for use in tubes with cathodes about 0.05 inch thick. The time to depletion of half the zirconium, for 0.1 percent by weight concentration, was predicted to be 20 years at 200 mA/cm² density and normal operating temperatures. However, with 0.003-inch-thick cathodes, the equivalent time is only about 2000 hours. Study of materials for this case resulted in the finding that a combination of tungsten and magnesium would give comparable predicted lifetimes.

Application of the new knowledge to production tubes rather than to laboratory tubes resulted in some surprises. Even though thick nickel could usually be used in microwave tubes, these devices required much internal glass and metal area as compared with the cathode area. In the carrier-

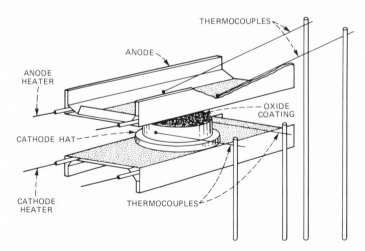

Fig. 3-24.   Diagram of a planar diode used in studies aimed at extending the life of cathodes.

type tubes, however, the cathode area is a much larger proportion of the total internal area. It was found that the former cathodes, operating at high current density, were much more easily disrupted by processing variations than were the carrier types. As a result, the pure nickel, single-additive technology was not generally adapted to microwave tubes, whereas the technology was successful in grid-controlled carrier types. Essentially all grid-controlled tubes made by Western Electric used the tungsten-magnesium alloy. The microwave tubes used an alloy containing several reducing elements.

In 1964, another contribution was made to the oxide cathode art, the coated powder cathode.[38] It was shown that under high-current emission, the current flow may cause a high field across the coating. This field will cause positive ion donors to be attracted toward the base metal, leaving a depletion at the cathode coating surface. If the gradient across the coating could be reduced, the surface depletion would also be less under high-current-density operation. To this end, the particles of carbonate that make up the coating were individually coated by a thin nickel layer. The carbonate was later converted to oxide during activation of the cathode, but the nickel coating remained. Experience proved that excellent life was obtained at a current density as high as 1 $A/cm^2$. The coated powder cathode was used in traveling wave tubes and PICTUREPHONE visual telephone service. An additional benefit from use of the cathode was its lessened sensitivity to process variables.

### 4.2 Klystrons and Microwave Triodes

Of the common carrier bands available for radio relay purposes, the one centered at 4 GHz was preferable from the point of view of directivity, antenna size, and rain attenuation. At the end of World War II, prior to the development of the traveling wave tube, only velocity variation tubes (klystrons) and close-spaced gridded planar tubes were available as satisfactory RF amplifiers at 4 GHz. At the time, the only proven reliable vacuum seals were metal to glass, the ceramic-to-metal vacuum seal technology having just been born. These considerations limited the design of multicavity klystrons to two cavities (rather than three- or four-cavity designs) and high beam voltages (1500 V), since low voltages require close spacings. To provide the gain, power output, and bandwidth required of a 4-GHz RF amplifier, four stages dissipating 180 W were needed.

If one could shrink the interelement spacings of the then current microwave triodes to less than 0.001 inch, it appeared that a simpler, low-voltage and lower-cost RF amplifier could be produced. Consequently, a close-spaced planar triode design, later coded the 416A, was begun by J. A. Morton and his associates with an objective of 0.0005-inch grid spacing, and a grid pitch of 1000 turns per inch using 0.0003-inch diameter wire.[39]

A drawing of the resultant structure is shown in Fig. 3-25. Nonconventional cathode construction was required to provide for the close spacing. The cathode was coated on a thick nickel base assembled on a ceramic reference ring. These elements were made coplanar by a grinding operation after assembly. A special technique was developed to coat the cathode with a precise application of high-density carbonate coating. The applied thickness was limited to 0.0005 inch, and the assembly was supported by spring loading the ceramic cathode reference ring to the grid frame with a precision spacer. The thickness of the spacer is the sum of the cathode coating thickness, the change in cathode position when heated to operating temperature, and the desired cathode-to-grid spacing.

The grid consisted of parallel wires wound on a molybdenum grid frame. Experience showed that the tungsten grid wire could be drawn to 0.0005 inch. The diameter was further reduced to 0.0003 inch by electrochemical etching in a continuous etcher that was feedback controlled from a resistivity measurement made on the completed wire. The grid frames were given an evaporated gold plating on one side and then mounted back to back in a winding machine chuck. The winding pitch was obtained by reducing the feed obtained from a precision lead screw by a factor of 10 using a sine bar lever. Winding tension was controlled by a dynamic system that overcame the "whip" caused by rotating the flat grid frame.

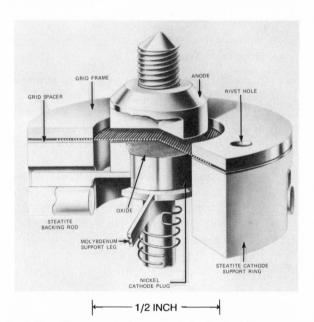

Fig. 3-25.   Close-spaced microwave triode, developed by a group led by J. A. Morton. The design objective was an improved, low-voltage, cost-effective RF amplifier for 4-GHz operation.

At an operating potential of 200 V and a current density of 0.2 A/cm², three of these tubes cascaded in an appropriate circuit produced 23-dB gain, 0.5 W of power output with greater than 30-MHz bandwidth. The extent to which dimensions were reduced can be appreciated from Fig. 3-26, which compares the 416A with a more conventional triode of that time.

With continued development work, the average life of triodes operating in low-level input stages eventually reached 40,000 hours; a life of 15,000 hours was typical at 0.5-W output. Further development allowed operation of the amplifier at 1-W output.

Tube development and production started by using nickel-iron-cobalt glass seals. As the metal-to-ceramic seal technology evolved, the glass-to-metal seals eventually were replaced by beryllia ceramic-to-metal seals. This resulted in reduced RF losses in the seals and increased RF power output.[34] The high thermal conduction of the beryllia seals allowed the total dc power input to be increased. With these changes, the power output of the 4-GHz amplifier eventually reached 5 W.

The 416 series of close-spaced triodes was used as transmitter amplifiers in the TD-2 and TD-3D radio systems, which started with 0.5-W output and, with a series of improvements, reached ever-increasing power levels until a level of 5 W was achieved. Starting with a route capacity of 2400 telephone message circuits in 1950, a later version of the 4-GHz system was capable of 19,800 circuits per route. In addition, these systems were used for network television and data transmission.[34] At one time, about

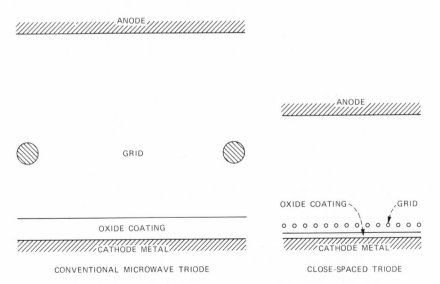

Fig. 3-26. Comparison of sizes and spacings of a 1940s triode (left) and the 416A microwave close-spaced design (right) with a grid spacing of 1/1000 inch.

70 percent of the long-distance circuit miles used radio systems with close-spaced triode amplifiers. About 100,000 tubes were used in amplifier circuits, and more than half that number were used in signal generators and frequency converters.*

### 4.3 Traveling Wave Tubes

A giant step in microwave-frequency gain-band figure of merit occurred with the invention by R. Kompfner, and the subsequent exploitation by Pierce, of the basic traveling wave interaction scheme.[40] Here was a new amplifier device that presented the promise of large bandwidth and high gain simultaneously. The reason for such capability is implied in the name. Modulation of an electron stream by information to be amplified occurs continuously during nearly synchronous electron beam and electromagnetic (EM) wave travel. There are no frequency limitations caused by transit time. The associated physical structure, together with traveling wave interaction, permits the input, amplification, and output of a wide range of frequencies. There is, however, a limitation on gain set both by requirements for stability and by practical limits on the amount of energy that can be transferred from the electron beam to the EM wave without losing the near synchronism between the motion of the electrons and the wave. (See *Communications Sciences (1925-1980)*, Chapter 4, section III.)

Like other microwave tubes, the traveling wave tube (TWT) operates on the principle of electrons giving up energy to an EM field. The interaction occurs between a linear electron beam and an EM wave that propagates along a slow-wave circuit. This circuit causes the EM wave to follow a route that results in its forward progress in the beam direction being slowed until the latter is the order of 0.1 c (the free space velocity of light). The slow-wave circuit typically surrounds the electron beam. Gain is obtained by adjusting the electron beam velocity to be just slightly faster than the velocity of the EM wave in the direction of the beam axis. The beam is then modulated and, on the average, slowed by its interaction with the EM wave; the kinetic energy lost by the electrons is transferred to the EM wave as the latter progresses toward the output circuit.

Many slow-wave circuits have been devised, but by far the most widely used is a simple wire helix. An EM wave traveling along the wire has a velocity along the helix axis of about $pc/\pi d$, where p is the helix turn-to-turn spacing, d its mean diameter, and c is, again, the free space velocity of light. Typically, the helix is wound to make the axial wave velocity

---

* Between 1979 and 1983, solid-state amplifiers using gallium arsenide field-effect transistors were installed in place of the electron tube transmitting amplifier. With the replacement of the close-spaced triode, the entire TD radio system was converted to solid-state devices exclusively.

equivalent to that of electrons accelerated through 1000 to 4000 V. Other slow-wave mechanical configurations are useful for higher powers; examples are a series of resonant cavities or a meandering waveguide. Shown in Fig. 3-27 is a schematic of a helix-type tube.

Early important contributions in the TWT field by Bell Laboratories were amplifier stability and quantification of tube design. Later contributions included noise reduction and beam focus. Bell Laboratories also offered the first mechanical design of precision slow wave structures, introduced the first mass-produced TWT, and built rugged TWTs for military and earth satellite use.

By the mid-1950s, several TWT developments were maturing for application. One of these was a 3-GHz low-noise tube, the 6784, for military use. A second, the 444A, a broadband output amplifier, was developed for use in the TH-1 long-haul radio system then being readied for service.[41] In addition to the basic contributions given earlier, the extended research and development program on these two tubes yielded most of the knowledge on TWTs that was applied to succeeding codes. Some of the most significant findings and developments are outlined below.

*Ion Damage to Cathodes.* Because the TWT is long, there is enhanced probability of positive ion formation from electron-gas molecule collisions even in the most optimally processed tube. The negative space charge of the electron beam tends to prevent the ions from leaving the region of the electron beam. Consequently, many of the ions produced in the region of the slow-wave circuit drift close to the beam axis toward the electron gun and then are accelerated toward the cathode once they enter the anode-cathode space, which provides an accelerating field. It was found that cathode life can be reduced by an order of magnitude from this bombardment. The solution is to design the tube in a way that requires the anode of the electron gun to be biased more positively than the slow wave circuit to effect a barrier to ion travel from the region of the slow wave circuit to the cathode. In all but those tubes designed for the very lowest noise, this practice has been followed.

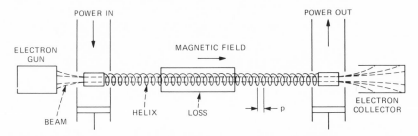

Fig. 3-27. Schematic of a helix-type traveling wave tube. Typically, the helix is wound to make the axial velocity of an RF wave along the helix equivalent to the velocity of electrons.

*Collector Depression.* By proper mechanical design, it is possible to operate the electron collector under RF drive as low as half the voltage of the slow-wave circuit; by this means overall TWT efficiency is significantly increased because, except for power to heat the electron-emitting cathode, most dissipation in a TWT occurs in the collector electrode.

*Noise.* In tubes designed to deliver moderate power, a converging electron beam is often used to avoid excessive current density from the cathode. The cathode has many times the area of the electron beam focused through the slow-wave structure. This beam compression tends to magnify the growing noise wave that is produced during beam flow. It was found that this wave can be greatly suppressed by introducing controlled magnetic flux at the cathode plane. Noise can often be reduced by an order of magnitude.

*Helix Accuracy.* There are adverse effects—instability, granularity of gain versus frequency, and intermodulations—that may be produced by periodic variations of pitch in the helical slow-wave circuit. Very accurate winding and means for maintaining accuracy are necessary. The first need was met by machine development, and the second by development of a technique of attaching helix wire to ceramic supporting members at each crossing of the wire. The result, widely used in later tubes, is sufficiently accurate that no evidence of periodicity can be detected.

*Magnetic Focusing.* Initially, dc solenoids were used for TWT focus. Investigation of materials and physical shapes resulted in a design procedure for attaining minimum material, permanent-magnet focus arrangements using a uniform straight field. For the 444A for the TH-1 radio system, such a circuit was adopted for production. Later in the 1950s, Bell Laboratories contributed to the development of a new scheme for focus, in which a series of ring magnets produces a sinusoidally varying magnetic field.[42,43] Because much less magnetic material is needed than for a uniform field, and because the periodic field produces much lower external leakage magnetic flux, most modern TWT amplifiers use magnetic focusing.

The 444A became the first TWT to be manufactured in significant quantities. Again, by cooperation between Bell Laboratories and Western Electric, assembly techniques peculiar to the TWT were successfully adopted in manufacture, and over 20,000 tubes had been shipped for customer use by the end of 1974. The device was used as the output amplifier and as a source of local oscillator power in TH-1 radio. It consistently delivered more than 7 W of power for periods of 3 to 5 years.

Figure 3-28 shows the 444A tube and focus circuit. The collector was cooled by air flow from centrally located blowers in the TH-1 station. The outer shell, much larger than the actual magnet, was provided to lower leakage fields that might affect nearby apparatus. Succeeding power TWT amplifiers were designed for TD-3 radio, TM radio, and TH-3 radio. All used periodic focus and incorporated the growing body of cathode knowl-

|← 5 INCHES →|

Fig. 3-28. The 444A tube and focus circuit. The outer shell, much larger than the actual magnet, was provided to lower leakage fields that might affect nearby apparatus.

edge to attain long life. One of these, the 461A shown in Fig. 3-29, consistently delivered an average operating life of eight years.

One additional TWT departed somewhat from the general designs, although it borrowed heavily from the technology and processing of them. This is the M4041 tube for the Telstar satellite.[44] To provide 3.5 W of power at 4.0 GHz in an extremely reliable package, this unit had a cathode current density of 80 mA/cm$^2$, compared to 200 mA/cm$^2$ in the power tubes. A cathode consisting of ultrapure nickel with zirconium additive, as discussed in section 4.1, was used. This satellite tube was constructed in the laboratory under ultraclean conditions. The focusing circuit was another innovation—a compromise between a heavy straight field system and a periodic system. It used a field reversal at the center. With this circuit, a favorable compromise was reached between total weight and efficiency; weight was cut by a factor of six from a single straight field magnet. Compared with use of a periodic field, a smoother beam profile was maintained, allowing lower collector voltage and therefore higher efficiency, as previously explained. Reliable performance in space was achieved by very careful processing and testing, and selection of tubes

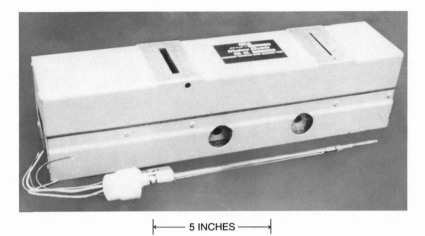

|← 5 INCHES →|

Fig. 3-29.   The 461A traveling wave tube consistently delivered an average operating life of eight years.

only after 2000 hours of aging and inspection. The mechanical design was made very rugged to withstand the rocket launch.

Project Telstar also required the largest and most powerful continuous wave (CW) communications TWT developed at Bell Laboratories, the M4040.[45] This tube, a schematic of which is shown in Fig. 3-30, developed under extreme time constraints by R. J. Collier, G. D. Helm, J. P. Laico, and K. M. Striny, was a 2-kW CW traveling wave amplifier, 4 feet long

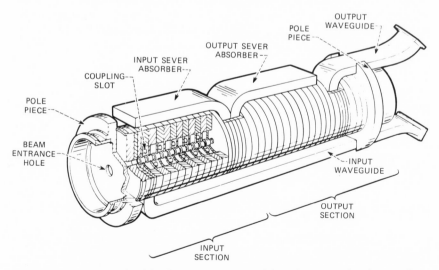

Fig. 3-30.   Internal details of the slow-wave structure of the Telstar ground station traveling wave tube.

and weighing 230 pounds. The only nonhelix Bell Laboratories TWT, it employed a coupled-cavity slow-wave structure with a 13-percent bandwidth centered at 6.15 GHz and a gain exceeding 27 dB. In early 1962, M4040s were installed high on the antenna feeds in Telstar ground transmitter stations at Andover, Maine and Pleumeur-Bodou, France; they functioned reliably as the final power amplifier of the signal beamed to the Telstar satellite.

One of the last TWT designs by Bell Laboratories was for the AR6A single-sideband microwave radio system. The performance objectives for this tube, coded 473A, were significantly different from those of its predecessors. Instead of being designed primarily for good efficiency at continuous operation levels near saturation, the 473A had to deliver high, short-period, peak-power output. However, its major requirements were specified at a power output level far below saturation; although it had a peak capability of about 70 W, the critical parameters were measured in the region of 0.5-W output, as shown in Table 3-1.

Many TWTs were investigated for the AR6A application and found lacking in required intermodulation performance. The intermodulation levels for the 473A tube were at least 3 dB lower than any other available tube. The flatness of the third-order intermodulation as a function of power output was superior in the 473A when compared to the other tubes under investigation. Further, the restricted gain-noise sum limit required departure from the conservative electron gun designs previously used. Major features of the 473A met all the above detailed requirements. After a short period of manufacture, it was found that 473A yields were rapidly increasing and shortly exceeded predicted values. Application in AR6A bays demonstrated that the design met all requirements. It was operated with a power supply designed for it and the combination of the tube and the power supply was stable with extended field use. (A total of 293 devices were shipped for field use before Western Electric terminated all TWT manufacture in December 1981.)

No Bell System application of the TWT has taken advantage of the potential instantaneous bandwidth of the device. This is because of statutory frequency assignments and also because modulation systems naturally divide available frequency space into channels. As a result, it has not been necessary to provide wideband input and output circuits.

| **Table 3-1.** Critical Parameters of the 473A TWT | |
| --- | --- |
| RF gain: | 40 to 46 dB |
| Gain plus noise figure: | 69.5 dB max. |
| Third-order intermodulation: | $-90.5 \pm 0.4$ dB ($P_0 = 19$ to 29 dBm) |
| Fifth-order intermodulation: | $-180$ dB ($P_0 = 22$ dBm) |
| | $-160$ dB ($P_0 = 18$ dBm) |

Mention should be made of a number of experimental electron tubes—both klystrons and TWTs—that were developed for use as oscillators and amplifiers in a millimeter-wave communication system (later referred to as the WT-4 system). Based on the use of the low-loss $TE_{01}$ mode in a circular waveguide, this system was the subject of continuing research at Bell Laboratories and elsewhere from the late 1940s into the 1960s because of its extremely wide transmission band.

A description of such innovative devices as spatial harmonic amplifiers, the backward-wave oscillator, double-stream amplifiers, and other devices using electron-wave interactions is to be found in *Communications Sciences (1925-1980)*, Chapter 4.

### 4.4 High-Figure-of-Merit Carrier Tubes

The need for circuits with greater bandwidth led to continuous evolution in the design and production of low-power, carrier-type grid control tubes. The most significant steps were in the application of known design principles to maximize the gain-bandwidth (GB) product figure of merit and application of new cathode technology, as previously described.

Investigation of the requirements for high-GB product revealed that it is largely a function of transconductance: $GB \sim g_m/(C_1 + C_2)$, where $C_1$ and $C_2$ are tube input and output capacitances. Since by definition, $g_m$ is the change in plate current divided by the corresponding change in grid-to-cathode voltage controlling the plate current, anything that will increase the influence of the grid voltage (signal input) on plate current flow is desirable. Moving the control grid closer to the cathode will do this. Unfortunately, it will also increase $C_1$; however, the improvement in $g_m$ much more than offsets this increase, particularly if a small grid wire is used. For a tetrode or pentode tube, a decrease in cathode-to-grid spacing requires that the screen grid also be moved closer to the cathode if the operating voltage is not to be unusually high. Similarly, the plate should be moved closer, both to operate at moderate voltage and to prevent influence on the electrical characteristics due to space charge.

The 6AK5/403A represented the state of the art at the end of World War II. The improvements in subsequent designs are shown in Table 3-2. These improvements were achieved by changes in mechanical design, precision piece parts, newer cathode technology, and statistical controls.

| Table 3-2. Improvements in the 6AK5 Series | | | | |
|---|---|---|---|---|
| Tube | 6AK5/408A | 404A | 435A | 436A |
| Figure of merit (gain-bandwidth product, MHz) | 72 | 123 | 146 | 165 |

Bell Laboratories contributions along these lines can be described by citing development of tubes for use in the L3 coaxial system.[46]

The 0.0035-inch cathode grid spacing of the 6AK5 is about the minimum for conventional construction techniques. Studies of designs for possible L3 amplifier tubes indicated that the cathode grid spacing should be 0.0025 inch. The smaller grid cathode spacing required a new grid design, as shown in Fig. 3-31. The grid frame consisted of two side rods supported

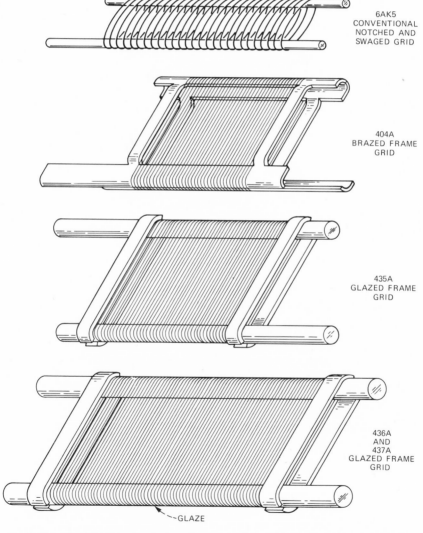

Fig. 3-31. The 0.0035-inch cathode grid spacing of the 6AK5 is about the minimum for conventional grid construction techniques (top). New grid designs were required for closer spacing; three examples are shown.

by straps at each end, leaving a space to be occupied by the grid laterals. The side rods were ground to precise dimensions to control the grid cathode spacing. The grid was wound on the frame and held under tension, and a glass frit glaze was fired on to maintain the tension. New machines were developed to obtain the precision required in winding these grids. Gold plating, important to inhibit thermionic emission from the grid, which was heated by the nearby cathode, was put on the grid to a thickness of 20 microinches. The 0.0003-inch diameter tungsten wire was wound at 410 turns per inch on the 436A and 380 turns per inch on the 435A. The tension in the lateral wire was equal to 200,000 pounds per square inch, an order of magnitude greater than the working stress of architectural steel. The screen grid was wound by conventional means, and the tube elements mounted on mica supports. The mica dimensions had tolerances tighter than any previous designs. The technique used was to refurbish dies as wear approached specified limits. Multiple sets of dies were made so that a usable die to specified tolerances was available at all times. Figure 3-32 is a cutaway view of the 436A, an L3 tetrode tube. Extensive testing

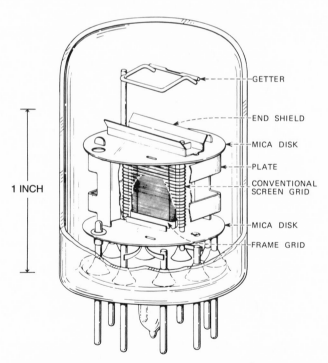

Fig. 3-32.   The 436A tetrode tube for the L3 coaxial cable transmission system. Grid wire was 0.0003 inch in diameter, wound 410 turns per inch.

proved the soundness of these construction methods, and the tube was manufactured at high yield. The L3 coaxial system employed careful statistical control methods to assure that minor manufacturing deviations did not pile up to cause out-of-specification system performance.[47]

### 4.5 Gas Tubes

From the 1930s through the 1950s (except for the World War II years), over 20 types of cold-cathode tubes were designed and produced for the Bell System, resulting in a design experience of providing unique structures and unique improvements in the normal characteristics of industrial-type applications. Some of the developments were as follows.

(1) An unusually stable voltage reference tube, the 423C, for the TD-2 radio relay automatic switching system and for the TH and TJ systems. [Fig. 3-33] This tube used parallel, flat, pure molybdenum electrodes that would provide an operating voltage stable to less than 0.1 V (typically 0.02 V) out of 100 V per 1000 hours of operation, an otherwise unavailable

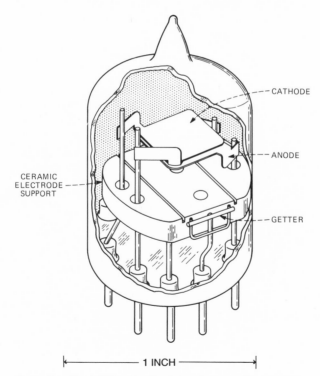

Fig. 3-33. The 423C cold cathode gas tube, used to provide a reference voltage for the TD-2, TH, and TJ microwave radio transmission systems. [Gewartowski and Watson, *Principles of Electron Tubes* (1956): 556.]

degree of stability. A later version, the 432B, provided the oscillation-free and surge-free reference voltage required by the A2A and A2B video transmission systems.

(2) A voltage regulator tube, the 427A, providing nominally 0.1-percent regulation from 5 to 40 mA, in contrast to the 5-percent regulation of commercial tubes. This tube also used parallel molybdenum electrodes to obtain needed stability.

(3) A counting or stepping tube, the 6167 and 439A, in the early 1950s,[48,49] which would transfer the glow discharge from cathode to cathode in response to incoming pulses by the action of the glow discharge moving from the location (on each cathode) of initial discharge to a preferred location in the direction of the stepping or counting movement. The arrangement of the electrodes is shown in Fig. 3-34 and again schematically in Fig. 3-35. The cathode design used the recently developed concept of a hollow region of a cathode being the preferred discharge location. In this case, the hollow was formed by coiling molybdenum wire, the wire also providing a mounting leg and the "pick-up" extension for the discharge. While counting tubes with simpler electrode structures later became available commercially, this tube provided manufacturing experience with the hollow cathode for later use.

|←————————— 1 INCH —————————→|

Fig. 3-34.    Arrangement of electrodes of the 6167 stepping (counting) tube.

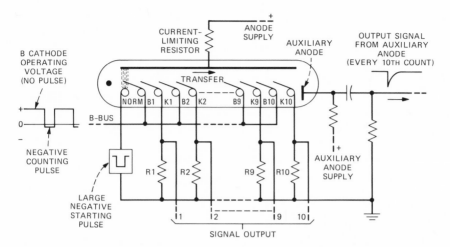

Fig. 3-35. Schematic of the 6167 counting action. Glow discharge stepped from position to position in response to input pulses.

## V. ELECTRON TUBES FOR SUBMARINE CABLE SYSTEMS

The electron tubes developed for submarine telephone cable systems were probably the most reliable tubes ever made. Because of the use of electron tubes in costly to access sea-bottom repeaters, the first transatlantic cable system, TAT-1, was recognized as a very ambitious project. As former Bell Laboratories president W. O. Baker stated in the *Bell Labs News* of January 29, 1979, "TAT-1, for its time, ranked in engineering difficulty with the later feat of putting an earth satellite into orbit."[50]

While all components of the system were required to be highly reliable, the tubes, with a known wear-out mechanism, were the key to the success of the project. Reliability was designed and built into the tubes. At Bell Laboratories, work on tubes for use in a proposed transatlantic system was started in 1933.[51] A short field trial cable from Key West, Florida to Havana, Cuba was laid in 1950, and the first major system was put into commercial operation across the Atlantic in 1956.

The suggested objective for submerged repeaters was that the tubes should not be responsible for a system failure for many (possibly 20) years. At that time, an average operating life of a few thousand hours (perhaps three to four months) was satisfactory for tubes in the home entertainment field. In land-based telephone equipment, an average life of a few years was considered reasonable. To meet the 20-year objective, no new or untried technologies were introduced; the emphasis was on conservative design and careful manufacture.

In design, three basic assumptions were made: (1) operation at the lowest practical cathode temperature would result in the longest thermionic life, (2) operating anode and screen grid voltages for the tubes should be kept low, and (3) the cathode current density should be kept as low as practicable.

In manufacture, care would be exercised in the selection and processing of materials, in fabrication procedures, and in detailed testing; long-term aging of all tubes would be carried out. Records of these items would be reviewed as part of the final selection of individual tubes for use in the repeaters. This selection concept was unique at that time.

The tube developed and manufactured by Bell Laboratories was the 175HQ shown in Fig. 3-36. The operating characteristics were cathode temperature, 670 degrees C (true); anode and screen voltages, 32 to 51 V; and cathode current density 0.7 mA/cm$^2$.

The cathode operating temperature was derived empirically. Life tests were run for many years, even during World War II when testing had to be squeezed in, and while the bulk of the work was done at about 710 degrees C, the results at 670 degrees C and 615 degrees C indicated that 670 degrees C would be appropriate. Normal cathode temperatures were approximately 750 degrees C.

|←— 2 INCHES —→|

Fig. 3-36.   Early model of the 175HQ (upright) and the final version (on side). In the first transatlantic system, 306 of these tubes operated for 22 years without failure.

Life tests were also used to study the effects of anode and screen grid voltages over the range of 40 to 60 V. No essential differences in performance were noted after eight years of operation, so the voltages listed above reflected the needs of the system. Voltages used in other vacuum tube amplifiers were typically 150 to 200 V.

The effect of cathode current density on thermionic life was also studied by life tests. After about 14 years, there was practically no difference over a 12-to-1 range of densities, i.e., 2.5 mA/cm² and 0.2 mA/cm². The final choice was to operate at 0.7 mA/cm², in contrast to normal current density of 50 mA/cm².

One common cause of tube deterioration with life is the formation of an interface layer on the surface of the cathode sleeve. As mentioned above, this depends in a complex way on the chemical composition of the cathode core material and, in effect, introduces a resistance in series with the cathode. This results in negative feedback and reduces the effective transconductance of the tube. The low transconductance of the 175HQ tube (1000 mS), the low cathode temperature, and the large cathode area were favorable factors to minimize the effects of this interface resistance. Again, as a result of life testing, one batch of nickel (melt 84) was selected for use in the production.

The tungsten heater for maintaining the cathode at the proper temperature was critical, since the heaters of all tubes in the system were connected in series. One open heater would disrupt the cable. Protection of the heaters from power surges was provided by a bypass gas tube across the three heaters of each repeater. Careful selection and control of the tungsten wire was instituted. The heaters were run at 1100 degrees C, which is well below that used in conventional tubes.

All 175HQ tubes used in submarine cables were made at Bell Laboratories, under the close supervision of many of the original development engineers. Fabrication was carried out with extreme care by operators specially selected for the job. Nylon smocks, acetate rubber gloves, and restricted areas were early steps toward modern ultraclean facilities. Thorough mechanical and electrical inspections over a 5000-hour aging period provided information for selection of individual tubes for cable use. By normal commercial test limits, the yield would have been about 98 percent. With the criteria used, only one tube in seven was accepted.

The original objective of a 20-year life was met by the first transatlantic cable, where 306 tubes operated for 22 years without a tube-related failure. This cable was retired in November 1978, for economic reasons. At that time, the 1608 175HQ tubes on sea bottom in seven cable systems had given 287 million tube-hours of trouble-free service.

The development of a new tube for the SD (1-MHz) cable system was started in 1955. The resulting 455A-F tubes[52] were designed by Bell Laboratories and manufactured by Western Electric in Allentown, Pennsylvania

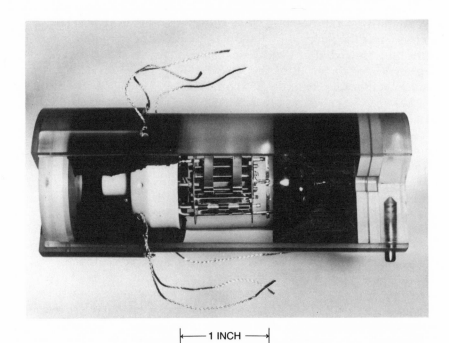

|←—— 1 INCH ——→|

Fig. 3-37. The 455A tube cushioned in a methacrylate housing, used in the 1-MHz SD submarine cable system. As of November 1978, 5874 such tubes in 10 cable systems had accumulated 738 million tube-hours of service.

with the close surveillance and cooperation of a resident group of Bell Laboratories engineering personnel. This tube is shown in Fig. 3-37.

To meet the same 20-year system reliability objective, developers of the new tube used an extension of the 175HQ design philosophy, updated in those areas where significant progress in basic knowledge had been made. For the broadband system, a higher-transconductance tube was needed, and a much closer grid-to-cathode spacing was required. The spacing for the 175HQ tube was 0.024 inch, compared to 0.0055 inch for the 455A-F. The use of frame-type grids aided in maintaining the closer spacing. Table 3-3 compares the two tubes.

| **Table 3-3.** Comparison of Submarine Cable Tubes | | |
|---|---|---|
| | **175HQ** | **455A-F** |
| Cathode temperature (true) | 670°C | 670°C |
| Cathode current density | 0.7 mA/cm$^2$ | 10 mA/cm$^2$ |
| Grid-cathode spacing | 0.024 in. | 0.0055 in. |
| Maximum element voltage | 51 V | 45 V |
| Transconductance | 1000 $\mu$s | 6000 $\mu$s |

With the closer spacing, concern for small particles of debris in the tube structure was increased. To improve microscopic inspection for such unwanted particles, development engineers designed the tube with an open structure. Minimization of particle generation was effected by assembling the tubes under laminar flow hoods located in segregated clean rooms and by other advances in environmental control. This was the showplace clean room of its day.

As seen in Table 3-3, the cathode temperature was kept at 670 degrees C (true) but the current density was increased from 0.7 to 10 mA/cm². This was necessary to achieve higher transconductance but was considered reasonable in the light of long-time life tests. The choice of cathode material was again of prime importance. By this time, work by physical chemists had advanced the knowledge of cathode emission phenomena to the point where the use of specific materials could be recommended. Thus cathodes were made from selected melts of high-purity nickel to which had been added 2 percent of magnesium.

Once more, close attention was paid to the tungsten used for heaters, since the heaters, protected in a more refined way by gas tubes, would again be connected in series for the entire system. Control of materials, assembly, processing, aging, testing, and inspection were pursued to a high degree, and again all information was scrutinized in selecting individual tubes for system use.

The performance record of the 455A-F tubes was most impressive. As of November 1978, 5874 tubes in 10 cable systems had accumulated 738 million tube-hours of service. Two tubes were considered "probable failures," although no service interruption could be directly attributed to either tube.

The gas-tube surge protectors were made with the same conservative care. While they were not actively part of the repeaters, they provided standby protection and were required to perform over the same time period in the same remote locations.

Gas-filled bypass tubes were needed in each repeater as a part of the fault-locating arrangement. They were designed to break down in the event of an open circuit within the repeater, thereby restoring circuit continuity so that the remainder of the repeaters could continue operating. The tubes were also necessary to protect against transient surges that might propagate down the cable in the event of external damage to the cable sheath. Requirements for the bypass tubes were (1) the breakdown voltage must be safely above the normal operating voltage drop across the repeater; (2) when breakdown occurs as a result of an open circuit within the repeater, the tube must be able to carry the full cable current of 0.25 A with a sustaining voltage less than 20 V to minimize heat generation in the repeater; and (3) for surge protection, the tube must safely carry short-duration current pulses of either polarity measured in hundreds of amperes. An argon-filled tube containing a cold cathode heated to emitting temperatures by ionic bombardment successfully met these requirements.[53,54]

## VI. PICTUREPHONE VISUAL TELEPHONE SERVICE TUBES

Two of the last tubes developed by Bell Laboratories were used with experimental visual telephone service. The display tube used for this system was not the usual cathode ray tube.[55] [Fig. 3-38] It was a plug-in device requiring no adjustment during installation. It was encased in plastic and had a glass shield across its face to serve as protection against accidental implosion. The best technology of the time was used to achieve long-life coated powder cathodes. The cathodes were operated continuously at a slightly reduced temperature to allow "instant on."

The camera tube, developed by a group led by E. I. Gordon, was based on the conventional vidicon tube used in the 1964 World's Fair PICTUREPHONE visual telephone service sets. However, the experience at the Fair showed how vulnerable the antimony trisulphide image-sensing target was to accidental exposure to bright light. The effects ranged from burn-in of ghost images to local destruction of the target. To overcome this problem and to take advantage of integrated circuit technology, de-

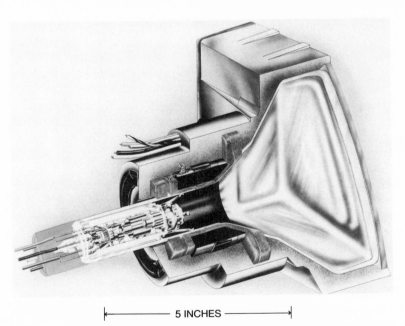

|← —————————— 5 INCHES —————————— →|

Fig. 3-38. The PICTUREPHONE visual telephone service display tube. Unlike entertainment television tubes, it required no adjustments. A glass shield in front and plastic encasement served as protection against accidental implosion.

velopers turned to the use of a silicon diode array containing upwards of 250,000 diodes on 20-micrometer centers. This concept was invented early in 1966 and by July 1966, feasibility had been established. The bright-light problem was solved; a camera tube with good sensitivity, especially in the near infrared, resulted.[56] [Fig. 3-39] An intense development effort involving Gordon's group at Murray Hill, New Jersey and a group under S. O. Ekstrand at Reading, Pennsylvania resulted in models for an experimental trial and Western Electric production during 1970.

In addition to use in PICTUREPHONE visual telephone service, this camera tube was used as the red channel of color cameras, and was used for surveillance in parking lots, apartment houses, etc. It also had important military applications. One version of it was used in the Apollo moon landing experiments. A storage tube version was used in luggage surveillance systems at airports.

### VII. TUBES FOR ELECTRONIC SWITCHING SYSTEMS

Initial systems planning for a high-speed electronic switching system involved a central computer-type control using the high-speed and random-access storage capabilities of electron-beam deflection tubes. A semipermanent memory called the flying spot store, incorporating a cathode ray tube to access information stored on photographic plates, was also developed. Parallel readouts of this memory were obtained by the simultaneous interrogation of many photographic plates arranged in front of a single cathode ray tube. The system plan involved using the low-noise negative resistive characteristics of the hollow cathode gaseous discharge tube that had been under development for some time for the talking paths. Development of production tubes for the initial switching system trial at Morris, Illinois was begun in the early 1950s. (See another volume in this series subtitled *Switching Technology (1925-1975)*, Chapter 9.)

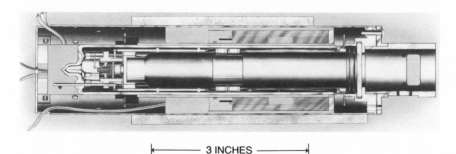

|← 3 INCHES →|

Fig. 3-39.   Camera tube for the PICTUREPHONE visual telephone service system. It used a silicon diode array as the light-sensitive target to solve the bright-light problem, and was especially sensitive in the near infrared.

### 7.1 Barrier Grid Storage Tube

The barrier grid storage tube, A4004, used an electrostatically deflected electron beam to deposit charge on a dielectric surface (writing) and the same beam at a later time to detect and remove the charge (reading).[57] The basic structure of this tube is shown in Fig. 3-40. The control of secondary electrons produced when the electron beam strikes the surface is an important function in the charge and discharge. As shown in Fig. 3-41, a fine grid, acting as a barrier and adjacent to the flat dielectric surface, prevents the redistribution of secondary electrons from disturbing the charge on adjacent spots during writing and reading on any particular spot. The A4004 could store binary information on a square array of 128 by 128 elements—in modern parlance, a 16K random-access memory (RAM). The tube could write and read in less than 1 microsecond ($\mu$s) and was used in the Morris system at full capacity at a 2.5-$\mu$s storage cycle for both reading and writing.[58] Reading destroyed the charge (erased) and rewriting was required if the charge information was to be retained for subsequent use.

### 7.2 Cathode Ray Tube for Flying Spot Store

The design of the flying spot store for the semipermanent memory was based on the availability of a suitable cathode ray tube under development commercially. Unfortunately, that development fell far short of meeting the specialized requirements and had to be abandoned. Development was then undertaken in the electron device department to meet the desired requirements in time for the systems trial. [Fig. 3-42] Tube requirements included (1) negligible beam distortion with electrostatic deflection, (2) a phosphor screen with adequate life and fast decay after excitation, and (3) a nearly optically flat tube face without appreciable imperfections in the glass, which would interfere with the imaging of the individual light

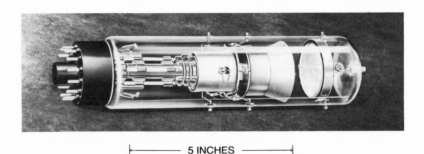

|←——————— 5 INCHES ———————|

Fig. 3-40.   The barrier grid storage tube used as a memory device in early field experiments with electronic telephone switching. An electron beam deposited information on a dielectric surface and also read information from the surface.

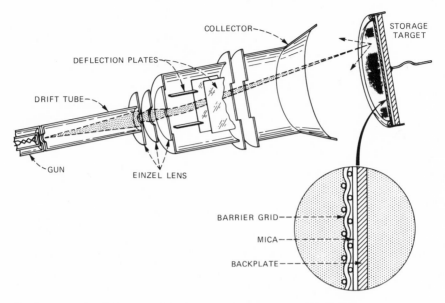

COLLECTOR

STORAGE
TARGET

DEFLECTION PLATES

DRIFT TUBE

GUN

EINZEL LENS

BARRIER GRID

MICA

BACKPLATE

Fig. 3-41.   Essential elements of the barrier grid tube. It was a 16K random access memory with a 2.5-$\mu$s read-write cycle time.

spots. Fortunately, production models of the cathode ray tube M2000 developed at the Murray Hill site met the necessary requirements and were installed in the Morris system trial.[59] The optical face plate for the tube was an interesting challenge. Glass manufacturers were not able to meet Bell Laboratories specifications for the 15-inch diameter, 3/4 inch-thick flat plate. The problem was occlusions in glass greater than 1 to 2 mils in diameter. It was solved by inspecting selected pieces of glass from two manufacturers until suitable areas could be found.

### 7.3 Talking Path Gas Tube

As discussed above in section 4.5, a hollow cathode structure was designed as part of a counting or stepping tube. Among the characteristics of interest were its low impedance to the transmission of audio frequency signals and its negative resistance for useful discharge currents.[60] These characteristics permitted the design of switching networks in which gas tubes replaced relay contacts in the talking path. [Fig. 3-43] These tubes took advantage of the stable discharge characteristics developed in the use of pure molybdenum cathodes.

The Morris field trial of an electronic switching system made extensive use of such unique gas diodes as electronic crosspoints in the switching matrix that provided connections through the system.[61] These tubes also

Fig. 3-42.   Laboratory model of the flying spot store. Design objectives included minimum beam distortion, a phosphor screen with very fast decay, and a nearly optically perfect tube face.

provided other switching control functions because of their stability and current capacity.[62,63] (See *Switching Technology (1925-1975)*, Chapter 9, section 4.6.)

The gas tube crosspoint provided the switching function as a result of the difference between the breakdown voltage of 200 V and the discharge-sustaining voltage of 110 V. The residual free electrons in the gas needed to permit rapid breakdown were provided by an auxiliary photoelectric surface in the form of barium getter evaporated on the glass wall of the tube, activated by the illumination of fluorescent lamps in the network cabinets. After breakdown, the tube provided the talking path, the negative resistance furnishing a small amount of voice frequency amplification to offset other loss in the voice path.

For all applications, about 70,000 of these crosspoint diodes were re-

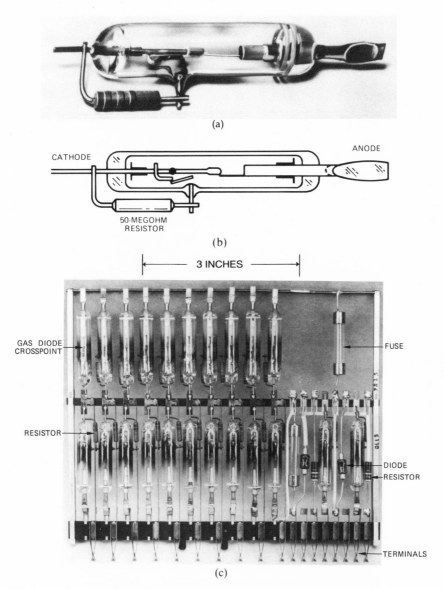

(a)

(b)

CATHODE

ANODE

50-MEGOHM
RESISTOR

3 INCHES

GAS DIODE
CROSSPOINT

FUSE

RESISTOR

DIODE
RESISTOR

TERMINALS

(c)

Fig. 3-43. Gas tubes for the talking path. (a) Gas tube crosspoint. (b) Schematic of gas tube crosspoint. (c) Plug-in network module for the Morris, Illinois field trial. In addition to its switching function, the device amplified voice signals. (Scale is for (c) only.)

quired for the Morris installation. Over 2000 other control tubes were also required to provide appropriate selection and latching of an available path through the network, and later to release the path. Most of these required the stability of the molybdenum cathode discharge for operation.

Thus the developed technology of the cold-cathode gas discharge provided the tools for initial study of the concepts of electronic switching, which would go forward from those years with the broadening technology of semiconductor devices.

## VIII. EPILOGUE

It has been possible here to cover only the highlights of Bell Laboratories contributions to the electron tube art and science. Because of space limitations, many areas are discussed more briefly than they merit. These include magnetron improvement for aircraft and missile detection systems, reliability improvement in these systems, and TWTs for application in the hostile environment of missile guidance systems.

As semiconductor technology advanced, application of resources to electron tubes steadily declined so that, except for TWTs and the microwave triode, development activity by the late 1970s was negligible. Yet in 1974, Western Electric manufactured nearly 1,200,000 tubes to support equipment designed before and during the semiconductor era.

## REFERENCES

1. M. J. Kelly, "The Manufacture of Vacuum Tubes," *Bell Lab. Rec.* **2** (June 1926), pp. 137-144.
2. W. Wilson, "Reducing the Cost of Electrons," *Bell Lab. Rec.* **3** (November 1926), pp. 69-71.
3. J. A. Becker, "The Role of Barium in Vacuum Tubes," *Bell Lab. Rec.* **9** (October 1930), pp. 54-58.
4. W. R. Ballard, "The High Vacuum Tube Comes Before the Supreme Court," *Bell Lab. Rec.* **9** (July 1931), pp. 513-516.
5. H. E. Mendenhall, "Radiation-Cooled Power Tubes for Radio Transmitters," *Bell Lab. Rec.* **11** (October 1932), pp. 30-36.
6. W. H. Doherty and O. W. Towner, "A 50 Kilowatt Broadcast Station Utilizing the Doherty Amplifier and Designed for Expansion to 500 Kilowatts," *Proc. IRE* **27** (September 1939), pp. 531-534.
7. Notes of an interview with J. O. McNally. [AT&T Bell Laboratories Archives Collection.]
8. J. W. Gewartowski and H. A. Watson, *Principles of Electron Tubes* (Princeton, New Jersey: D. Van Nostrand Co., 1965), p. 216.
9. C. E. Fay, "A New Vacuum Tube for Ultra High Frequencies," *Bell Lab. Rec.* **13** (August 1935), pp. 379-382.
10. M. J. Kelly and A. L. Samuel, "Vacuum Tubes as High Frequency Oscillators," *Bell Syst. Tech. J.* **14** (January 1935), pp. 97-134.
11. M. S. Glass, U. S. Patent No. 1,980,196; filed October 10, 1931; issued November 13, 1934.
12. M. S. Glass, U. S. Patent No. 2,139,678; filed October 10, 1936; issued December 13, 1938.
13. M. S. Glass, U. S. Patent No. 2,170,944; filed May 29, 1937; issued August 29, 1939.
14. M. S. Glass and D. Hale, U. S. Patent No. 2,268,194; filed December 2, 1939; issued December 30, 1941.

    S. O. Ekstrand, U. S. Patent No. 2,245,581; filed December 2, 1939; issued June 7, 1941.
15. J. B. Johnson, "A Movable-Screen Cathode Ray Tube," *Bell Lab. Rec.* **26** (May 1948), pp. 219-221.

16. C. H. Prescott, Jr. and M. J. Kelly, "The Caesium-Oxygen-Silver Photoelectric Cell," *Bell Syst. Tech. J.* **11** (July 1932), pp. 334-367.
17. M. J. Kelly, "The Caesium-Oxygen-Silver Photoelectric Cell," *Bell Lab. Rec.* **12** (October 1933), pp. 34-39.
18. M. Glass, U. S. Patent No. 2,414,099; filed April 2, 1942; issued January 14, 1947.
19. S. B. Ingram, "Cold-Cathode Gas-Filled Tubes as Circuit Elements," *Elec. Eng.* **58** (July 1939), pp. 342-346.
20. G. H. Rockwood, "Current Rating and Life of Cold-Cathode Tubes," *Elec. Eng.* **60** (September 1941), pp. 901-902.
21. J. R. Pierce, "Rectilinear Electron Flow in Beams," *J. Appl. Phys.* **11** (August 1940), pp. 548-554.
22. J. R. Pierce, *Theory and Design of Electron Beams* (New York: D. Van Nostrand Co., 1954).
23. G. T. Ford, "Midget Tubes for High Frequencies," *Bell Lab. Rec.* **22** (November 1944), pp. 605-609.
24. G. C. Dalman, "A New Miniature Double Triode," *Bell Lab. Rec.* **25** (September 1947), pp. 325-329.
25. R. H. Varian and S. F. Varian, "A High Frequency Oscillator and Amplifier," *J. Appl. Phys.* **10** (May 1939), pp. 321-327.
26. J. R. Pierce and W. G. Shepherd, "Reflex Oscillators," *Bell Syst. Tech. J.* **26** (July 1947), pp. 460-681; see also Bell Telephone Laboratories, *Radar Systems and Components* (New York: D. Van Nostrand Co., 1949), pp. 488-709.
27. J. B. Fisk, H. D. Hagstrum, and P. L. Hartman, "The Magnetron as a Generator of Centimeter Waves," *Bell Syst. Tech. J.* **25** (April 1946), pp. 167-348; see also Bell Telephone Laboratories, *Radar Systems and Components* (New York: D. Van Nostrand Co., 1949), pp. 56-152.
28. G. B. Collins, ed., *Radiation Laboratory Series: Microwave Magnetrons*, Vol. 6 (New York: McGraw-Hill, 1948), p. 497.
29. J. Feinstein and R. J. Collier, "A Magnetron Controlled by a Symmetrically-Coupled $TE_{011}$ Mode Cavity," *Le Vide* **70** (July/August 1957), pp. 247-254.
30. R. J. Collier and J. Feinstein, U. S. Patent No. 2,854,603; filed May 23, 1955; issued September 30, 1958.
31. A. L. Samuel, J. W. Clark, and W. W. Mumford, "The Gas-Discharge Transmit-Receive Switch," in Bell Telephone Laboratories, *Radar Systems and Components* (New York: D. Van Nostrand Co., 1949), pp. 310-363.
32. K. C. Black, "Stevens Point and Minneapolis Linked by Coaxial System," *Bell Lab. Rec.* **20** (January 1942), pp. 127-132.
33. H. T. Friis, "Microwave Repeater Research," *Bell Syst. Tech. J.* **27** (April 1948), pp. 183-246.
34. A. C. Dickieson, "The TD-2 Story, From Research to Field Trial," *Bell Lab. Rec.* **45** (October 1967), pp. 283-289.
    A. C. Dickieson, "The TD-2 Story, Vacuum Tubes and Systems Engineering," *Bell Lab. Rec.* **45** (November 1967), pp. 325-331.
    A. C. Dickieson, "The TD-2 Story, Changing for the Future," *Bell Lab. Rec.* **45** (December 1967), pp. 356-363.
35. K. M. Olsen, "High-Purity Nickel," *Bell Lab. Rec.* **38** (February 1960), pp. 54-58.
36. J. J. Lander, H. E. Kern, and A. L. Beech, "Solubility and Diffusion Coefficient of Carbon in Nickel: Reaction Rates of Nickel-Carbon Alloys With Barium Oxide," *J. Appl. Phys.* **23** (December 1953), pp. 1305-1309.
37. H. E. Kern, "Research on Oxide-Coated Cathodes," *Bell Lab. Rec.* **38** (December 1960), pp. 451-456.
38. D. W. Maurer and C. M. Pleass, "The CPC: A Medium Current Density, High Reliability Cathode," *Bell Syst. Tech. J.* **46** (December 1967), pp. 2375-2404.
39. J. A. Morton, "A Microwave Triode For Radio Relay," *Bell Lab. Rec.* **27** (May 1949), pp. 166-170.

40. R. Kompfner, "The Traveling Wave-Tube as Amplifier at Microwaves," *Proc. IRE* **35** (February 1947), pp. 124-127.
41. J. P. Laico, H. L. McDowell, and C. R. Moster, "A Medium Power Traveling-Wave Tube For 6000-Mc Radio Relay," *Bell Syst. Tech. J.* **35** (November 1956), pp. 1285-1346.
42. J. E. Sterrett and H. Heffner, "The Design of Periodic Magnetic Focusing Structures," *IRE Trans. Electron Dev.* **ED-5** (January 1958), pp. 35-42.
43. C. E. Bradford and C. J. Waldron, "TH-3 Microwave Radio System: The Traveling-Wave Tube Amplifier," *Bell Syst. Tech. J.* **50** (September 1971), pp. 2223-2234.
44. M. G. Bodmer, J. P. Laico, E. G. Olsen, and A. T. Ross, "The Satellite Traveling-Wave Tube," *Bell Syst. Tech. J.* **42** (July 1963), pp. 1703-1748.
45. R. J. Collier, G. D. Helm, J. P. Laico, and K. M. Striny, "The Ground Station High-Power Traveling-Wave Tube," *Bell Syst. Tech. J.* **42** (July 1963), pp. 1829-1861.
46. G. T. Ford and E. J. Walsh, "The Development of Electron Tubes for a New Coaxial Transmission System," *Bell Syst. Tech. J.* **30** (October 1951), pp. 1103-1128.
47. R. F. Garrett, T. L. Tuffnell, and R. A. Waddell, "The L3 Coaxial System," *Bell Syst. Tech. J.* **32** (July 1953), pp. 969-1005.
48. M. A. Townsend, "Construction of Cold-Cathode Counting or Stepping Tubes," *Elec. Eng.* **69** (September 1950), pp. 810-813.
49. D. S. Peck, "Cold-Cathode Counting Tube," *Bell Lab. Rec.* **31** (April 1953), pp. 127-129.
50. "Two-Decade TAT-1 Record Called 'A Major Triumph,' " *Bell Labs News* (January 29, 1979), p. 3.
51. J. O. McNally, G. H. Metson, E. A. Veazie, and M. F. Holmes, "Electron Tubes for the Transatlantic Cable System," *Bell Syst. Tech. J.* **36** (January 1957), pp. 163-188.
52. V. L. Holdaway, W. Van Haste, and E. J. Walsh, "Electron Tubes for the SD Submarine Cable System," *Bell Syst. Tech. J.* **43** (July 1964), pp. 1311-1338.
53. T. F. Gleichmann, A. H. Lince, M. C. Wooley, and F. J. Braga, "Repeater Design for the North Atlantic Link," *Bell Syst. Tech. J.* **36** (January 1957), pp. 69-101.
54. J. O. McNally, "Early Tube and Cathode Work Leading to Submarine Cable Tubes." [AT&T Bell Laboratories Archives Collection: *Vacuum Tubes.*]
55. S. O. Ekstrand, "Devices—The Hardware of Progress," *Bell Lab. Rec.* **39** (May/June 1969), pp. 174-180.
56. E. I. Gordon, "A 'Solid-State' Electron Tube for the PICTUREPHONE Set," *Bell Lab. Rec.* **45** (June 1967), pp. 174-179.
57. M. E. Hines, M. Chruney, and J. A. McCarthy, "Digital Memory in Barrier-Grid Storage Tubes," *Bell Syst. Tech. J.* **34** (November 1955), pp. 1241-1264.
58. T. S. Greenwood and R. E. Staehler, "A High-Speed Barrier Grid Store," *Bell Syst. Tech. J.* **37** (September 1958), pp. 1195-1220.
59. C. W. Hoover, Jr., "An Experimental Flying-Spot-Store for Electronic Switching," *Bell Lab. Rec.* **37** (October 1959), pp. 366-372.
    C. W. Hoover, Jr., G. Haugk, and D. R. Herriott, "System Design of the Flying Spot Store," *Bell Syst. Tech. J.* **38** (March 1959), pp. 365-401.
    C. W. Hoover, Jr., R. E. Staehler, and R. W. Ketchledge, "Fundamental Concepts in the Design of the Flying Spot Store," *Bell Syst. Tech. J.* **37** (September 1958), pp. 1161-1194.
    M. B. Purvis, G. V. Deverall, and D. R. Herriott, "Optics and Photography in the Flying Spot Store," *Bell Syst. Tech. J.* **38** (March 1959), pp. 403-424.
60. M. A. Townsend and W. A. Depp, "Cold-Cathode Tubes for Transmission of Audio Frequency Signals," *Bell Syst. Tech. J.* **32** (November 1953), pp. 1371-1391.
61. A. D. White, "An Experimental Gas Diode Switch," *Bell Lab. Rec.* **36** (December 1958), pp. 446-449.
62. M. A. Townsend, "Cold Cathode Gas Tubes for Telephone Switching Systems," *Bell Syst. Tech. J.* **36** (May 1957), pp. 755-768.
63. K. S. Dunlap and R. L. Simms, Jr., "A Cold-Cathode Gas Tube Space-Division Network for an Electronic Telephone Switching System," paper presented at AIEE meeting, Morris, Illinois (October 1960).

# Chapter 4

# Optical Devices

*In the 1920s, the Bell System developed optical devices for sound motion pictures, telephoto transmission, and television. Optical work regained prominence in the more modern era through the classic 1958 paper by A. L. Schawlow and C. H. Townes, which pointed the way to coherent, unidirectional sources of light (lasers). Subsequent developments in laser technology at Bell Laboratories included lasers of many wavelengths and power levels. Of special importance to communications were solid-state lasers operating at room temperature and avalanche photodiodes that provided very sensitive light detection. Other areas in optics to which Bell Laboratories has made contributions include optoisolators that are significant improvements over their electromechanical counterparts, a silicon target television camera tube, a solid-state television camera, and light-modulation and -deflection devices.*

## I. INTRODUCTION

The history of optical devices in the Bell System dates back to experiments by Alexander Graham Bell. Bell transmitted speech over a beam of light using a device he called the photophone, for which he was granted a patent in 1880. Practical lightwave communications, as it turned out, had to await the development of lasers and low-loss optical fibers in the 1960s and 1970s, developments to which Bell Laboratories scientists made major contributions. Although not directly related to telephony, the importance of transmitting visual images was recognized very early in the Bell System and led to pioneering work in the 1920s, first in telephotography and then in television. The vacuum tube photocell was essential to these first demonstrations and is an example of an early Bell Laboratories contribution to optical devices. Optical device development in the Bell System over the years has produced advances in television, sound recording on motion picture film, high-speed motion picture photography, and information display and storage. With the advent of lasers in the early 1960s, work on solid-state optical devices came into prominence, providing the foundation for modern optical communications.

Principal authors: L. A. D'Asaro and L. K. Anderson

## II. LASERS AND LIGHT-EMITTING DIODES

The first light sources utilized in the Bell System were incandescent arc lamps and filament lamps. (For more information on these devices, see a companion volume in this series subtitled *The Early Years (1875-1925)*, Chapter 6, section 3.4.5.) Filament lamps were used as early as 1890 in a telephone trunk signaling circuit and in 1894 for switchboard line signaling. The development of improved tungsten-filament indicator lamps was pursued at Bell Laboratories until 1932, and these lamps were widely used even up to the 1970s, when they began to be replaced by light-emitting diodes (LEDs), discussed later in this section.

Filament and arc lamps radiate energy in all directions and over a wide range of wavelengths. Although directionality and monochromaticity can be achieved by spatial and wavelength filtering of a filament lamp or arc lamp, the filtering process produces an output beam of extremely low intensity and hence of limited practical usefulness. It was only with the development of the laser that a light source with the directionality, monochromaticity, and high intensity needed for optical communications became available. The special characteristics of the laser are directly attributable to the optical gain or net stimulated emission of the laser medium and the feedback provided by the laser cavity mirrors.

While the theoretical basis for stimulated emission was given by Einstein[1] as early as 1917, it was not until 1954 that the first practical device based on stimulated emission, the microwave maser (Microwave Amplification by Stimulated Emission of Radiation), was invented by J. P. Gordon, H. J. Zeiger, and C. H. Townes of Columbia University.[2] This device formed the basis for extremely low-noise microwave receivers that were used in the early Telstar satellite experiments and in radio astronomy. (Early maser work is described in a companion volume in this series subtitled *Physical Sciences (1925-1980)*, Chapter 5, section I.) The maser set the stage for the invention of the laser (Light Amplification by Stimulated Emission of Radiation), which was proposed in 1958 by A. L. Schawlow of Bell Laboratories in collaboration with Townes, who was at that time a Bell Laboratories consultant.[3] As described in some detail in Chapter 5 of *Physical Sciences (1925-1980)*, these scientists described an optical oscillator, pumped by radiation, and introduced an important new concept, the open resonator, consisting of an amplifying medium between two parallel, opposing reflectors.

The initial publication of the Schawlow-Townes article sparked a flurry of attempts to produce a workable solid-state laser following their prescription. In 1960, T. H. Maiman of the Hughes Research Laboratories used a chromium-doped ruby crystal illuminated by a pulsed flash lamp to create the first successful device.[4] It operated in the red portion of the visible spectrum at a wavelength of 0.6943 micrometer ($\mu$m). Publication of Maiman's results was followed by a paper from Bell Laboratories, which was the first to show the unusual directionality and coherence of laser light.[5]

Another form of laser, the helium-neon laser, was proposed by A. Javan in mid-1959 and first demonstrated by Javan, W. R. Bennett, Jr., and D. R. Herriott at the end of 1960.[6] The gain medium in this case was an electrically excited gas discharge. It operated continuously at low power at a wavelength of 1.15 $\mu$m in the near infrared portion of the spectrum. This was the first continuously operating laser and produced enormous interest and excitement.

In 1962, A. D. White and J. D. Rigden demonstrated the first red-emitting (0.63-$\mu$m), continuously operating helium-neon laser.[7] The helium-neon system was explored in great detail and resulted in gas lasers as small as one inch long. For the first time, an optical oscillator having all the characteristics of microwave devices was available for communications experiments. Another important gas laser emitting visible radiation was the argon ion laser,[8] which, because of its combination of radiation in the short-wavelength blue and green portions of the spectrum and its increased power capability, opened up broad new areas of applications, such as eye surgery, holography, and fluorescence.

Continued work on solid-state lasers soon led in quick succession to the demonstration of the first continuously operating solid-state laser, utilizing a new laser material, $CaWO_4$:Nd, by L. F. Johnson, G. D. Boyd, K. Nassau, and R. R. Soden,[9] and of the first continuously operating ruby laser at liquid-nitrogen temperatures by D. F. Nelson and W. S. Boyle.[10]

The most important and widely used continuous dielectric solid-state laser, utilizing a crystal of yttrium aluminum garnet (YAG) doped with neodymium, was invented by J. E. Geusic, H. M. Marcos, and L. G. Van Uitert in 1964.[11] This YAG laser emitted up to several hundred watts of continuous light power in the infrared region (1.06 $\mu$m) and saw wide commercial application in machining and drilling. [Fig. 4-1]

The most efficient of all dielectric solid-state lasers, a holmium laser known as Alphabet YAG (because it had so many elements added as dopants during crystal growth), operated at liquid-nitrogen temperatures and emitted radiation at 2.1 $\mu$m. It was introduced by Johnson, Geusic, and Van Uitert in 1966.[12]

While the power capability of these solid-state lasers was quite impressive, the $CO_2$ gas-discharge laser of C. K. N. Patel, radiating at 10.6 $\mu$m, was, at the time, the most dramatic in terms of power capability.[13,14] (See *Physical Sciences (1925-1980)*, Chapter 5, section IV.)

All dielectric lasers were pumped by external light sources such as gas-discharge lamps. For improved reliability for applications in optical communications and data storage systems, an all-solid-state laser was desired in which not only the laser material but also the light source that pumps

Fig. 4-1.   A high-power Nd:YAG laser burning a hole in a block of wood. This 1964 innovation emitted up to several hundred watts of continuous light power in the infrared range.

the laser would be solid state. F. W. Ostermeyer succeeded in designing the first such device, a continuously operating, room-temperature YAG laser pumped by LEDs.[15] However, this arrangement was later supplanted by the structurally simpler semiconductor laser, discussed below.

While different lasers could be made to operate at a variety of discrete wavelengths, a number of new techniques were developed to obtain tunable coherent radiation with the qualities of laser light. These techniques involved nonlinear optical effects and included optical second harmonic generation, optical parametric oscillation, and stimulated Raman oscillation. Among Bell Laboratories contributions were (1) the introduction of the concept of phase matching, which makes possible practical nonlinear interactions;[16,17] (2) the first successful operation of pulsed[18] and continuous[19] optical parametric oscillators; and (3) the concept and implementation of the spin-flip Raman laser, which was tunable over wide regions of the infrared portion of the spectrum.[20] The tunability of coherent optical sources was perceived at the time to be of potential interest for a broad range of applications, including optical communications and high-resolution spectroscopy. Tunable lasers were actually used in air pollution monitoring,[21] but otherwise did not find commercial applications.

Geusic, H. J. Levinstein, S. Singh, R. G. Smith, and Van Uitert[22] succeeded in developing the first practical solid-state laser source in the visible region by second harmonic generation of the YAG laser. [Fig. 4-2] Conversion to visible output was based on the development and understanding of nonlinear optical materials such as barium sodium niobate and lithium niobate.[23,24]

For electronic applications, a compact p-n junction device converting electrical energy directly into optical energy is most desirable. In the mid-

Fig. 4-2.   S. Singh observing green light generated in barium sodium niobate from the invisible light output of an Nd:YAG laser.

1950s, it was recognized that compounds made from elements in columns III and V of the periodic chart (III-V compounds) have similar structural and chemical properties to those of germanium and silicon, with the added capability of light generation by radiative electron-hole recombination. Attention was focused especially on the compound gallium phosphide (GaP), since it was known that this material could potentially generate light in the wavelength range corresponding to the red and green parts of the visible spectrum. From the late 1950s and through the 1960s, the scientists of the research area thoroughly investigated this material and laid the foundation of an LED device technology. This work is detailed in other volumes in this series subtitled *Physical Sciences (1925-1980)*, pp. 87-90 and 428-430, and *Communications Sciences (1925-1980)*, Chapter 7, section 2.4.

Significant advances were made in the mid-1960s in the understanding of the physics of radiative recombination[25] and in the liquid-phase epitaxial growth of efficient light-emitting junctions in GaP crystals.[26,27] P-n junction epitaxial crystals were made to emit both red and green light at intensities suitable for display application under normal illumination. In the course of this work, it was demonstrated that the radiative mechanism responsible for light generation was tied to a special type of defect center at which

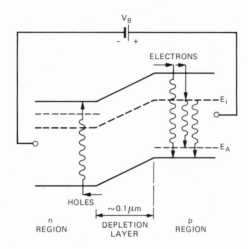

Fig. 4-3.   The electroluminescence process: band diagram of a forward-biased p-n junction showing both full-gap radiation and trap-dominated radiation. [Bergh and Dean, *Light-Emitting Diodes* (1976): 36.]

minority carriers injected across the p-n junction recombine. [Fig. 4-3] These centers are isoelectronic (having the same number of valence electrons as the host atom or atoms that they replace), and they act as traps for mobile charge carriers, providing a center for radiative recombination. The most important centers are: (1) a nitrogen impurity on a phosphorus site, which gives rise to green luminescence;[28] and (2) a nearest neighbor pair of zinc and oxygen atoms on gallium and phosphorus sites respectively, which produces red luminescence.[29]

In the late 1960s, the work on LEDs gradually moved from research into the development area, where the efficiency of both red and green devices was substantially improved, and where a mass production technology was developed.[30,31] Large-area, single-crystal substrates grown by a high-pressure liquid encapsulation Czochralski method developed by S. J. Bass and P. E. Oliver in England[32] were used to form efficient junctions by a new high-capacity liquid-phase epitaxial process,[33] the first commercial application of this technique.

At about the same time, workers at Monsanto Laboratories found that the nitrogen doping that gives rise to green luminescence in GaP crystals also enhances red luminescent efficiency in ternary $GaAs_{1-x}P_x$ epitaxial layers grown by gas-phase epitaxy.[34] This technology became widely used for red and yellow LEDs, while nitrogen-doped liquid-phase epitaxy GaP remained the best choice for green.

Single indicators [Fig. 4-4], illuminators, numeric and alphanumeric displays, and optically coupled isolators were all placed in Western Electric manufacture. These devices found wide application in the Bell System as reliable, low-power lamps and displays for consoles, test panels, and station apparatus, and as efficient low-noise couplers and isolators in various transmission and switching systems. (See also section IV.)

Another important member of the III-V family of compounds is gallium arsenide (GaAs). This material gives more efficient electroluminescence than GaP but in the near infrared portion of the spectrum, i.e., at 0.9 μm. The radiative mechanism responsible for light generation in this case does not involve the use of special types of impurities as with GaP, but is based on direct recombination of electrons and holes.

In 1962, the process of semiconductor junction laser action in GaAs p-n junctions was first realized at General Electric Research Laboratories.[35] This laser could operate only in a pulsed mode to achieve the necessary current density. To increase the performance of such lasers, the stripe geometry p-n junction laser was invented by D. K. Wilson for mode control and better thermal dissipation.[36] In the following years, considerable effort was expended on improving the performance of these lasers by better heat sinking and by better minority carrier confinement.

A major breakthrough was made by I. Hayashi, M. B. Panish, P. W. Foy, and S. Sumski, who used advances made in liquid-phase epitaxial growth of lattice-matched layers of gallium aluminum arsenide on GaAs to produce a double-heterostructure laser in which injected carriers from a forward-biased p-n junction and light from the electron-hole recombi-

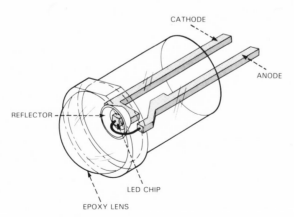

Fig. 4-4. Schematic of a typical LED indicator. The LED chip is mounted to one of the metal leads. [Bergh and Dean, *Light-Emitting Diodes* (1976): 531.]

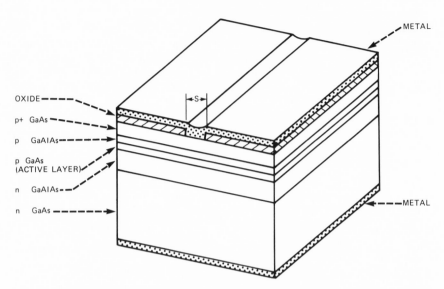

Fig. 4-5. Stripe geometry double-heterostructure laser, showing the p-type GaAs active layer ($\sim 1$ μm thick) sandwiched between p- and n-type layers of gallium aluminum arsenide. The width of the active region in which lasing occurs is controlled by the voltage drop in these layers and the stripe width S is usually chosen to be comparable to a filament width (5-10 μm). The top of the laser is bonded to a heat sink.

nation were both confined in the same thin active layer.[37] The result was a much lower current, called threshold current, at which laser action would commence; reduced temperature dependence; and higher lasing efficiency than in junction lasers used earlier. By using an active stripe 12 μm wide and by bonding the laser to a type IIA diamond heat sink, reproducible continuous lasing at and above room temperature was achieved in 1970.[38] Later, improved stripe geometry, double-heterostructure lasers operated with a dc input current of 100 milliamperes (mA) at 2 volts (V) and produced an output of many milliwatts. The structure of this laser is shown in Fig. 4-5 and the threshold characteristic in Fig. 4-6.

A very bothersome feature of these early junction lasers was degradation of their light output during use to a point where the devices became useless. Progress towards long-term reliability was obtained by eliminating strain, which induced clusters of dislocations during lasing.[39] Operation for 100,000 hours at room temperature was thus achieved in lasers intended for optical communications.[40]

Continued development of this technology in a related material system (indium phosphide) resulted in other double-heterostructure lasers that operated at very low currents (less than 25 mA) and produced power up to 30 milliwatts (mW). These devices were capable of operating for up to one million hours at 10 degrees C (sea-bottom temperature), a requirement for their use in submarine cables.

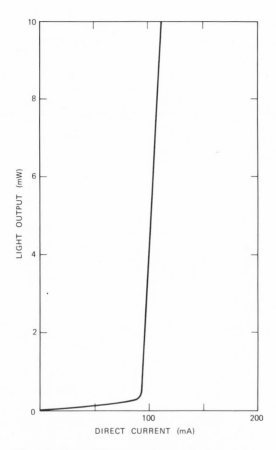

Fig. 4-6. Laser light output from one mirror face as a function of direct current, showing the sharp threshold of lasing.

In 1971, a high-pulse-power, single-transverse mode design was demonstrated, in which the laser cavity was several micrometers thick, and which contained critical thicknesses of regions having controlled gain and loss to suppress high-order modes.[41] These lasers produced an output of 0.5 watt (W) in a 100-nanosecond pulse. This power was used for micromachining thin films in high-resolution image-recording systems, as described in section VI.

### III. LIGHT-DEFLECTION AND -MODULATION DEVICES

While work on optical sources at Bell Laboratories started only in the late 1950s, work on devices to modulate light beams dates from the mid-1920s, when a novel light valve was invented that could vary the intensity of a light beam in direct proportion to the strength of a modulated electrical current. This light valve was the key component necessary for motion picture technology to emerge from silence into the era of the talkies. The operating principle was based on the mechanical motion of an electrical conductor in a magnetic field. An audio signal was first converted by a microphone to a modulated electrical current. The space between a pair of parallel conducting wires connected in series served as an optical slit that opened and closed as the electrical current changed. The light valve, driven by this current, in turn modulated the intensity of a light beam that illuminated a track on a moving photographic film. Upon development of the film, the sound track formed a permanent recording of the original audio signal.

Improvements to the design of this light valve during the late 1920s and 1930s resulted in a device capable of modulating light at frequencies covering the full audio spectrum. Advantage was taken of this high-fidelity capability in the development of a stereophonic-sound film system that made it possible for the first time to produce a sound facsimile with concert-hall quality. A detailed description of this four-track sound film system was presented by Bell Laboratories in a series of seven papers at the joint meeting of the Acoustical Society of America and the Society of Motion Picture Engineers in the spring of 1941.[42]

The success of the light valve's performance in this system appeared to mark the end of development in light-valve technology. However, the end was only temporary, for with the invention of the laser, new systems concepts emerged in the 1960s for using a laser beam in information storage systems, graphics and visual display techniques, and broadband communications. All of these applications required extremely high-speed light valves (modulators) and light-beam deflectors. The speeds required were, in fact, so high that they could not be achieved with any device, such as the original light valve, that operated by mechanical motion of its parts.

By 1962, experiments were under way at Bell Laboratories that laid the foundations for modern electro-optic modulators and deflectors (see the companion volume subtitled *Physical Sciences (1925-1980)*, p. 191). These modulators are based on the application of an electric voltage to some types of transparent crystals. In some cases, this causes the velocity of light that is linearly polarized along a certain axis within the crystal to be changed, while light that is polarized orthogonally remains unaffected. This application can be used to build a light deflector, since the polarization of light at 45 degrees to the axis can be changed by 90 degrees. With standard optical components, it is possible to deflect the rotated component of the laser beam into a different direction from the unrotated component, and make a two-position beam deflector or high-speed light valve. [Fig. 4-7]

This concept was extended by T. J. Nelson,[43] who, by properly connecting and directing a laser beam through a series of $n$ simple two-position beam deflectors, produced a composite deflector having $2^n$ resolvable output

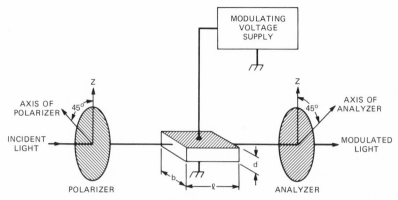

Fig. 4-7. Optic amplitude modulation. Application of an electric voltage to certain types of transparent crystals allows 90-degree rotation of the plane of polarization of the incident laser beam at a rate limited only by voltage application. [Denton in *Laser Handbook,* ed. Arecchi and Shulz-Dubois (1972): 706.]

positions. Models of such a composite scanner were made with 20 stages that simulated an output array of 1024-by-1024 resolvable beam positions. As the number of stages of the composite electro-optic deflector increased, so did the electrical drive power requirements and the optical quality requirements of the electro-optic crystals. S. K. Kurtz[44] showed that the ultimate limit in such an optical addressing system can be described by a capacity-speed product, the magnitude of which involves material constants and the electrical drive power. Since there are practical limits to the power that can be dissipated in such a system, in-depth materials studies and developments were conducted that were aimed at producing a high-quality crystal material that exhibited an unusually large electro-optic effect. Lithium niobate and lithium tantalate, two very useful man-made crystalline materials, evolved from this effort.[45,46]

An alternative approach to light deflection is acousto-optic deflection, in which light is diffracted, as in a grating, by the sinusoidal refractive index pattern produced by an ultrasonic wave.[47] [Fig. 4-8] The deflection angle is changed by altering the frequency of the ultrasound. In this case, the capacity-speed product is related to the frequency bandwidth over which the deflector can be operated. Interestingly, the lithium niobate crystal also exhibited a very large piezoelectric effect as well as large acousto-optic interaction, making the material very useful in acousto-optic devices. Much of the initial work in acousto-optic theory and material improvements is summarized by R. W. Dixon,[48] while the material properties required for optimum performance of acousto-optic deflectors were described by D. A. Pinnow.[49]

In 1966, R. C. LeCraw introduced another class of high-speed light modulators that operated on the magneto-optic effect.[50] (See also *Communications Sciences (1925-1980),* Chapter 7, section 4.2.)

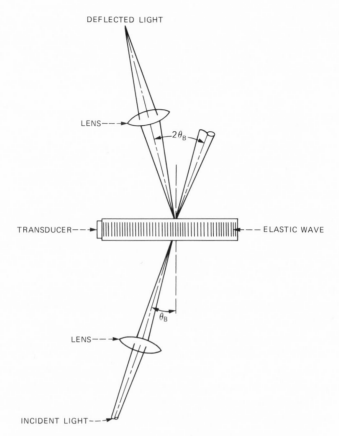

Fig. 4-8. Acousto-optic deflection. The light beam is diffracted, as in a grating, by the sinusoidal modulation of the refractive index produced by an ultrasonic wave. The deflection angle is changed by altering the frequency of the ultrasound. [Denton in *Laser Handbook*, ed. Arecchi and Shulz-Dubois (1972): 720.]

Acousto-optic light deflectors found applications in production systems that scanned laser beams to accomplish various useful functions, such as the generation of pattern masks for silicon integrated circuits[51] and the inspection of these masks for defects.[52] In addition, exploratory electro-optic and acousto-optic devices were developed for optical communications. These devices were capable of modulating light beams with bandwidths up to several gigahertz.[53]

## IV. LIGHT-DETECTION DEVICES

Light-detection devices fulfill a number of needs. Where simple detectors are needed (for example, in lightwave communication), an individual point detector, such as the photodiode, is used. In applications such as optical memories, arrays of photodetectors of limited spatial resolution are adequate. For conversion of images into electrical signals, high-resolution photodetection devices are required.

Even before the invention of the transistor, it was noticed that reverse-biased metal-semiconductor junctions were quite sensitive to optical illumination. The advent of semiconductor p-n junctions made high-sensitivity photodiodes possible. With proper design, essentially all incident optical radiation can be absorbed within or close to the space charge layer of these reverse-biased junctions. The photogenerated electron-hole pairs are then separated and collected as photocurrent across the high-electric-field region of the junction.

Interest in the detection of high-speed laser signals led R. P. Riesz to develop fast photodiodes in the early 1960s.[54] Small-area semiconductor p-n and p-i-n junction photodiodes as well as metal-semiconductor junction and point-contact photodiodes were fabricated from both silicon and germanium from 1962 to 1966. These high-speed silicon and germanium photodiodes had high light-collection efficiency throughout the visible and near infrared regions of the spectrum to 1.1 and 1.6 $\mu$m, respectively. [Fig. 4-9] The fastest of these photodiodes reached rise and fall times shorter than 100 picoseconds (psec).

As an outgrowth of investigations of the reverse breakdown characteristics of junction diodes, it was observed that high internal current gain could be obtained in photodiodes operated near breakdown. Under these conditions, the photogenerated electrons and holes gain sufficient energy by moving through the high-field region of the junction to release an avalanche of electron-hole pairs through impact ionization. In many aspects, these avalanche photodiodes constitute a solid-state analog to the photomultiplier tube.

A serious obstacle to the use of avalanche photodiodes was the presence of tiny localized regions of breakdown, called microplasmas, in the p-n junction. These regions fluctuated off and on in a random way, producing excess noise. The first low-noise, microplasma-free avalanche photodiodes suitable for use as high-gain photodectors in an optical transmission system were made in 1964.[55] Subsequently, avalanche photodiodes found wide application as high-speed, low-noise detectors in optical transmission systems.[56]

An alternate approach to photodetectors with gain is the phototransistor. Its origin dates back to 1949, when J. N. Shive noticed the influence of light on the electrical characteristics of the early germanium transistors.[57] In phototransistors, absorbed optical radiation generates electron-hole pairs, which are then separated at the collector-base junction, thereby causing a current flow into the base region. This base current is then amplified by the current gain of the transistor.

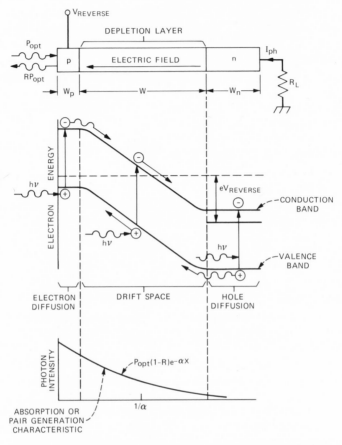

Fig. 4-9. Operation of the solid-state photodiode. High-speed silicon and germanium devices have high collection efficiency throughout the visible and near infrared range. Photogenerated electron-hole pairs are separated and collected as photocurrent across the high-electric-field junction region. [Melchior in *Laser Handbook*, ed. Arecchi and Shulz-Dubois (1972): 771.]

A very early application of an array of phototransistors to a card translator[58] is described in Chapter 1, section 2.1. Arrays of phototransistors have continued in applications as readout devices in systems such as the experimental holographic optical memory system developed in the late 1960s.[59]

The devices described so far are either individual detectors or array detectors of fairly limited resolution. The well-known vidicon used in television is an example of a large-area, high-resolution device required for the transmission of images. Conventional vidicons operate only over a limited range of light levels and cannot tolerate direct exposure to sunlight.

To overcome these shortcomings for the PICTUREPHONE* visual telephone service, in 1967, a team under E. I. Gordon's leadership developed
a camera tube whose image-sensing target consisted of an array of planar
silicon photodiodes addressed by a low-energy scanning electron beam.[60]
The target of such a silicon vidicon consisted typically of a silicon wafer
2 centimeters (cm) in diameter containing about one-half million photodiodes on 15-$\mu$m center-to-center spacings. The wafer thickness was 15
$\mu$m. Image light incident on this target lowered the voltage of the individual
photodiodes, which were reverse biased by the scanning electron beam.
The current needed for the recharging of each photodiode during the
electron beam scanning cycle was a measure of the incident-light intensity.
Successful fabrication of these silicon targets required extraordinarily clean
processing facilities and proper gettering procedures to eliminate defects
and to reduce dark currents. In addition, a resistive layer had to be developed to prevent space charge buildup from interfering with the scanning
electron beam (see also Chapter 3, section VI).

While the silicon camera tube combined the reliability and sensitivity
of a silicon integrated diode array with the low cost and simplicity of
electron beam scanning, its operation still required a glass envelope. This
limitation is avoided in the charge-coupled device (CCD) invented by
Boyle and G. E. Smith in 1970, as described in detail in Chapter 2, section
6.5 of this volume. This device led to many applications, including filters,
card readers, pattern recognition devices, and a compact all-solid-state
camera. In a CCD imaging device, the carriers, which are excited optically
in a silicon wafer, are stored as packets of charge in potential wells located
underneath biased electrodes, and moved from one electrode to adjacent
electrodes towards a readout terminal by appropriate pulsing of the electrode potentials. Both linear and two-dimensional CCDs can be used as
image sensors. While early CCDs held 8 bits of information, later line
scanners had up to 1700 elements. Two-dimensional CCD arrays progressed
from 8 by 16 elements for the first feasibility demonstration to a 500-
by-500 storage element area. CCD arrays, combined with color-splitting
prisms, made possible compact, lightweight, hand-held color television cameras with a sensitivity comparable to commercial vidicons.
[Fig. 4-10]

All detectors discussed thus far respond to a limited range of optical
wavelengths. In particular, they are insensitive in the far infrared range.
For some applications, a detector that responds to all incident radiation,
particularly in the infrared portion of the spectrum, is of interest. A detector
based on the pyroelectric effect fills this need. Although this effect has
been known since ancient times, it was not until 1956, when A. G.

---

* Service mark of AT&T.

├─ 4 centimeters ─┤

Fig. 4-10. Television camera with CCD imaging array. The CCD array consists of 500 by 500 picture elements.

Chynoweth investigated it in detail, that the foundations were laid for its use in light detectors.[61] A strong pyroelectric effect is observed in certain ferroelectric crystals, such as barium titanate, strontium barium niobate, and triglycene sulfate, when these crystals are operated close to their Curie temperatures. In pyroelectric detectors, optical radiation is absorbed, producing a change in temperature, which, in turn, induces a change in polarization that is sensed as an electrical signal. Although pyroelectric detectors do not reach the detection sensitivities of semiconductor detectors, they have a broad spectral response, extending from the visible throughout the infrared region, and provide high-speed response.

### V. OPTOISOLATORS

The availability of efficient LEDs in the early 1960s led to a new possibility for electronic devices; a combination of an LED and an efficient photodetector could provide a means of isolating dc potentials from ac signals. This function, which is frequently needed in electronic circuitry, had traditionally been performed with relays or transformers. The solid-state optoisolator is smaller, lighter, and more reliable, and ultimately became less costly than the traditional devices.

The first solid-state optoisolators were made by coupling a GaP LED to a silicon solar cell[62] using a low-melting-point glass. Subsequent devices used LEDs made of GaP, GaAs, $GaAs_{1-x}P_x$, or $Ga_{1-x}Al_xAs$, but all optoisolators of modern design use silicon photoconductors, junction photo-

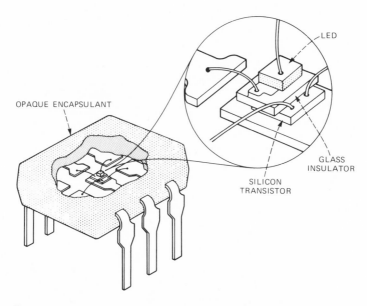

OPAQUE ENCAPSULANT

LED

GLASS
INSULATOR

SILICON
TRANSISTOR

Fig. 4-11. Diagram of an optoisolator in a plastic-encapsulated dual in-line package. [Bergh and Dean, *Light-Emitting Diodes* (1976): 563.]

diodes, or phototransistors as detectors. [Fig. 4-11] The coupling material is typically a transparent plastic. In all cases, the optoisolator must be followed by an amplifier stage, usually integrated into a single package with the LED and photodetector, to compensate for the attenuation between the LED input and the photodetector output.

The availability of optoisolators led to new freedom in circuit designs. Optoisolators can be used to avoid ground loops in complex systems and to reduce noise pickup. They resulted in lighter and more compact equipment by eliminating the need for bulky transformers and relays.

## VI. IMAGE DISPLAY DEVICES

For the period of this history, the most important and durable information display device for high-density displays was the cathode ray tube in its many manifestations. These include oscilloscope displays, computer alphanumeric displays, and television displays in both monochrome and color. The awkward bulk of the cathode ray tube has led to numerous attempts to replace it with solid-state or other more compact display devices, but as of 1975, none had been successful in commercial application for high-density displays.

Section II described the use of LEDs as indicators and in alphanumeric displays. The development of sophisticated laser technology and related light-control devices in the 1960s led to the idea of using laser-addressed

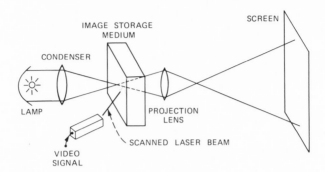

Fig. 4-12.   Display system for the projection of an image written by a low-power laser beam onto a two-dimensional storage medium used as a slide or transparency.

light deflectors for graphics display applications. Attempts to use visible lasers directly for display—for example, by scanning a modulated laser beam over a screen to produce a video image—while technically successful, proved economically unattractive and potentially hazardous because of the high laser power required (on the order of 1 W for each square meter of screen area).[63] As a result, attention turned to a laser-addressed display technology in which a deflected and modulated low-power laser beam was used to write an image onto a two-dimensional storage medium that could then serve as a slide or transparency in a more or less conventional optical projection system.[64] [Fig. 4-12] This approach takes advantage of the high resolution and fast writing capability of a finely focused, rapidly deflected laser beam, while relying on an efficient, bright, and economical conventional lamp to provide basic screen illumination.

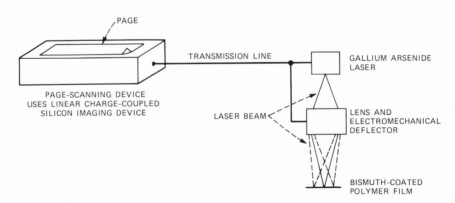

Fig. 4-13.   Details of image recording, using a bismuth-coated, polymer film. A GaAs laser is used to machine a microfilm transparency in the bismuth layer.

The key to the practicality of such systems was the storage medium, and several possibilities were investigated. In one system, a very thin coating of bismuth on a transparent polymer film was used.[65,66] Images were written onto the film using a scanning pulsed YAG or GaAs laser that locally melted a succession of circular spots in the bismuth. Surface tension in the liquid metal caused this liquid to pile up at the edge of the melted spot, leaving a transparent hole in the bismuth. Microfilm pictures built up by machining a succession of such spots are permanent, and suitable for high-speed facsimile and related applications. [Fig. 4-13] Subsequently, a number of researchers, primarily at the Philips Laboratories Division of North American Philips Corporation, have made this laser hole-burning technique the basis of a high-density data storage technology (optical disk recording), in which the storage medium is a rotating disk covered with a thin laser-machinable coating.

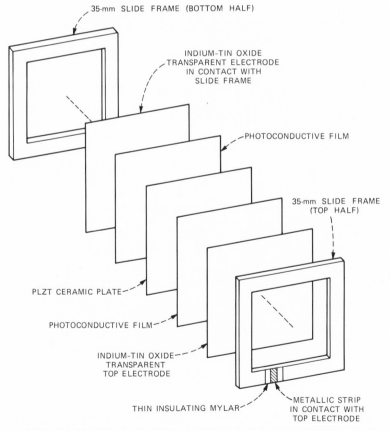

Fig. 4-14. Exploded view of a ferpic-type image storage system, which provides erasable storage on a photoconductive film on a thin ceramic plate. [Maldonado and Fraser, *Proc. IEEE* **61** (1973): 975.]

Another approach was explored in the early 1970s, based on the unique properties of certain ferroelectric ceramic materials such as Pb-based lanthanum-doped zirconate titanates (PLZTs). In these materials, a thin plate of ceramic was coated with a thin photoconductive film and transparent electrodes to form a sandwich structure called a ferpic, whose optical properties could be varied by a combination of voltage and light.[67] [Fig. 4-14] In particular, if the ferpic was scanned with a low-power blue laser beam while a potential was applied across the transparent electrodes, the addressed regions ferroelectrically switched to a different, reversible optical state, thus providing erasable image storage viewable in projection.

In yet another approach, a thin liquid-crystal layer was sandwiched between transparent electrode substrates.[68] [Fig. 4-15] The liquid-crystal layer could be locally thermally switched from a stable, transparent state to an equally stable but erasable scattering state by the heat from a finely focused, low-power infrared laser beam. Of the two technologies, the

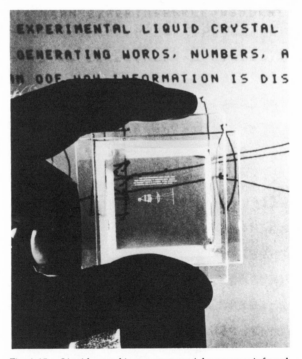

Fig. 4-15. Liquid-crystal image storage. A low-power infrared laser switches the liquid-crystal film from a stable transparent state to a stable (but erasable) scattering state to form an image. In reflected light, the image is white on black. However, in a projected image that uses transmitted light, the image is black on white, as shown in the background. [Maydan, *Proc. IEEE* **61** (1973): 1011.]

liquid crystal offered the greatest long-term promise. In particular, liquid-crystal imaging has attractive applications in slow-scan graphics display systems in which high-resolution pictures are reconstructed for temporary viewing from data sent over ordinary voice-grade telephone channels. Possible applications for such systems include remote blackboards for educational purposes and graphics terminals for computer applications.

## VII. DEVICES FOR LIGHTWAVE COMMUNICATIONS

Much of the work on optical devices, from the appearance of the laser in the early 1960s through the early 1970s, was directed toward optical communications applications. The various types of lasers, modulators, and photodetectors described so far are key elements of such systems. However, what was needed for practical applications of this new device technology was a reliable transmission medium. It was not until 1970, when glass optical fibers reached an attenuation of 20 dB per kilometer (dB/km), that optical communications systems were thought to be on the verge of practicality. Enormous progress was made in the early 1970s in reducing the attenuation in glass fiberguides to a point where losses of a few dB/km— a degree of transparency unheard of heretofore—became routine. With continued development, losses were reduced to only a fraction of a dB/km, thus opening the way for repeaterless transmission spans of many tens of kilometers. These achievements in fiberguide technology, coupled with equally impressive achievements in improving the performance and extending the reliability of GaAs junction laser sources, led to the recognition that practical optical communications or lightwave communications systems were technically feasible. For the research aspects of lightwave communication, see Chapter 7 of *Communications Sciences (1925-1980)*, Chapter 13, section II of *Physical Sciences (1925-1980)*, and the review of early work on lightwave technology by S. E. Miller, E. A. J. Marcatili, and T. Li.[69]

To test the applicability of lightwave communications systems in interoffice digital trunking in metropolitan areas, a high-capacity, experimental system operating at 45 megabits per second (Mb/sec) was designed and set up in an environment approaching field conditions at the Bell Laboratories facility in Atlanta, Georgia in early 1976.[70] The basic components of the system were: (1) a 2000-foot sheathed and protected cable, which was 1.3 cm in diameter and contained 144 fibers, pulled through standard underground ducts [Fig. 4-16], (2) a feedback-stabilized GaAlAs junction laser transmitter [Fig. 4-17], and (3) a high-sensitivity silicon avalanche photodiode receiver. [Fig. 4-18] With these self-contained, packaged modules that could be mounted on circuit boards together with standard regenerator electronics, 50 dB of transmission loss could be tolerated in an operating lightwave system. With low-loss fiberguides in the cable, this design translated into repeater spacings of 7 km—enough to interconnect

Fig. 4-16.   Experimental fiberguide cable. Messages were transmitted via pulses of light in 144 fibers in the 1976 field trial in Atlanta, Georgia. [*Physics Today* **29** (May 1976): cover.]

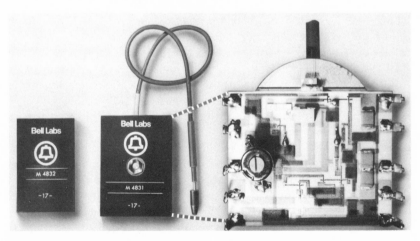

Fig. 4-17.   Directly modulated GaAlAs laser transmitter operating at a wavelength of 0.825 $\mu$m for lightwave communications systems. The average optical power was 0.5 mW in the fiberguide.

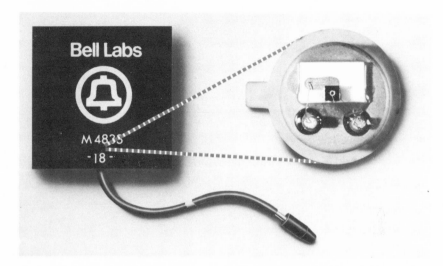

Fig. 4-18.   Silicon avalanche photodiode receiver for lightwave communications systems. With average optical input power of $0.5 \times 10^{-5}$ mW, this subsystem had a bit error rate of only $10^{-9}$.

central offices in major metropolitan areas with very few, and in most cases no, manhole repeaters.

This experiment demonstrated that the pioneering work covered in this chapter did indeed lay the foundation of a lightwave technology suitable for practical transmission systems. Since then, this technology has continued to evolve at a very rapid rate. At the time of the publication of this book, single-mode systems have been installed that offer repeaterless spans of more than 30 km at data rates in excess of 400 Mb/sec, marking the beginning of a new era in transmission technology.

## REFERENCES

1. A. Einstein, "Zur Quantentheorie der Strahlung," *Phys. Zeit.* **18** (1917), pp. 121-128.
2. J. P. Gordon, H. J. Zeiger, and C. H. Townes, "Molecular Microwave Oscillator and New Hyperfine Structure in the Microwave Spectrum of $NH_3$," *Phys. Rev.* **95** (July 1, 1954), pp. 282-284.
3. A. L. Schawlow and C. H. Townes, "Infrared and Optical Masers," *Phys. Rev.* **112** (December 1958), pp. 1940-1949.
4. T. H. Maiman, "Stimulated Optical Radiation in Ruby," *Nature* **187** (August 6, 1960), pp. 493-494.
5. R. J. Collins, D. F. Nelson, A. L. Schawlow, W. Bond, C. G. B. Garrett, and W. Kaiser, "Coherence, Narrowing, Directionality, and Relaxation Oscillations in the Light Emission from Ruby," *Phys. Rev. Lett.* **5** (October 1, 1960), pp. 303-305.
6. A. Javan, "Possibility of Production of Negative Temperature in Gas Discharges," *Phys. Rev. Lett.* **3** (July 15, 1959), pp. 87-89.
   A. Javan, W. R. Bennett, Jr., and D. R. Herriott, "Population Inversion and Continuous Optical Maser Oscillation in a Gas Discharge Containing a He-Ne Mixture," *Phys. Rev. Lett.* **6** (February 1, 1961), pp. 106-110.

7. A. D. White and J. D. Rigden, "Continuous Gas Maser Operation in the Visible," *Proc. IRE* **50** (July 1962), p. 1697.

8. E. I. Gordon, E. F. Labuda, and W. B. Bridges, "Continuous Visible Laser Action in Singly Ionized Argon, Krypton, and Xenon," *Appl. Phys. Lett.* **4** (May 15, 1964), pp. 178-180.

9. L. F. Johnson, G. D. Boyd, K. Nassau, and R. R. Soden, "Continuous Operation of the $CaWO_4:Nd^{+3}$ Optical Maser," *Proc. IRE* **50** (February 1962), p. 213.

10. D. F. Nelson and W. S. Boyle, "A Continuously Operating Ruby Optical Maser," *Appl. Opt.* **1** (March 1962), pp. 181-183.

11. J. E. Geusic, H. M. Marcos, and L. G. Van Uitert, "Laser Oscillations in Nd-Doped Yttrium Aluminum, Yttrium Gallium and Gadolinium Garnets," *Appl. Phys. Lett.* **4** (May 15, 1964), pp. 182-184.

12. L. F. Johnson, J. E. Geusic, and L. G. Van Uitert, "Efficient, High-Power Coherent Emission from $HO^{3+}$ Ions in Yttrium Aluminum Garnet, Assisted by Energy Transfer," *Appl. Phys. Lett.* **8** (April 15, 1966), pp. 200-202.

13. C. K. N. Patel, "Continuous-Wave Laser Action on Vibrational-Rotational Transitions of $CO_2$," *Phys. Rev.* **136** (1964), p. A1187.

14. C. K. N. Patel, "CW High Power $N_2$-$CO_2$ Laser," *Appl. Phys. Lett.* **7** (July 1, 1965), pp. 15-17.

15. F. W. Ostermayer, Jr., R. B. Allen, and E. G. Dierschke, "Room-Temperature cw Operation of a $GaAs_{1-x}P_x$ Diode-Pumped YAG:Nd Laser," *Appl. Phys. Lett.* **19** (October 15, 1971), pp. 289-292.

16. J. A. Giordmaine, "Mixing of Light Beams in Crystals," *Phys. Rev. Lett.* **8** (January 1, 1962), pp. 19-20.

17. A. Ashkin, G. D. Boyd, and J. M. Dziedzic, "Observation of Continuous Optical Harmonic Generation with Gas Masers," *Phys. Rev. Lett.* **11** (July 1, 1963), pp. 14-16.

18. J. A. Giordmaine and R. C. Miller, "Tunable Coherent Parametric Oscillation $LiNbO_3$ at Optical Frequencies," *Phys. Rev. Lett.* **14** (June 14, 1965), pp. 973-976.

19. R. G. Smith, J. E. Geusic, H. J. Levinstein, J. J. Rubin, S. Singh, and L. G. Van Uitert, "Continuous Optical Parametric Oscillation in $Ba_2NaNb_5O_{15}$," *Appl. Phys. Lett.* **12** (May 1, 1968), pp. 308-310.

20. C. K. N. Patel, E. D. Shaw, and R. J. Kerl, "Tunable Spin-Flip Laser and Infrared Spectroscopy," *Phys. Rev. Lett.* **25** (July 6, 1970), pp. 8-11.

21. L. B. Kreuzer and C. K. N. Patel, "Nitric Oxide Air Pollution Detection by Optoacoustic Spectroscopy," *Science* **173** (July 2, 1971), pp. 45-47.

22. J. E. Geusic, H. J. Levinstein, S. Singh, R. G. Smith, and L. G. Van Uitert, "Continuous $0.532$-$\mu$ Solid-State Source Using $Ba_2NaNb_5O_{15}$," *Appl. Phys. Lett.* **12** (May 1, 1968), pp. 306-308.

23. J. E. Geusic, H. J. Levinstein, J. J. Rubin, S. Singh, and L. G. Van Uitert, "The Nonlinear Optical Properties of $Ba_2NaNb_5O_{15}$," *Appl. Phys. Lett.* **11** (November 1, 1967), pp. 269-271.

24. S. Singh, "Non-Linear Optical Materials," in *Handbook of Lasers*, ed. R. J. Pressley (Cleveland, Ohio: The Chemical Rubber Co., 1971), pp. 489-525.

25. M. Gershenzon, R. A. Logan, and D. F. Nelson, "Electrical and Electroluminescent Properties of Gallium Phosphide Diffused $p$-$n$ Junctions," *Phys. Rev.* **149** (September 16, 1966), pp. 580-597.

26. R. A. Logan, H. G. White, and F. A. Trumbore, "P-N Junctions in GaP with External Electroluminescence Efficiency $\sim 2\%$ at $25°C$," *Appl. Phys. Lett.* **10** (April 1, 1967), pp. 206-208.

27. R. A. Logan, H. G. White, and W. Wiegmann, "Efficient Green Electroluminescence in Nitrogen-Doped GaP $p$-$n$ Junctions," *Appl. Phys. Lett.* **13** (August 15, 1968), pp. 139-141.

28. D. G. Thomas, J. J. Hopfield, and C. J. Frosch, "Isoelectronic Traps Due to Nitrogen in Gallium Phosphide," *Phys. Rev. Lett.* **15** (November 29, 1965), pp. 857-860.

29. C. H. Henry, P. J. Dean, and J. D. Cuthbert, "New Red Pair Luminescence from GaP," *Phys. Rev. Lett.* **166** (February 15, 1968), pp. 754-756.

30. R. H. Saul, J. Armstrong, and W. H. Hackett, Jr., "GaP Red Electroluminescent Diodes with an External Quantum Efficiency of 7%," *Appl. Phys. Lett.* **15** (October 1, 1969), pp. 229-231.

31. R. H. Saul and O. G. Lorimer, "Liquid Phase Epitaxy Processes for GaP LED's," *J. Cryst. Grow.* **27** (December 1974), pp. 183-192.

32. S. J. Bass and P. E. Oliver, "Pulling of Gallium Phosphide Crystals by Liquid Encapsulation," *J. Cryst. Grow.* **3** (1968), pp. 286-290.

33. R. H. Saul and D. D. Roccasecca, "Vapor-Doped Multislice LPE for Efficient GaP Green LED's," *J. Electrochem. Soc.* **120** (August 1973), pp. 1128-1131.

34. M. G. Craford, D. L. Keune, W. O. Groves, and A. H. Herzog, "The Luminescent Properties of Nitrogen Doped GaAsP Light Emitting Diodes," *J. Electron. Mater.* **2** (February 1973), pp. 137-158.

35. R. N. Hall, G. E. Fenner, J. D. Kingsley, T. J. Soltys, and R. O. Carlson, "Coherent Light Emission from GaAs Junctions," *Phys. Rev. Lett.* **9** (November 1, 1962), pp. 366-368.

36. D. K. Wilson, "Mode Control in *PN* Junction Lasers," in *Radiative Recombination in Semiconductors* (New York: Academic Press, 1965), pp. 171-176.

37. I. Hayashi, M. B. Panish, P. W. Foy, and S. Sumski, "Junction Lasers Which Operate Continuously at Room Temperature," *Appl. Phys. Lett.* **17** (August 1, 1970), pp. 109-111.

38. J. E. Ripper, J. C. Dyment, L. A. D'Asaro, and T. L. Paoli, "Stripe-Geometry Double Heterostructure Junction Lasers: Mode Structure and cw Operation Above Room Temperature," *Appl. Phys. Lett.* **18** (February 15, 1971), pp. 155-157.

39. B. C. De Loach, Jr., B. W. Hakki, R. L. Hartman, and L. A. D'Asaro, "Degradation of CW GaAs Double-Heterojunction Lasers at 300 K," *Proc. IEEE* **61** (July 1973), pp. 1042-1044.

40. W. B. Joyce, R. W. Dixon, and R. L. Hartman, "Statistical Characterization of the Lifetimes of Continuously Operated (Al,Ga)As Double-Heterostructure Lasers," *Appl. Phys. Lett.* **28** (June 1, 1976), pp. 684-686.

41. T. L. Paoli, B. W. Hakki, and B. I. Miller, "Zero-Order Transverse Mode Operation of GaAs Double-Heterostructure Lasers with Thick Waveguides," *J. Appl. Phys.* **44** (March 1973), pp. 1276-1280.

42. H. Fletcher, "The Stereophonic Sound-Film System—General Theory," *J. Acoust. Soc. Amer.* **13** (October 1941), pp. 89-99.

E. C. Wente, R. Biddulph, L. A. Elmer, and A. B. Anderson, "Mechanical and Optical Equipment for the Stereophonic Sound Film System," *J. Acoust. Soc. Amer.* **13** (October 1941), pp. 100-106.

J. C. Steinberg, "The Stereophonic Sound Film System—Pre- and Post-Equalization of Compandor Systems," *J. Acoust. Soc. Amer.* **13** (October 1941), pp. 107-114.

W. B. Snow and A. R. Soffel, "Electrical Equipment for the Stereophonic Sound-Film System," *J. Soc. Motion Pict. Eng.* **37** (October 1941), pp. 380-396.

E. C. Wente and R. Biddulph, "A Light-Valve for the Stereophonic Sound-Film System," *J. Soc. Motion Pict. Eng.* **37** (October 1941), pp. 397-405.

E. C. Wente and A. H. Müller, "Internally Damped Rollers," *J. Soc. Motion Pict. Eng.* **37** (October 1941), pp. 406-417.

L. A. Elmer, "A Non-Cinching Film Rewind Machine," *J. Soc. Motion Pict. Eng.* **37** (October 1941), pp. 418-426.

43. T. J. Nelson, "Digital Light Deflection," *Bell Syst. Tech. J.* **43** (May 1964), pp. 821-845.

44. S. K. Kurtz, "Design of an Electro-Optic Polarization Switch for a High-Capacity High-Speed Digital Light Deflection System," *Bell Syst. Tech. J.* **45** (October 1966), pp. 1209-1246.

45. P. V. Lenzo, E. H. Turner, E. G. Spencer, and A. A. Ballman, "Electrooptic Coefficients

and Elastic-Wave Propagation in Single-Domain Ferroelectric Lithium Tantalate," *Appl. Phys. Lett.* **8** (February 15, 1966), pp. 81-82.

46. E. H. Turner, "High-Frequency Electro-optic Coefficients of Lithium Niobate," *Appl. Phys. Lett.* **8** (June 1, 1966), pp. 303-304.
47. E. I. Gordon, "A Review of Acoustooptical Deflection and Modulation Devices," *Appl. Opt.* **5** (October 1966), pp. 1629-1639.
48. R. W. Dixon, "Acoustooptic Interactions and Devices," *IEEE Trans. Electron Dev.* **ED-17** (March 1970), pp. 229-235.
49. D. A. Pinnow, "Guide Lines for the Selection of Acoustooptic Materials," *IEEE J. Quantum Electron.* **QE-6** (April 1970), pp. 223-238.
50. R. C. LeCraw, "Wide-Band Infrared Magneto-Optic Modulation," *IEEE Trans. Magn.* **MAG-2** (September 1966), p. 304.
51. See special issue on Device Photolithography, *Bell Syst. Tech. J.* **49** (November 1970), pp. 1995-2220.
52. J. H. Bruning, M. Feldman, T. S. Kinsel, E. K. Sittig, and R. L. Townsend, "An Automated Mask Inspection System—AMIS," *IEEE Trans. Electron Dev.* **ED-22** (July 1975), pp. 487-495.
53. R. T. Denton, F. S. Chen, and A. A. Ballman, "Lithium Tantalate Light Modulators," *J. Appl. Phys.* **38** (March 15, 1967), pp. 1611-1617.
54. R. P. Riesz, "High-Speed Semiconductor Photodiodes," *Rev. Sci. Instrum.* **33** (September 15, 1962), pp. 994-998.
55. L. K. Anderson, P. G. McMullin, L. A. D'Asaro, and A. Goetzberger, "Microwave Photodiodes Exhibiting Microplasma-Free Carrier Multiplication," *Appl. Phys. Lett.* **6** (February 15, 1965), pp. 62-64.
56. H. Melchior, "Demodulation and Photodetection Techniques," in *Laser Handbook*, Vol. 1, ed. F. T. Arecchi and E. O. Shulz-DuBois (Amsterdam: North-Holland Publishing Co., 1972), pp. 725-835.
57. W. Shockley, M. Sparks, and G. K. Teal, "*p-n* Junction Transistors," *Phys. Rev.* **83** (July 1, 1951), pp. 151-162.
58. J. N. Shive, "A New Germanium Photo-Resistance Cell," *Phys. Rev.* **76** (August 15, 1969), p. 575.
59. L. K. Anderson, "Holographic Optical Memory For Bulk Data Storage," *Bell Lab. Rec.* **46** (November 1968), pp. 318-325.
60. M. H. Crowell, T. M. Buck, E. F. Labuda, J. V. Dalton, and E. J. Walsh, "A Camera Tube with a Silicon Diode Array Target," *Bell Syst. Tech. J.* **46** (February 1967), pp. 491-495.
61. A. G. Chynoweth, "Pyroelectricity, Internal Domains, and Interface Charges in Triglycine Sulfate," *Phys. Rev.* **117** (March 1, 1960), pp. 1235-1243.
62. J. E. Iwersen, L. A. D'Asaro, E. E. LaBate, and P. P. Peron, "An Optoelectronic Relay," *IRE Trans. Electron Dev.* **ED-9**, Abstract (November 1962), p. 503.
63. E. I. Gordon and L. K. Anderson, "New Display Technologies—An Editorial Viewpoint," *Proc. IEEE* **61** (July 1973), pp. 807-813.
64. D. Maydan, "Infrared Laser Addressing of Media for Recording and Displaying of High-Resolution Graphic Information," *Proc. IEEE* **61** (July 1973), pp. 1007-1013.
65. D. Maydan, "Micromachining and Image Recording on Thin Films by Laser Beams," *Bell Syst. Tech. J.* **50** (July/August 1971), pp. 1761-1789.
66. R. C. Miller, R. H. Willens, H. A. Watson, L. A. D'Asaro, and M. Feldman, "A Gallium-Arsenide Laser Facsimile Printer," *Bell Syst. Tech. J.* **58** (November 1979), pp. 1909-1998.
67. J. R. Maldonado and D. B. Fraser, "PLZT Ceramic Display Devices for Slow-Scan Graphic Projection Displays," *Proc. IEEE* **61** (July 1973), pp. 975-981.
68. L. K. Anderson, "Projecting Images With Liquid Crystals," *Bell Lab. Rec.* **52** (July/August 1974), pp. 223-229.
69. S. E. Miller, E. A. J. Marcatili, and T. Li, "Research Toward Optical-Fiber Transmission Systems," *Proc. IEEE* **61** (December 1973), pp. 1703-1751.
70. I. Jacobs, "Lightwave Communications Passes Its First Test," *Bell Lab. Rec.* **54** (December 1976), pp. 290-297.

# Chapter 5

# Magnetic Memories

*Magnetic memory devices, developed in the 1950s and 1960s, were the mainstay of the large-capacity memory technology for the Bell System into the 1970s. The random-access ferrite core and ferrite sheet memories became important elements in early electronic switching systems. A Bell Laboratories innovation, the twistor, improved memory integrity by storing information on permanent-magnet memory cards. Operating problems related to card loading were resolved in the piggyback twistor by wrapping the equivalent of permanent magnets directly on twistor wire, producing a less bulky memory with on-line write capability. During the early 1970s, the arrival of magnetic bubble technology, also developed at Bell Laboratories, brought a sequential readout memory of high packing density without moving parts.*

## I. INTRODUCTION

The memory function is not new to the telephone plant. In the form of relay registers and wired translators, memory was used as early as the development of the panel central office in the 1920s. The electromechanical card translator used in the No. 4 toll crossbar, developed in the late 1940s, was one of the first large-capacity memories. The need for large memories increased dramatically when solid-state digital computers became available through rapid advances in semiconductor technology in the latter half of the 1950s. A flood of memory devices quickly emerged from laboratories throughout the world. Quartz acoustic delay lines, ferrite cores, ferrite sheets, barrier-grid cathode-ray-tube-based stores, and magnetic flat-film arrays were candidates for read-write data storage, while photographic stores based on cathode ray tubes, magnetic drums, permanent magnet twistor memories, and hard-wired ferrite core stores were candidates for read-only memories (ROMs). For telephone switching applications, these developments opened the door to stored program control, an approach

Principal authors: L. W. Stammerjohn and A. H. Bobeck

permitting many new service offerings. (See another volume in this series subtitled *Switching Technology (1925-1975)*, for a comprehensive discussion of switching applications.)

The foremost operational requirements for telephone switching systems are extremely high system integrity and availability. Permissible downtimes are measured in hours per 40 years, a far more stringent requirement than that for any general-purpose computer. Recovery from any system failure must be swift, and absolute integrity of the stored program must be maintained. These factors led to choices of memory technologies that often differed from those taken by the emerging computer industry, which emphasized performance over reliability.

The general-purpose computer systems that evolved in the late 1950s relied primarily on magnetic core or magnetic thin-film memories for "scratchpad" storage, on capacitor read-only control stores, and on rotating magnetic disk and/or tape memories for backup storage. Although the electrical characteristics of early magnetic disk memories were adequate, the reliability of these memories did not meet the data integrity required by electronic switching systems, so a different memory strategy was pursued.

Stored program control of telephone switching systems requires three classes of data to be retained: (1) The master control program consists of data that needs to be modified only to accommodate changes of the system resulting from growth, different types of services, or correction of errors. (2) The translation data is made up of subscriber information, such as translation from directory number to equipment number, or other special information associated with customer services; a small percentage of the translation data is changed each day. (3) Information about each call must be retained temporarily while a call is in progress.

Early on, it was decided that only two types of memories would be needed to handle the three classes of data: a semipermanent memory to store the systems program and the bulk of the translation information, and a scratchpad memory to handle call-progress and recent-update translation data. The recent-update portion of the scratchpad memory contained the daily changes, such as changes of subscriber location or types of service. The capacity allocated to recent-change data depended on the frequency at which the contents were transferred to the semipermanent memory.

The detailed choices of memory technology, module size, etc. were strongly influenced by estimates of the required program size and change frequency. For offices in the 10,000- to 50,000-line range, the size of the program memory was initially estimated at $1.0 \times 10^6$ to $2.5 \times 10^6$ bits with an additional $10^6$ to $10^7$ bits of semipermanent memory needed for the translation data. Furthermore, 40,000 to 500,000 bits of scratchpad memory could handle the call-in-progress information as well as any recent changes in translation data. Revisions to the program were projected perhaps every 5 to 20 years. As system development progressed, however, it soon was realized that much larger program memories were needed and that the changes were much more frequent than anticipated.

Although other material systems were studied as possible means of providing the memory function,* the simplicity and low cost of magnetic memories allowed them to dominate the field for an extended period of time. The developments at Bell Laboratories as described below fall into three categories: (1) the core and ferrite sheet memories that served as scratchpad memories; (2) the twistor memories that were primarily used as semipermanent memories; and (3) magnetic bubbles, with serial access to data, which were functionally similar to disk memories.

## II. THE FERRITE CORE AND FERRITE SHEET

J. W. Forrester of the Massachusetts Institute of Technology is generally credited with the concept of using magnetic cores with square loop magnetization-drive characteristics in coincident-selection arrays to store binary information.[1] Devised in 1949, the concept was reduced to practice with metallic cores; however, these were quickly replaced by ferrite cores.

The square loop characteristic, shown in Fig. 5-1, implies that the magnetization B switches from one state to the other when a magnetic field H exceeding a critical value is applied. Below this critical field, the original state remains unaffected. In a matrix of magnetic cores threaded by con-

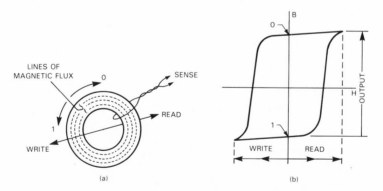

Fig. 5-1. Information storage in ferrite cores. (a) The two stable states of magnetization (clockwise and counterclockwise) are associated with logical 1s and 0s, respectively. (b) The square loop characteristic implies that the magnetization B switches from one state to the other state only when the magnetic field H exceeds a critical value. This permits coincident-current selection, as shown in Figure 5-2.

---

* Noteworthy alternatives were the barrier-grid memory (a vacuum-tube-based temporary memory) and the flying spot store (a memory in which information is stored on photographic film); see Chapter 9 of *Switching Technology (1925-1975)* and Chapter 3 of this volume.

ductors, as shown in Fig. 5-2, an individual core within the matrix can be selected by applying currents somewhat below the critical value to each of the two wires threading that core. The resulting magnetic field will exceed the critical value only in that core and drive the magnetization in one direction—for example, clockwise. The selected core will switch if it was previously magnetized counterclockwise or remain unaffected if the prior magnetization was clockwise. A sense wire that threads all the cores of the matrix detects the presence or absence of switching. Since, during a read operation, a core is driven to the 0 state, the readout is destructive. Of course, had a 1 been read, that core had to be rewritten later if the information was to be retained.

Each of the selection wires requires appropriate decode and drive circuitry to locate an element from a memory address. The matrix organization minimizes the amount of control circuitry, so that many magnetic storage elements can be served by a few drive and selection circuits. Early magnetic memories used even more complex wiring than that shown in Fig. 5-2 to minimize the control circuitry, which at the time was relatively expensive. Generally, four wires threaded each core.

During the period of the early ferrite core work, the potential of a fully transistorized digital computer was recognized, and in 1951 the Air Force funded the transistorized digital computer (TRADIC) project at Bell Lab-

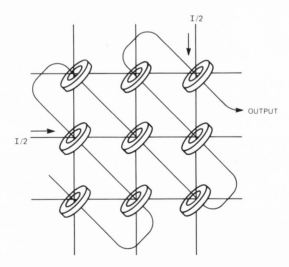

Fig. 5-2.   Schematic wiring pattern of a coincident-current core matrix. A current I creates a magnetic field in excess of the threshold. Applying a current I/2 to an X-oriented and a Y-oriented wire will result in the threshold magnetic field being exceeded only in the core at the intersection of those two wires.

oratories under the direction of J. H. Felker. (See another volume in this series subtitled *National Service in War and Peace (1925-1975)*, Chapter 13, section 2.2.3 and Chapter 1, section 2.1 of this volume.) The first TRADIC computer was completed in 1954. It used quartz multifaceted delay lines as its memory.

Because of the more desirable characteristics of random-access memories (RAMs), in which the access time does not depend on the address of the data element being accessed, and following an article by J. A. Rajchman of RCA in June 1952,[2] a ferrite core memory development was included in the TRADIC project.

In 1954, A. H. Bobeck[3] and E. F. Sartori[4] completed a fully transistorized ferrite core memory system based on the General Ceramics S-3 80-mil outside diameter, 50-mil inside diameter core. This system, believed to be the first all-solid-state memory constructed, used type 2008 high-power germanium switching transistors provided by the semiconductor device development group.

The memory system was word organized; that is, information was transmitted as 16-bit words to and from the 256-bit core plane. [Fig. 5-3] Word

Fig. 5-3. Photograph of an early coincident-current core matrix plane, used in the early TRADIC military computer.

currents of 0.175 ampere (A) were derived from a transistor-driven magnetic stepping switch invented by M. Karnaugh.[5] The time required to complete a read-write cycle was 32 microseconds ($\mu$s). All logic operations were performed using magnetic techniques. Besides the 256-bit cores used for the scratchpad memory, a source of constants requiring 160 additional memory cores was also provided. Fifty master program steps and ten subroutine steps were provided by the magnetic switch. An average power of 15 watts was consumed by the system.

Next, Bobeck and E. L. Younker, working with International Telemeter Magnetics of California, developed a 1024-word, 16-bit/word transistor-driven core memory for the next-generation computer, which was completed in 1957.[6]

Meanwhile, in a magnetic device group under D. H. Looney, attention was being directed toward the development of the ferrite sheet. The principal objective of the ferrite sheet was to permit the fabrication of many memory cells in a single part to simplify the mechanics of assembling memory modules.

It might seem to have been a straightforward process to replace individual cores by batch fabrication, but this was not the case. Numerous approaches were tried to integrate ferrite core memories. In addition to the multi-apertured ferrite sheet adopted at Bell Laboratories and Western Electric as well as at RCA, there were also attempts to embed the access and sense conductors within the ferrite itself. Examples were the ferrite bead of Looney[7] and the ferrite sandwich of R. Shahbender at RCA,[8] neither of which proved to be feasible.

The ferrite sheet was described by R. H. Meinken in 1956.[9] As seen in Fig. 5-4, it was a thin sheet of manganese-magnesium ferrite (a square loop ferrimagnetic material) with 256 holes arranged in a 16-by-16 matrix. The ferrite immediately surrounding a hole formed the equivalent of a single core. Four conductors passed through each hole to form what was known as a three-dimensional memory array. One of the conductors, plated on the sheet, connected the 256 cells in series. The other three conductors were wires threaded through sheets stacked to form a memory module. [Fig. 5-5]

Various approaches were explored for making ferrite sheets. At first, ultrasonically drilling the necessary 256 holes in a postage-stamp-size ferrite slab seemed easiest. This process, however, did not allow the properties of the magnesium-manganese ferrite to be locally tempered, so that the switching could be limited to the ferrite immediately surrounding the hole. This tempering was needed to isolate the magnetic reversals of one hole and prevent it from affecting the magnetic state of another. Eventually, ferrite powder was pressed into final form in a multipost die, and the interaction between neighboring holes was overcome both by sheet processing and array wiring configurations.

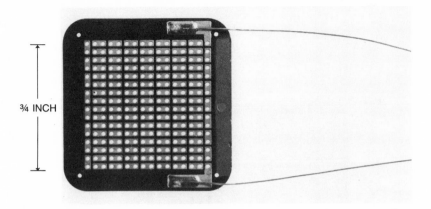

¾ INCH

Fig. 5-4.    The ferrite sheet, the building block of the first magnetic memory used extensively in the Bell System. Each ferrite sheet contained 256 holes and a printed conductor. It was electrically equivalent to a matrix of 256 individual ferrite cores with a conductor threading all cores in series.

The ferrite sheet memory was read by current pulses of about 250 milliamperes (mA) on each of two drive conductors (the X and Y address lines). The output from a coincidently accessed location storing a 1 was roughly a triangular voltage pulse with a peak of 75 millivolts (mV) and a base of 0.9 $\mu$s. A 1 was written by applying current pulses of the opposite polarity to the same pair of conductors. To write a 0, the same procedure as writing a 1 was followed, except that an inhibit pulse was applied to a third conductor. The cycle time for this read-write sequence was 5.5 $\mu$s. A broad operating temperature range of 0 to 55 degrees C was achieved. However, because the characteristics of ferrites change with temperature, temperature-sensing circuits were needed to track drive current levels with temperature variation.

Soon after Vice President J. A. Morton committed the ferrite sheet (rather than cores) to the evolving electronic switching systems, the development was transferred to the Allentown, Pennsylvania location where a joint Bell Laboratories-Western Electric team successfully introduced the ferrite sheet into manufacture. The memory was first used in the 1ESS* and the 101ESS electronic switches. For the smallest office, the 1ESS electronic switch call store contained four ferrite sheet memory modules organized to store 8192 words, each 24 bits long.[10] Multiple units were used in larger offices. Later, a different organization of this memory was also used in the early installations of the 2ESS electronic switch.

Production of the ferrite sheet memory started at Western Electric in 1963. In the same year, it was introduced in the first office using 101ESS

---

* Trademark of AT&T Technologies, Inc.

Fig. 5-5.   The ferrite sheet memory, as used in 1ESS switching
equipment. One module consisted of 12 blocks of 16 sheets and
a spare sheet. Each block stored 4096 bits. Four modules composed
a call store organized as 8192 words of 24 bits.

switching equipment in Cocoa Beach, Florida, followed in 1965 by the
first office with 1ESS switching equipment in Succasunna, New Jersey.[11]
It was used in all electronic switching system offices until 1973, when it
was replaced by the core memory described later in this section. Nevertheless, as of 1975, the ferrite sheet memory was still in manufacture,
since it was needed for growth of older offices and other systems.

The intrinsic switching time of the ferrite in the ferrite sheet was about
1 $\mu$s. Here, and throughout the industry, a more basic understanding of
the switching process led to other material compositions with superior
switching properties. However, it proved too difficult to introduce these
new compositions into the ongoing ferrite sheet production.

By 1972, a new ferrite core store architecture made possible the development of a call store with significantly improved performance and
lower per-bit cost than the ferrite sheet store could offer. In this memory
organization, the address selection and coincident-current read and write

operations made use of considerably more semiconductor circuitry than the ferrite sheet memory, allowing an array construction that required only two wires to thread each memory core. This array geometry, together with developments in memory core manufacture, handling, and placement, made it economical to develop a call store with a ferrite core with an outer diameter smaller than the 25-mil hole in the ferrite sheet.

The memory designed for the more advanced 1A processor consisted of two 832-by-512 matrices of magnesium-manganese-zinc ferrite cores.[12] A novel accessing scheme, employing integrated circuit (IC) diode arrays, was used to interface the IC logic with the core array, selecting and energizing the appropriate wires, and giving a 32,768-word by 26-bit/word organization. [Fig. 5-6] Used in a 24-bit/word format, each of the new memories superseded four of the 8192-word ferrite sheet call stores in new offices with 1ESS switching equipment.

The coincident drive currents required for reading and writing the core memory were 300 mA in each of two windings. The 1 output was a voltage pulse of 30 mV, and the minimum read-write cycle time was 1.2 $\mu$s. Read access time was 0.45 $\mu$s. The core memory was designed to operate in ambient temperatures from 0 to 50 degrees C.

The key advantages of this memory were its high speed, small size, reduced cost, and reduced power. The store was first placed in operation in the Bell System in the Alton, Illinois office with 1ESS switching equipment in 1972. [Fig. 5-7] The first offices with 4ESS switching equipment in 1976 used the memory as part of the 1A processors employed in these systems.

This memory was in use for only a short time, however, before advances

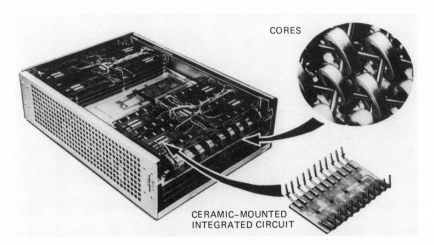

CORES

CERAMIC–MOUNTED
INTEGRATED CIRCUIT

Fig. 5-6.    The 20A memory module, which contained 851,968 cores with two wires threaded through each core. For increased functional density, several integrated circuits were mounted on ceramic units, which in turn were assembled on a printed circuit board. The module was organized as 32,768 words of 26 bits.

Fig. 5-7.    The Alton, Illinois office of 1ESS switching equipment, the first site using core memory modules. Five modules can be seen mounted vertically on the first three equipment frames.

in ICs resulted in a situation in which the cost of interfacing silicon circuitry with the magnetic storage medium exceeded the cost of embedding semiconductor storage elements within the ICs. Accordingly, magnetic RAMs in later-generation switching systems were replaced by semiconductor memories.

## III. THE TWISTOR

The third magnetic memory of major importance to electronic switching systems was the twistor, first described by Bobeck in 1957.[13] The original twistor concept made use of the Wiedemann effect (the creation of magnetic anisotropy by mechanical strain) to create helical magnetization in magnetic wire. The twisted wires were supported within glass tubes. [Fig. 5-8] An equivalent geometry, a thin tape of Permalloy* (a magnetically

---

* Trademark of Allegheny Ludlum Steel Corp.

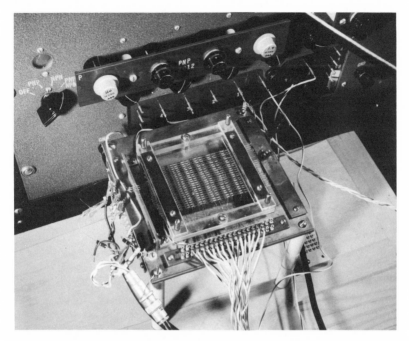

Fig. 5-8. Experimental twistor memory array with magnetic wires under torsion housed in glass tubes.

permeable iron-nickel alloy) wrapped helically around a copper core wire—the twistor wire—was more readily reduced to practice. [Fig. 5-9] (See also another volume in this series subtitled *Physical Sciences (1925-1980)*, Chapter 12, section 1.1.2.) Small regions of the Permalloy tape served as memory cells, and the copper core wire acted as both a sense wire and an access wire. A single-turn strip conductor solenoid or a multiturn solenoid encircling the twistor wires in an orthogonal orientation completed the selection matrix. The twistor was first operated as a word-organized, destructive read-out memory. Known as the variable twistor memory, it was used in military systems.

Twistor memory planes were word organized. Generally the twistor wires functioned as the bit lines and the orthogonal drive conductors as the word lines. To write a word into the variable twistor memory, the addressed word line and selected twistor bit lines were driven in coincidence. To read a word, however, the word line was simply overdriven with a current pulse of the opposite polarity. Those sections of the twistor wires that did switch (i.e., storing 1s) induced voltage signals that traveled the length of the twistor wire to sense circuitry. [Fig. 5-10] Following a read operation, the magnetization of segments of all the twistor wires in proximity to the energized word line was left in the 0 state.

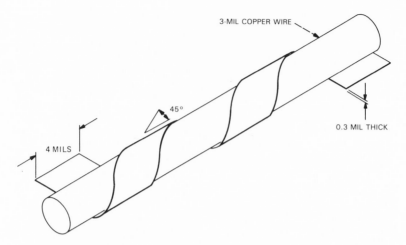

Fig. 5-9.   The twistor wire, consisting of a copper wire helically wrapped with a thin Permalloy tape.

An example of a variable twistor memory module is the 512-word, 24-bit/word GS-59556 storage register designed for Nike Zeus computers. The memory unit contained 30 twistor wires, paired with copper wires of the same diameter encased in Mylar* to form a belt. A. J. Munn designed a machine to fabricate this belt in a continuous operation, actually building the machine in a shop he had set up in his garage. This timely contribution was a crucial element in making twistor memories cost effective. Included as part of the memory unit was a plane of 512 large-diameter ferrite cores, each of which was coupled to a word access winding that wrapped the belt. The cores constituted a coincident-current biased-core switch by which access to the word locations on the twistor belt was provided. The entire unit was encased in a Mumetal™ box to shield it from stray magnetic fields.

Since the readout in all memories described so far is destructive, each read cycle must be followed by a write cycle. Because this mode of operation could not assure the long-term data integrity required of a program store, memories with nondestructive readout were preferred as program stores.

The semipermanent memory developed for electronic switching systems used twistor wires to sense the presence or absence of magnetization of tiny permanent magnets. The concept of storing information in an array of permanent magnets was advanced by S. M. Shackell, a professor at Stevens Institute of Technology who worked part-time at Bell Laboratories.[14] The permanent magnet twistor configuration was suggested by J. Janik, Jr.,[15] who was familiar with the Shackell scheme.

---

* Trademark of E. I. DuPont de Nemours Co., Inc.

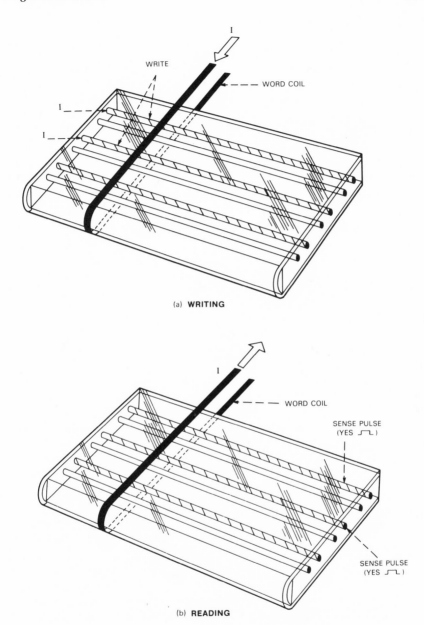

Fig. 5-10. The twistor memory. (a) A coincidence of currents in selected twistor wires and word coil writes data bits in the word line at selected twistor bit lines. (b) Data is read by energizing the word coil to reverse the magnetization of twistor wires previously written. The induced signals travel to sense electronics via a twistor wire/copper wire balanced transmission line pair that rejects extraneous noise signals.

The structure of the memory is shown schematically in Fig. 5-11. The small permanent magnets storing the information are carried on aluminum cards inserted into the memory.[11,16] [Fig. 5-12] An unmagnetized magnet allows switching of the nearby twistor wire and represents a stored 1, whereas a magnetized magnet inhibits switching and thus represents a stored 0. The condition of a magnet is interrogated by a current pulse in the appropriate copper solenoid. The memory is word organized, and current applied to any solenoid will read out in parallel the information stored in the magnets over each of the twistor wires associated with the selected solenoid. Random access to words is achieved with a biased–core switch matrix that selects a single solenoid for any combination of X and Y drives.

Construction of a module required a belt of 44 twistor wires and 44 copper return wires embedded between sheets of Mylar. After the copper strip solenoid tapes were cemented to 64 glass-bonded mica support planes, the twistor belt, aligned perpendicular to the solenoids, was attached. The planes attached to the belt were then folded into a module, leaving space between the planes for the insertion of cards that carried the permanent

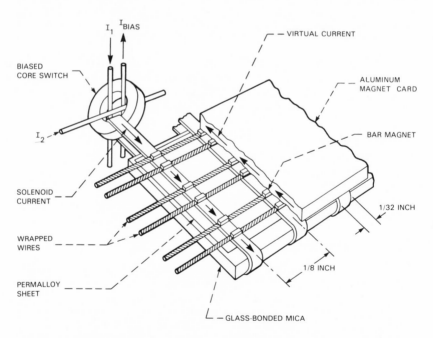

Fig. 5-11.   Schematic of the semipermanent twistor memory. An unmagnetized magnet represents a stored 0, and a magnetized magnet represents a stored 1. The condition of the magnet is sensed by a current in the copper solenoid. This memory can be read an unlimited number of times without destroying the stored information. [Stammerjohn, *IEEE Trans. Commun. Electron.* **83** (1964): 817.]

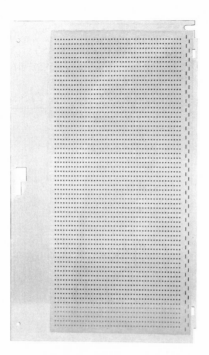

Fig. 5-12.  The aluminum magnet card, which carried an array of 2816 information bar magnets and a double stripe of initializing magnets. A twistor store consisted of 16 modules of 128 cards each; an electronic switching system central office contained two to six stores.

magnets containing the information. Because over 1525 feet of twistor wire was wrapped in one operation, and belts over 700 feet long (enough for seven memory modules) were laminated in one run, the result was a low-cost memory.

The biased-core access array that provided the word selection currents for the twistor memory used a dc bias current to maintain all cores at magnetic saturation. Two orthogonal drive windings, when energized, switched the core at the intersection of those windings, thereby inducing current in the word line. When the drives were removed, the dc bias current would reswitch the selected core. These drive windings each required 2.5 ampere-turns of drive to overcome the dc bias and switch the selected core.

The nominal output from the twistor bit lines was 2.5 mV across a 300-ohm near-end terminating resistor. The maximum signal transmission delay from the shorted far end of the bit line pair to the near end, a distance of some 70 feet, was about 200 nanoseconds (ns). The read cycle time was 5.5 $\mu$s.

Each permanent-magnet twistor module contained 360,448 bits organized as 4096 words of 88 bits each, requiring 128 magnet cards. In 1ESS electronic switches, the program store contained 16 modules arranged in a 4-by-4 array. The average office contained seven program stores, for a total capacity of about 40 million bits.

The cards containing the permanent magnets had to be removed from the memory to be written. Since the energy required to write the card—i.e., to change the magnetization of the permanent magnets—was large compared to the energy that could be provided by the memory-module access circuitry, the stored information was secure, as required for the program memory.

To write and load cards expeditiously into a program store required two additional pieces of equipment—a card loader and a card writer. The card loader was an electrically driven magazine that could extract or insert 128 cards simultaneously in a memory module. This operation took about one minute. The card writer was operated either from magnetic tape or from a remote location via data link. It magnetized or demagnetized the small magnets on the cards.

Permanent magnet twistor memories were first installed in the Cocoa Beach office with 101ESS electronic switching equipment in 1963 and in the Succasunna office with 1ESS electronic switching equipment in 1965. [Fig. 5-13] The first memories were manufactured at the Columbus, Ohio location of Western Electric in 1963 and were still being manufactured for new offices with 1ESS electronic switching equipment in 1975. More than 100,000 units were produced.

It became apparent that the program memory required far more frequent changes than initially anticipated, and the operating complexities arising from card handling prompted the development of an electrically changeable program store. This store, the piggyback twistor, was first described in a patent disclosure by W. A. Barrett in 1962.[17] In this memory, a high-coercivity magnetic storage tape was wrapped barber-pole fashion on a central conductor along with a second, low-coercivity sensor tape.[18,19] [Fig. 5-14] The name piggyback twistor derives from this geometry. The high-coercivity tape provided the function of the magnet card but with an important distinction. Data could be entered using much the same access circuitry already in place to provide the sensing (read) operation. A major problem solved in reducing this idea to practice was that of simultaneously wrapping two dissimilar magnetic tapes on the copper core wire at high speed. A second major problem solved was the development of a non-magnetostrictive magnetic alloy for the storage tape in this memory. An iron-cobalt-gold alloy with a square loop characteristic was found to be suitable.

In the piggyback twistor memories, the twistor wires were embedded in pairs in a plastic belt. Again, copper solenoid straps were folded around

Fig. 5-13.  Program store of the Succasunna, New Jersey electronic switching system office. More than 5 million bits of information were stored in 16 modules of permanent-magnet twistor memory.

the twistor belt. The belt and solenoid arrays were folded in an accordion fashion, and the solenoids were connected to a biased-core access switch matrix. Belt lamination and assembly were far less critical than in the permanent magnet twistor, since there was no requirement to support and register a magnet card.

The memory was written by the coincidence of currents in the word solenoid and the twistor wire. During reading, the word solenoid was energized at a low current level. Consequently, read operations did not affect the stored information. However, a write operation, if generated in error, could still disturb the stored information. A major contribution in store design for the piggyback twistor was the development of an effective method for preventing unwanted write operations in certain protected areas of the store.

The piggyback twistor memory had a biased-core access matrix similar to the permanent-magnet twistor memory. The single-turn bias winding carried 3.9 A of continuous current. For reading, each two-turn access winding was driven with 1.45 A, inducing a read current of 1.8 A in the

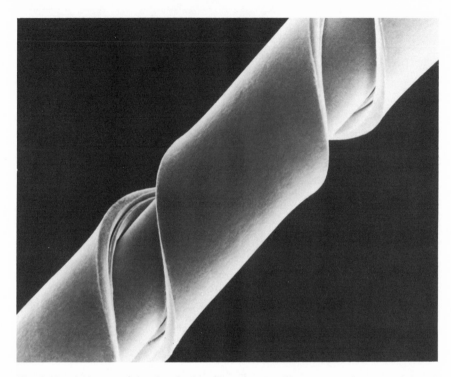

Fig. 5-14.   A closeup of the piggyback twistor, showing the two tapes that were wrapped on the copper core wire. Permalloy was used for the inner (sensing) tape and an iron-cobalt-gold alloy for the outer (storage) tape. This latter alloy was tailored to retain a square loop characteristic through the wrapping process.

solenoid word line. For writing, the access currents were increased to 2 A, and a coincident current of 180 mA was impressed on the bit lines. The bit line current polarity distinguished between writing a 0 or a 1.

Unlike the permanent magnet twistor, both wires that constituted a bit line were wrapped with magnetic tape, significantly increasing the transmission delay of a readout and causing an output signal to take 720 ns to travel 70 feet from one end of the bit line to the other. For this reason, the readout circuitry timing was keyed to the memory address to compensate for the distance that the output had to travel along the bit line before it reached the sensing circuit.

The 1 and 0 output signals from a piggyback twistor memory were of opposite polarity, in contrast to those of other magnetic memories that differ only in output level. These bipolar outputs simplified signal detection. A worst-case output could be as low as 1 mV for a 200-ns duration, although it had to be the correct polarity for the preceding 75 ns. The piggyback twistor memory had a 6.3-$\mu$s read cycle time. A write cycle, which occurred infrequently, took less than 20 $\mu$s. An attribute of the piggyback twistor was the incorporation of redundancy. The memory modules were built with 54 bits per word for an application requiring only 47 bits per word. This meant that defective bit lines could be replaced by spares either at the factory or in the field by a simple rewiring of external terminals.

The piggyback twistor was used in the stored program control unit in the No. 1 traffic service position system (TSPS) and in the electronic translator system for the No. 4 toll crossbar. This store functioned as both a program and a call store. The first memories were installed in the Morristown, New Jersey TSPS office in January 1969.

Before the piggyback twistor could find widespread application, however, the reliability of rotating magnetic disks had improved, and they could be used in electronic switching systems as backup stores. This now permitted programs to be stored in the scratchpad memory, since they could be rapidly reloaded from disk after a system failure. Thus, by the end of the period covered by this history, all magnetic memories discussed so far were being phased out of electronic switching equipment in favor of semiconductor memories, augmented by magnetic disk memories.

## IV. BUBBLE TECHNOLOGY

When magnetic tape is used to record audio or video signals, those signals correspond to the magnetization of particles on the tape. Information is imprinted (stored) on the tape as it moves at high velocity by a write-head, and is retrieved in much the same way. By contrast, in a magnetic bubble memory chip, regions of reversed magnetization (domains) are nucleated within a thin, single-crystal magnetic layer, and these regions are moved about within the magnetic layer.[20]

Magnetic bubble domains, when viewed with a polarizing microscope, appear as tiny circular dots. [Fig. 5-15] They are actually cylindrical island domains that are found in most thin, single-crystal magnetic films that support magnetization perpendicular to the plane of the film. Magnetization direction in the bubble is opposite to that of the surrounding area of the magnetic film in which it resides. A transition region, the domain wall, separates the magnetization of the bubble from that of its surroundings. Within the domain wall, of course, the magnetization must rotate through 180 degrees. Domain walls, particularly those found in defect-free, single-crystal magnetic materials, readily move in response to applied magnetic fields; this is the foundation of magnetic bubble technology.

Bubbles can be made to move rapidly along precisely defined paths in response to electrical stimuli. The pattern of stored bubbles represents the information. Generally the presence of a bubble in the moving data stream signifies a 1 and the absence of a bubble signifies a 0. The fact that magnetic

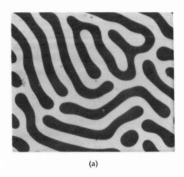 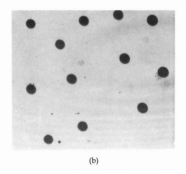

(a)                                                                      (b)

Fig. 5-15.   Magnetic bubbles. (a) Strip domains, seen by Faraday rotation of polarized light. They represent regions magnetized alternately inwardly and outwardly in a 55-micrometer ($\mu$m) thick platelet of terbium orthoferrite. Domains are typically 90 $\mu$m wide. (b) When a 50-oersted bias field is applied perpendicular to the orthoferrite sheet, a cylindrical domain (bubble) 35 $\mu$m in diameter is formed for each properly oriented, single-walled strip domain.

bubbles can be made visible is a characteristic that has aided immensely in the development of this technology. Visual observation of bubble manipulations in magnetic bubble chips is possible at quasi-static speeds using Faraday rotation, and stroboscopic techniques can be added to observe high-speed phenomena.

Bubble devices are the result of a long-term effort at Bell Laboratories to find a solid-state equivalent of the magnetic disk. The first attempt, in the late 1950s, used ferrite structures shaped like ladders.[21] Following a literature report on a shift register based on domain-wall motion,[22] various embodiments of such shift registers were explored at Bell Laboratories.[23,24] But these shift registers had a serious shortcoming: they were one dimensional and did not permit such operations as closed-loop operations without detection and regeneration. The solution came from the work of P. C. Michaelis[25] on magnetically anisotropic thin-film structures that supported domains that could be moved in two dimensions, and thus held the potential for closed loops and other functional arrangements.

However, in these films the isolated domains were large (0.1 by 0.4 millimeters [mm]), and they moved slowly when traveling along the hard axis. Recognizing this limitation, Bobeck pointed out in late 1965 that movement of domains with equal ease in any direction would not be possible as long as the magnetization direction is in the plane of the magnetic medium—it must align *perpendicular* to the surface. This requires a material with uniaxial anisotropy with the axis aligned perpendicular to the surface of the material.

At about the same time, Morton had induced W. Shockley, one of the inventors of the transistor (see Chapter 1), to return to Bell Laboratories on a part-time basis. Shockley developed an interest in domain-wall devices and soon became a central figure in the bubble story when he organized a gathering of mathematicians, magnetic materials experts, device engineers, and a patent attorney to discuss manganese bismuth (MnBi), a material known to support magnetization perpendicular to its surface.

At the meeting held on March 22, 1966 in Shockley's office it was soon realized that MnBi was not the material needed for a two-dimensional shift register. But R. C. Sherwood, who had previously collaborated with J. P. Remeika and H. J. Williams on magnetic materials research, proposed the use of orthoferrites, canted antiferromagnetic materials. The previous research had established that these materials have a low (approximately 150-gauss) magnetization and domain walls that were easy to move.[26] (See *Physical Sciences (1925-1980)*, Chapter 12, section 3.1.) Sherwood supplied Bobeck with several orthoferrite platelets grown by Remeika, and one of these, $ErFeO_3$, was used to demonstrate the first two-dimensional bubble shift register.

Subsequently, under the guidance of U. F. Gianola, a practical two-dimensional bubble-shift register was realized, and A. A. Thiele, an applied mathematician, provided an in-depth treatment of the theory of the stability of bubble domains.[27,28] With these efforts, the feasibility of bubble technology was established. Some of the principal contributors are shown in Fig. 5-16.

Much of the research and development that followed the initial effort and that eventually led to significantly improved magnetic bubble devices was conducted under the direction of H. E. D. Scovil. Early experiments were performed to determine the static and dynamic nature of bubbles in orthoferrites, semitransparent materials with an orthorhombic crystal structure. Unfortunately, the bubbles to be found in orthoferrites were large, typically 0.1 mm in diameter, and the crystals were small; only rudimentary experiments demonstrating controlled bubble motion, bubble generation, and detection could be done. [Fig. 5-17] Nonetheless, Morton showed enthusiasm for magnetic bubbles; in 1969, he conducted a press conference at which this technology was disclosed to the public. At the time, Morton predicted that bubble devices with millions of bits would eventually be realized.

Morton's prediction was a very daring one, for no one knew better than he the magnitude of the material problems to be solved. It was very clear, for example, that single-crystal platelets could never serve as a viable materials vehicle. The only hope lay in the heteroepitaxy of a uniaxial magnetic film on a nonmagnetic substrate crystal. However, not only was there no known magnetic material system having suitable properties but there was no demonstration of any kind of heteroepitaxy in *any* material system that did not have extensive defects, especially dislocations known to be incompatible with bubble propagation.

Fig. 5-16.   A photograph taken in 1971 of some of the principal contributors to
magnetic bubbles. Standing (left to right): L. G. Van Uitert, R. C. Sherwood, J. P.
Remeika, and P. C. Michaelis; seated: A. H. Bobeck (left) and A. A. Thiele (right).

To supplement the support available from the materials research area,
L. J. Varnerin organized a materials development effort that included, as
one of the first tasks by L. K. Shick and J. W. Nielsen, the heteroepitaxial
growth of orthoferrites on nonmagnetic substrates, using a tipping method
of flux epitaxy.[29] Since this approach did not prove feasible for orthoferrites,
material suitable for device development was provided by S. L. Blank,[30]
using the seeded Bridgman method.

Meanwhile, an unexpected development occurred that had major sig-
nificance. Bobeck noticed that garnet platelets cut from multifaceted crystals
provided by L. G. Van Uitert of the materials research area had regions
supporting bubble domains, which implies uniaxial anisotropy. [Fig. 5-18]
(See also *Physical Sciences (1925-1980)*, Chapter 19, section 1.1.) This discovery was surprising, since the symmetry of a perfect cubic crystal does
not allow for uniaxial anisotropy. It was demonstrated that anisotropy was
an intrinsic property of the garnet and was not strain induced by devising
a strain-insensitive garnet composition, $Tb_1Er_2Fe_5O_{12}$, using data on the
stress sensitivity of garnets generated by S. Iida.[31] Platelets cut from this

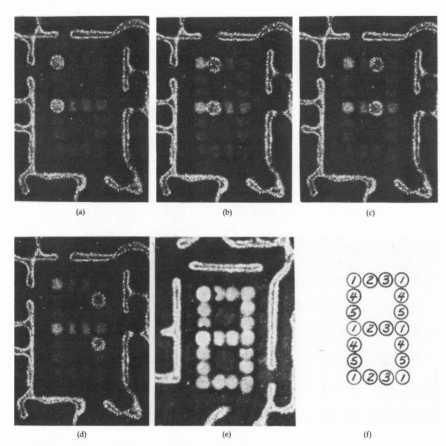

Fig. 5-17.   Sequence of photographs illustrating two-dimensional shifting of cylindrical magnetic domains in HoFeO₃ orthoferrite. Domains are moved through an array of drive loops connected to a five-phase pulser, in a manner shown in (f). (a) Starting position of two domains with phase 1 active. (b) Phase 2 active. (c) Phase 3 active. (d) Phase 1 followed by phase 4. (e) Residual pattern after continuous 1-2-3-1-4-5-1-3-2-1-5-4 sequence. Domains are made visible in a colloid viewer by incident light reflecting from tiny magnetic platelets that become attracted to the stray field of the domain.

crystal possessed highly uniaxial regions (associated with particular facets) and supported regular bubble domains.

A large joint effort of both research and development investigators showed that this unexpected uniaxial anisotropy found in some flux-grown garnets was indeed not related to strain but was, in fact, growth induced;[32] this discovery of growth-induced anisotropy was an entirely unexpected result.

This property implied that the garnet structure could be tailored by site substitutions to provide a wide range of magnetic properties well adapted

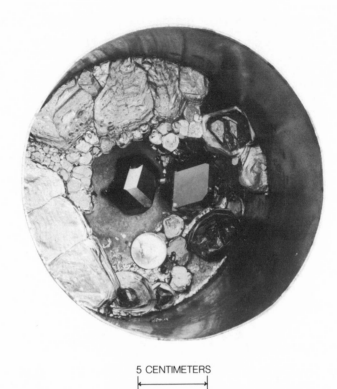

5 CENTIMETERS

Fig. 5-18.   Platinum crucible containing dozens of garnet crystals grown from a molten flux solution by L. G. Van Uitert. Many of the larger crystals were sliced and polished for device evaluation.

to bubbles. Further, and more importantly, garnets offered the possibility of heteroepitaxy. Not only were a wide range of garnets suitable for growing on substrates, but lattice constant matching was possible through appropriate substitutions. Thus, the possibility of dislocation-free heteroepitaxy needed for defect-free films could now be envisioned.

Shick and Nielsen used the tipping method of flux epitaxy for heteroepitaxial garnet growth and achieved films of gallium-substituted erbium and europium garnets on gadolinium-gallium garnet (GGG) of sufficient quality that bubble shift registers could be made.[33] H. J. Levinstein, R. W. Landorf, and S. J. Licht, utilizing similar flux compositions but employing a dipping configuration, were able to grow films with fewer defects, and more importantly, observed that growth occurred under supercooled conditions.[34] This permitted growth under isothermal conditions rather than under the inherently nonuniform conditions of slow cooling growth. Thus it was possible to achieve the high degrees of uniformity of the multi-component compositions essential for device use.

The engineering of bubble compositions and optimization of growth processes proceeded at a rapid rate, thanks to the work of Blank, W. A. Bonner, C. D. Brandle, J. E. Geusic, F. B. Hagedorn, Nielsen, and Van Uitert among others.[35-41] This included efforts in analyzing defect formation in the growth of GGG substrate materials that resulted in dislocation-free boules[39] as well as low-defect films required for large bubble circuits.[40] Studies by Blank and Nielsen[35,38] on the phase properties of the garnet system established methods for growing accurately controlled compositions, providing the basis for the engineering of garnets. Ultimately, suitable garnets were developed for devices with a storage density as great as $10^7$ bits per square centimeter and a temperature range of operation from $-54$ degrees C to $+140$ degrees C.

The success of bubble domain devices was critically dependent on establishing a viable manufacturing technology for these new materials. J. E. Kunzler, who had overall responsibility for electronic materials development, recognized in 1971 that only a "materials capability line" would provide meaningful early experience under pilot line conditions. This was rapidly implemented, and the technology that resulted became the forerunner of the production technology used throughout the world.[42] It provided materials with an exceedingly tight control of bubble film compositions while maintaining the throughput essential for volume production.

The bubbles generated in the orthoferrites were well behaved in that they did not show any erratic motions in uniform gradient fields, nor did their velocity versus drive characteristics show any nonlinearities. Such was not the case with bubbles produced in garnet materials—especially the high-quality garnet films grown by epitaxial means. These showed a continuum of unusual static and dynamic properties caused by differing domain-wall structures.[43]

The simplest bubble wall structure is the Bloch wall, in which the magnetization rotates in a plane parallel to the wall. In a "hard" bubble, the rotation within the wall is in both clockwise and counterclockwise segments. Ion implantation of ion species such as $H_2$ or Ne was found to be the most practical approach to avoiding hard bubbles, and all wafers are now so treated.[44] (For details on ion implantation, see Chapter 2, section 4.4.)

Methods were developed to perform functions needed in bubble chips: propagation, generation, detection, transfer, replication, annihilation, and swap. Of particular interest is the most basic operation, propagation. Conductor propagation was the method first used to move bubbles. It was replaced by field access, in which inductive power is coupled through patterned Permalloy features to produce bubble motion, as initially suggested by A. J. Perneski.[45] When a rotating magnetic field is applied in the plane of the magnetic material, the track of Permalloy features placed on or near the surface of the garnet guides the bubbles around the surface of the magnetic material in a controlled fashion. Initially Y-bar structures were used,[46] followed by T-bar structures. [Fig. 5-19] Many variants in chip organization are possible. For example, Fig. 5-20 illustrates schematically a 68-kilobit, endless loop shift-register serial chip and a 68-kilobit, major-minor parallel chip architecture.[47]

Bubble memory chips are packaged with two mutually orthogonal wire-wound solenoids driven to provide a rotating in-plane drive field, and with a permanent magnet that provides the bias field necessary to maintain stable bubbles. Because the field of the permanent magnet tracks the temperature characteristics of the bias field, the temperature range of operation is wide. Data retention is completely nonvolatile.

The first Bell System application of bubble memories was in the 13A voice announcement system, introduced in late 1976 in the No. 5 crossbar system in Detroit, Michigan. [Fig. 5-21] A pair of 1/4 megabit (Mbit) packages mounted on a printed circuit board with associated electronics stored 24 seconds of prerecorded messages. Low cost, ease of maintenance,

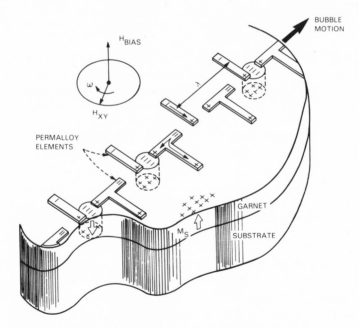

Fig. 5-19.   Isometric view of Permalloy T-bar pattern in contact with the surface of a garnet film. A rotating in-plane magnetic field generates poles in the Permalloy elements that cause the bubble domain to move. The bias field maintains the stable bubble domains.

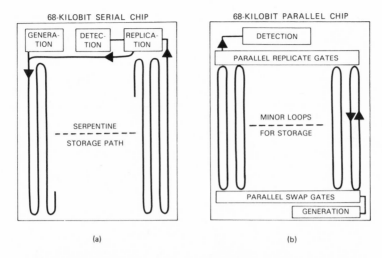

Fig. 5-20.   Architecture of (a) a serial magnetic bubble chip, and (b) a parallel magnetic bubble chip. In the serial chip, bubbles travel a single reentrant path. Such chips are simple to operate; however, the time to access data is relatively long in contrast to the parallel chip organization, which is composed of many minor loops.

and reduced size made the 13A a popular choice with the operating companies.

The large-capacity BELLSTORE* magnetic bubble memory subsystem was developed at Western Electric in Guilford Center, North Carolina and at Bell Laboratories in Morristown for military and other government applications. The memory, coded T250-2200, was an all-solid-state modular design expandable from 4 to 244 Mbit in 4-Mbit increments. Features were nonvolatility; bulk erase capability; nonrecoverable bit error rate better than $10^{-12}$; access time, 14 milliseconds (ms) to write and 8 ms to read; operating temperature range, 0 to 50 degrees C; and nonoperating memory retention temperature range, −40 to +80 degrees C. Although based on a bubble stepping rate of 0.1 megahertz, the system organization permitted data transfers at 1 megaword per second. Special-purpose memory boards were also designed and manufactured with one-, two-, or four-bubble packages.[48] [Fig. 5-22]

Bubble devices in manufacture at Western Electric in Reading, Pennsylvania by 1977 were based on a 3-micrometer ($\mu$m) bubble technology that allowed data storage on a 16-$\mu$m grid spacing. All processing was in

---

* Trademark of AT&T Technologies, Inc.

Fig. 5-21.    Photograph of the 13A voice announcement system. The system consists of eight
bubble memory boards and three special boards containing the controller, power supplies,
field drivers, etc.

a clean-room environment. Bubble circuits were processed in arrays of
106 chips on 3-inch-diameter wafers. There were two patterned metalization layers, one a good electrical conductor for carrying control pulses
to generate, transfer, and replicate bubbles, and the other a Permalloy
layer to propagate and detect them. Fabrication required working at the
limits of the photolithographic facilities then available. The Permalloy level
was the most difficult one to pattern, since it was necessary to open 1.5-
$\mu$m gaps between Permalloy elements and to maintain the edge definition
of elements within 0.15 $\mu$m. The processing making this possible was
developed under the direction of R. S. Wagner.[49]

Thus, at the end of the time period covered by this history, sequential
memories based on magnetic bubbles were in manufacture and had found
a variety of specialized applications. Since then, because of the continued
exponential growth of the semiconductor memory technology, the use of
bubble memory devices has not become as widespread as was anticipated

Fig. 5-22. Special-purpose printed circuit board with four 29B 1/4 Mbit storage capacity magnetic bubble packages and peripheral electronics.

at the beginning of the development program. Nevertheless, there continue to emerge specialized applications requiring some or all of the unique attributes of bubble memories: nonvolatility of the data, insensitivity to radiation, and the absence of moving parts.

## V. CONCLUSION

From the storing of a simple telephone number to the accumulation of huge data bases, telecommunications and other information-age systems require memories of many sizes and types. It should be apparent from this account that for over two decades, magnetic memories played a central role in meeting these needs. The continuously evolving technology presented real challenges to device designers, and Bell Laboratories people met these challenges with a series of inventions and developments that kept the Bell System in the forefront of memory technology. Although at the end of the period covered by this history, semiconductor memory devices had started to replace magnetic RAMs, the manufacture of serial bubble memories was just beginning. At the time of publication of this book, Western Electric—now AT&T Technology Systems—in Reading was producing a significant volume of bubble devices for a variety of uses.

## REFERENCES

1. J. W. Forrester, "Digital Information Storage in Three Dimensions Using Magnetic Cores," *J. Appl. Phys.* **22** (January 1951), pp. 44-48.
2. J. A. Rajchman, "Static Magnetic Matrix Memory and Switching Circuits," *RCA Rev.* **13** (June 1952), pp. 183-201.
3. A. H. Bobeck, "A 256-Bit Transistor-Driven Magnetic Memory System," *TRADIC Computer Research Program, Report for the Fifth Quarter* (July 1, 1955), pp. 185-209.
4. E. F. Sartori, "Input-Output Circuit for TRADIC Magnetic Memory System," *TRADIC Computer Research Program, Report for the Fifth Quarter* (July 1, 1955).
5. M. Karnaugh, "Pulse-Switching Circuits Using Magnetic Cores," *Proc. IRE* **43** (May 1955), pp. 570-584.
6. E. L. Younker, "Memory for the LEPRECHAUN Computer," *TRADIC Computer Research Program, Report for the Eighth Quarter* (April 1, 1956), pp. 53-75.
7. D. H. Looney, "The Ferrite Bead—A New Memory Device," *Proc. 2nd Conf. Magnetism and Magnetic Materials,* Boston, Massachusetts (October 16-18, 1956), p. 673.
8. R. Shahbender, K. Li, C. Wentworth, S. Hotchkiss, and J. Rajchman, "Laminated Ferrite Memory," *RCA Rev.* **24** (December 1963), pp. 705-729.
9. R. H. Meinken, "A Memory Array in a Sheet of Ferrite," *Proc. 2nd Conf. Magnetism and Magnetic Materials,* Boston, Massachusetts (October 16-18, 1956), p. 674.
10. R. M. Genke, P. A. Harding, and R. E. Staehler, "No. 1 ESS Call Store—A 0.2 Megabit Ferrite Sheet Memory," *Bell Syst. Tech. J.* **43** (September 1964), pp. 2147-2191.
11. R. H. Meinken and L. W. Stammerjohn, "Memory Devices," *Bell Lab. Rec.* **43** (June 1965), pp. 228-235.
12. J. G. Chevalier and M. W. Rolund, "New Memory Reduces No. 1 ESS Cost and Size," *Bell Lab. Rec.* **50** (April 1972), pp. 120-123.
13. A. H. Bobeck, "A New Storage Element Suitable for Large-Sized Memory Arrays—the Twistor," *Bell Syst. Tech. J.* **36** (November 1957), pp. 1319-1340.
14. S. M. Shackell, U.S. Patent No. 3,566,373; filed January 10, 1958; issued February 23, 1971.
15. J. J. DeBuske, J. Janik, Jr., and B. H. Simons, "A Card Changeable Nondestructive Readout Twistor Store," *Proc. Western Joint Computer Conf.,* San Francisco (March 3-5, 1959), pp. 41-46.
16. C. F. Ault, L. E. Gallaher, T. S. Greenwood, and D. C. Koehler, "No. 1 ESS Program Store," *Bell Syst. Tech. J.* **43** (September 1964), pp. 2097-2146.
17. W. A. Barrett, U.S. Patent No. 3,067,408; filed November 4, 1958; issued December 4, 1962.
18. W. A. Baker, G. A. Culp, G. W. Kinder, and F. H. Myers, "Stored Program Control No. 1A Store," *Bell Syst. Tech. J.* **49** (December 1970), pp. 2509-2560.
19. W. A. Baker, "The Piggyback Twistor—An Electrically Alterable Nondestructive Readout Twistor Memory," *IEEE Trans. Commun. Electron.* **83** (November 1964), pp. 829-833.
20. A. H. Bobeck, "Properties and Device Applications of Magnetic Domains in Orthoferrites," *Bell Syst. Tech. J.* **46** (October 1967), pp. 1901-1925.
21. U. F. Gianola and T. H. Crowley, "The Laddic—A Magnetic Device for Performing Logic," *Bell Syst. Tech. J.* **38** (January 1959), pp. 45-72.

22. K. D. Broadbent, "A Thin Magnetic Film Shift Register," *IRE Trans. Electron. Comput.* **EC-9** (September 1960), pp. 321-323.

23. A. H. Bobeck and R. F. Fischer, "Reversible, Diodeless, Twistor Shift Register," *J. Appl. Phys.* **30** (April 1959), pp. 43S-44S.

24. U. F. Gianola and D. H. Smith, "The Propagation of Magnetic Domains," *Bell Lab. Rec.* **44** (December 1966), pp. 364-370.

25. P. C. Michaelis, "A New Method of Propagating Domains in Thin Ferromagnetic Films," *J. Appl. Phys.* **39** (February 1, 1968), pp. 1224-1226.

26. R. C. Sherwood, J. P. Remeika, and H. J. Williams, "Domain Behavior in Some Transparent Magnetic Oxides," *J. Appl. Phys.* **30** (February 1959), pp. 217-225.

27. A. A. Thiele and R. H. Morrow, "Properties of the Magneto-Optical Garnet Memory," *IEEE Trans. Magn.* **MAG-3** (September 1967), p. 458.

28. A. A. Thiele, "The Theory of Cylindrical Magnetic Domains," *Bell Syst. Tech. J.* **48** (December 1969), pp. 3287-3335.

29. L. K. Shick and J. W. Nielsen, "Liquid-Phase Homoepitaxial Growth of Rare-Earth Orthoferrites," *J. Appl. Phys.* **42** (March 15, 1971), pp. 1554-1556.

30. S. L. Blank, L. K. Shick, and J. W. Nielsen, "Single Crystal Growth of Yttrium Orthoferrite by a Seeded Bridgman Technique," *J. Appl. Phys.* **42** (March 15, 1971), pp. 1556-1558.

31. S. Iida, "Magnetostriction Constants of Rare Earth Iron Garnets," *J. Phys. Soc. Japan* **22** (May 1967), pp. 1201-1209.

32. A. H. Bobeck, E. G. Spencer, L. G. Van Uitert, S. C. Abrahams, R. L. Barns, W. H. Grodkiewicz, R. C. Sherwood, P. H. Schmidt, D. H. Smith, and E. M. Walters, "Uniaxial Magnetic Garnets for Domain Wall 'Bubble' Devices," *Appl. Phys. Lett.* **17** (August 1, 1970), pp. 131-134.

33. L. K. Shick, J. W. Nielsen, A. H. Bobeck, A. J. Kurtzig, P. C. Michaelis, and J. P. Reekstin, "Liquid Phase Epitaxial Growth of Uniaxial Garnet Films; Circuit Deposition and Bubble Propagation," *Appl. Phys. Lett.* **18** (February 1, 1971), pp. 89-91.

34. H. J. Levinstein, R. W. Landorf, and S. J. Licht, "Rapid Technique for the Heteroepitaxial Growth of Thin Magnetic Garnet Films," *IEEE Trans. Magn.* **MAG-7** (September 1971), p. 470.

35. S. L. Blank and J. W. Nielsen, "The Growth of Magnetic Garnets by Liquid Phase Epitaxy," *J. Cryst. Grow.* **17** (December 1972), pp. 302-311.

36. W. A. Bonner, J. E. Geusic, D. H. Smith, L. G. Van Uitert, and G. P. Vella-Coleiro, "Growth and Characteristics of High Mobility Bubble Domain Garnets with Improved Temperature Stability," *Mater. Res. Bull.* **8** (October 1973), pp. 1223-1229.

37. B. S. Hewitt, R. D. Pierce, S. L. Blank, and S. Knight, "Technique for Controlling the Properties of Magnetic Garnet Films," *IEEE Trans. Magn.* **MAG-9** (September 1973), pp. 366-372.

38. S. L. Blank, J. W. Nielsen, and W. A. Biolsi, "Preparation and Properties of Magnetic Garnet Films Containing Divalent and Tetravalent Ions," *J. Electrochem. Soc.* **123** (June 1976), pp. 856-863.

39. C. D. Brandle, D. C. Miller, and J. W. Nielsen, "The Elimination of Defects in Czochralski Grown Rare-Earth Gallium Garnets," *J. Cryst. Grow.* **12** (March 1972), pp. 195-200.

40. J. E. Geusic, H. J. Levinstein, S. J. Licht, L. K. Shick, and C. D. Brandle, "Cylindrical Magnetic Domain Epitaxial Films with Low Defect Density," *Appl. Phys. Lett.* **19** (August 15, 1971), pp. 93-95.

41. F. B. Hagedorn, W. J. Tabor, J. E. Geusic, H. J. Levinstein, S. J. Licht, and L. K. Schick, "Cylindrical Magnetic Domain Epitaxial Film Characterization: Device and Growth Implications," *Appl. Phys. Lett.* **19** (August 15, 1971), pp. 95-98.

42. S. L. Blank and S. J. Licht, "The Simultaneous Multiple Dipping of Magnetic Bubble Garnets," *IEEE Trans. Magn.* **MAG-16** (September 1980), pp. 604-609.

43. W. J. Tabor, A. H. Bobeck, G. P. Vella-Coleiro, and A. Rosencwaig, "A New Type of Cylindrical Magnetic Domain (Bubble Isomers)," *Bell Syst. Tech. J.* **51** (July/August 1972), pp. 1427-1431.

44. R. Wolfe and J. C. North, "Suppression of Hard Bubbles in Magnetic Garnet Films by Ion Implantation," *Bell Syst. Tech J.* **51** (July/August 1972), pp. 1436-1440.

45. A. J. Perneski, "Propagation of Cylindrical Magnetic Domains in Orthoferrites," *IEEE Trans. Magn.* **MAG-5** (September 1969), pp. 554-557.

46. I. Danylchuk, "Operational Characteristics of $10^3$-Bit Garnet Y-Bar Shift Register," *J. Appl. Phys.* **42** (March 15, 1971), pp. 1358-1359.

47. P. I. Bonyhard, I. Danylchuk, D. E. Kish, and J. L. Smith, "Applications of Bubble Devices," *IEEE Trans. Magn.* **MAG-6** (September 1970), pp. 447-451.

48. J. E. Geusic, "Magnetic Bubble Devices: Moving from Lab to Factory," *Bell Lab. Rec.* **54** (November 1976), pp. 262-267.

49. T. W. Hou, C. J. Mogab, and R. S. Wagner, "Ion Milling Planarization for Magnetic Bubble Devices," *J. Vacuum Sci. Technol. A* **1** (October-December 1983), pp. 1801-1805.

# Chapter 6

# Piezoelectric Devices

*Throughout the 50-year period of this history, the quartz crystal unit was the basis for carrier telephony and radio transmission. In the 1920s, radio carrier frequencies began to be controlled by quartz crystal units. The following decade brought the first successful application of quartz resonators to channel filters for telephone transmission systems. Later, in World War II, the need for better control in FM transmission led to a significant expansion of production of quartz devices, while postwar ultrasonic delay lines using barium titanate piezoelectric ceramics as transducer materials significantly improved the capability of radar systems. Solder-sealed quartz devices gave way to the more stable cold-welded units. Subsequently, the plated crystal precision units of the late 1950s increased frequency stability and accuracy dramatically. Moreover, mathematical modeling allowed the development of the trapped energy concept, leading to the elimination of unwanted modes, which culminated in the development of multiresonator monolithic crystal filters.*

## I. EARLY QUARTZ RESONATORS AND FILTERS

For long-distance telephone transmission to be economical, multiple telephone calls must be multiplexed onto a single transmission line. The widely used frequency multiplexing requires carrier frequencies controlled to a high degree of accuracy as well as precise filters to separate the various channels from the composite signal at the receiving end of a system. For both of these purposes—frequency control and filtering—quartz devices based on the piezoelectric effect had no rival during the period of this history.

The piezoelectric effect is observed in certain dielectric crystals that, when deformed by pressure, generate an electrical charge. The name derives from the Greek *piezein*, which means to press. These crystals contain positively and negatively charged ions that separate when a stress is applied to the crystal. If the crystal structure has no center of symmetry, then the stress-induced separation of charges is nonsymmetric and a net electric

Principal authors: R. A. Sykes and T. R. Meeker

dipole moment is generated. If the crystal is isolated, this electric dipole moment appears as a voltage across the crystal. An electric current will flow when the crystal is short-circuited.

Conversely, an electric field applied to the crystal causes a mechanical strain or deformation in the crystal. Specifically, a piezoelectric crystal vibrates in response to the application of an alternating electric field. Since a quartz crystal has excellent elastic properties and little mechanical loss, it has well-defined mechanical resonances at which the electrical impedance also becomes resonant. Therefore, such a quartz resonator with its very stable frequency of vibration can be used to control or to select frequencies in electrical circuits.

A measure of the progress made from 1938 to 1972 is best illustrated by a comparison of the A1 and A6 channel bank filters used in telephone carrier systems, shown in Fig. 6-1. The A1 channel filter followed a design using discrete components in a lattice filter structure, whereas the A6

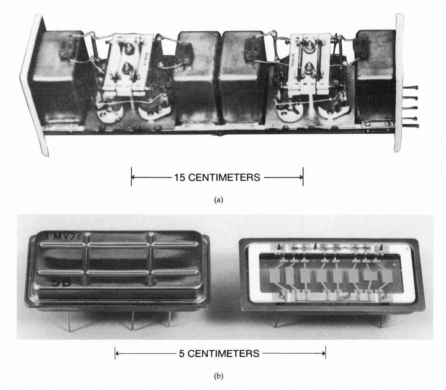

|←——— 15 CENTIMETERS ———→|

(a)

|←——— 5 CENTIMETERS ———→|

(b)

Fig. 6-1.   Comparison of two channel bank filters: (a) A1 channel filter, designed in 1938, used discrete components in a lattice filter structure; (b) the later A6 channel filter design, introduced in 1972, based on acoustically coupled trapped-energy resonators on a single quartz plate.

channel filter design, a monolithic crystal filter, is based on six acoustically coupled trapped energy resonators on a single quartz plate.

Radio broadcasting began in the early 1920s, and, following the investigations of piezoelectric crystal-controlled oscillators by A. M. Nicolson of Western Electric,[1,2] W. G. Cady of Wesleyan University,[3] and G. W. Pierce of Harvard University,[4] quartz crystals were used in 1924 to stabilize the carrier of the Bell System's pioneering radio station WEAF, then at 32 Sixth Avenue in New York City.

The application of quartz resonators in filters required considerable additional development. Filter design without quartz elements had its origin with the investigations of G. A. Campbell,[5,6] O. J. Zobel,[7] and K. S. Johnson,[8] which led to the invention, theory, and design equations for electric wave filters. This work led to bandpass filters in the frequency range of 8 to 30 kilohertz (kHz), enabling the production of carrier telephone systems in service as early as 1918.[9] For additional detail on this early work, see another volume in this series subtitled *The Early Years (1875-1925)*, Chapter 10, section II.

However, it was not until 1927 that L. Espenschied observed that the quartz crystal, with its high inherent stability and low dissipation (high Q), might be used in a filter to permit closer channel spacing and higher frequencies, which at the time were limited by the inductors used in channel filters.[10]

In 1925, W. P. Mason had begun work on quartz crystals for broadcasting. He found that the quartz resonators used for radio broadcast services contained so many spurious modes it was not possible to use them in bandpass filters, but noted that the lower-frequency modes (such as the extensional mode of a bar) did offer possibilities.[11] He showed that an inductor used in series with a quartz crystal unit in a filter yielded bandwidths as high as 13 percent. Crystal units alone limited the bandwidth to 0.3 percent.

The introduction of inductors, however, increased the insertion loss to undesirable levels. Mason showed that, with some network transformations, the resistive part of the inductors could be moved to form part of the terminating resistors. Figure 6-2 shows this equivalence. This innovation

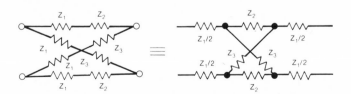

Fig. 6-2.   Equivalent lattice networks, as devised by W. P. Mason. $Z_1$ represents an inductor, and $Z_2$ and $Z_3$ are the crystal units.

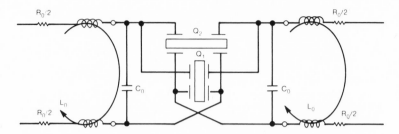

Fig. 6-3.   A crystal bandpass filter using divided plated crystal units and series inductors. $Q_1$ and $Q_2$ represent the divided plated crystal units of the lattice network, $C_0$ represents the shunt capacitances removed from the lattice, and $L_0$ represents the inductors producing a wideband channel filter. Over 40 years of channel filters were based on this design.

was followed by a further simplification, in which the number of inductors and crystal units was reduced by one-half by using coupled coils and divided plating on the surface of the crystal plates. With the dissipative part of the inductors forming part of the termination to produce a constant insertion loss at all frequencies, it was then possible to design channel filters with low loss in the passband. This design was one of the most significant theoretical contributions to crystal units in carrier telephone transmission systems. Designs for channel filters for the next 40 years were based essentially upon that shown in Fig. 6-3. More than two million were produced.

With this work, Mason began a long and productive career that spanned more than three decades. His work is also discussed in two companion volumes in this series subtitled *Communications Sciences (1925-1980)*, Chapter 1, section 3.1, and *Physical Sciences (1925-1980)*, Chapters 9, 14, 16, and 19.

## II. RESONATOR TECHNOLOGY REFINEMENTS

Initial theoretical work on the vibration of bars and plates applied mainly to long thin bars for extensional and flexure modes, and to plates of large dimensions for thickness-shear or transverse modes, configurations not characteristic of actual devices. Practical quartz crystal units for frequency control and selection were developed strictly experimentally with only limited theoretical confirmation and extension.

The properties of resonators depend both on the dimensions of the plates or bars and on their orientation with respect to the crystallographic axes. Figure 6-4 illustrates a few orientations of particular interest. Physical properties of a plate, such as the velocity of acoustic waves and the frequency-temperature coefficient, can be selected through the choice of a plane within the crystal and the rotational orientation of the plate or bar within that plane.

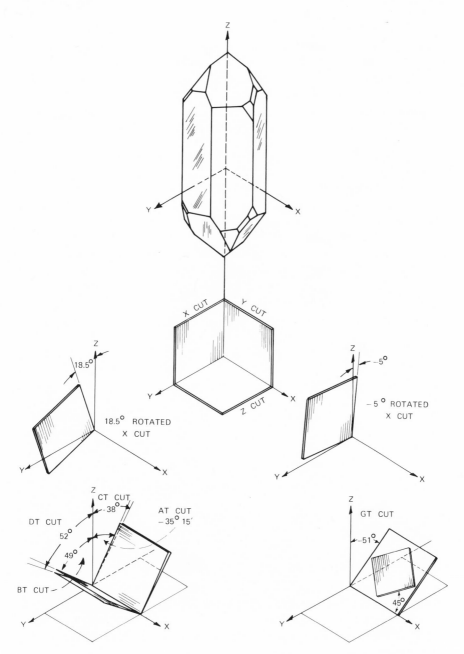

Fig. 6-4.  Schematic of a quartz crystal showing the orientation of selected plates for frequency control and selection.

Experimental crystal resonators and channel filters were produced from 34 kHz to 359 kHz. Unwanted responses in both filter crystals and the higher-frequency shear units for oscillators were the major problem. Studies had been made to determine the effects of rotated bars and plates as well as the resonance spectra of low-frequency crystal resonators as a function of dimensional ratios.[8] The purpose of these studies was to find the rotations for reduced or zero coupling between modes (required for single-mode operations) as well as those to produce zero frequency-temperature coefficients.

Figure 6-5 shows the effect of plate geometry on the resonant frequencies of a 0-degree, X-cut plate and a +18.5 degree, X-cut plate. When only the width is changed, the frequency of the length extensional mode is not affected in the case of the +18.5 degree, X-cut plate over a wide range of plate geometries.

Through the use of +18.5 degree, X-cut resonators, successful channel filters for telephone carrier systems were developed.[12,13] Twelve channels spaced 4 kHz apart from 60 to 108 kHz were chosen as a compromise between systems considerations and crystal plates of reasonable size to form the basic bank of filters for future carrier systems. As improvements were made in inductor core materials and mounting systems for the quartz plates, the −5 degree, X-cut crystal plate was substituted. A lower temperature coefficient was then possible, because higher impedances could be used with improved inductors.

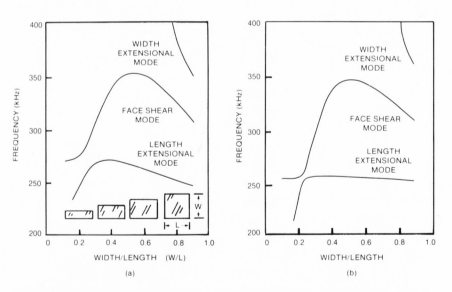

Fig. 6-5.  Effect of plate geometry on the resonant frequencies for plates of constant length, L, of 10 millimeters. (a) A 0-degree, X-cut plate. (b) A +18.5 degree, X-cut plate. When only the width is changed, the frequency of the length extensional mode is not affected over a wide range of plate geometries for the +18.5 degree cut. [Heising, *Quartz Crystals for Electrical Circuits* (1952): 223.]

The high-frequency plates, the AT and BT cuts, materially improved characteristics for the control of broadcast and shortwave radio transmitters because of small frequency-temperature coefficients and simpler unwanted mode spectra. By careful selection of dimensions, high-frequency units could be made to operate from 500 kHz to 15 megahertz (MHz). For many years, the BT cut was preferred because of its higher yield in manufacture. A great deal of information regarding surface contour and dimensional ratios was gathered empirically, since no reasonably simple theory could explain the resonance behavior. In contrast with the lower-frequency filter crystals whose electrodes were thin metallic films plated directly on the major surfaces, the high-frequency units were formed by placing the quartz plate between steel electrodes, as shown in Fig. 6-6(a). For plates of large dimensional ratios (e.g., both length and width greater than 55 times the thickness), the corners can be clamped to obtain a more stable unit, as seen in Fig. 6-6(b). These mounting methods permitted frequency adjustment by successively grinding and etching the major surfaces.

The introduction of coaxial carrier systems made it necessary to develop carrier and pilot selection filters for frequencies from 1 to 3 MHz. The resonator designs used in channel filters, if applied to the higher frequencies, would have been unsuitable because of the small plate size that would have resulted. A great deal of effort was expended, therefore, to obtain filters with satisfactory performance, and many types of resonator configurations were tried, including harmonic versions of the low-frequency

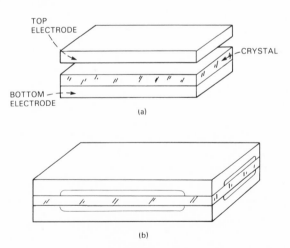

Fig. 6-6. High-frequency shear-type mountings: (a) crystal plate free with air gap; (b) plate clamped at corners. These mountings permitted frequency adjustment by successive grinding and etching. Only plates with large dimensional ratios can be clamped at the corners to achieve more stable units.

Fig. 6-7. R. A. Sykes, who made many contributions to piezoelectric
crystal units.

types as well as the high-frequency shear types, such as the AT and BT
cuts mentioned before. It was evident that a detailed study was required
of the resonance spectra of the high-frequency thickness-shear plate if
filter applications were to advance. This study comprised detailed observations of the changes in the resonance spectra near the principal mode
as lateral dimensions of the crystal plate were reduced by very small
increments. The same method was later applied to the filter and the oscillator
control units. For example, successful crystal filters for 2- and 3-MHz pilot
selection were developed as early as 1935. Later, R. A. Sykes [Fig. 6-7]
postulated the many types of modes existing in the high-frequency shear
plates and gave empirical formulas to predict their behavior.[14]

### III. MILITARY APPLICATIONS DURING WORLD WAR II

After the successful development of low-frequency crystal units for
filters, similar units were used for the basic control oscillator in frequency-
modulated transmitters for military applications. These units made use of
the CT and DT plates shown in Fig. 6-4. By late 1941, the Western Electric
Hawthorne Works in Chicago, Illinois was able to produce crystal units
for these devices at a rate 1000 times greater than any previous production
volume. To assist in the initial production, engineers and scientists selected

from the research, development, and systems areas of Bell Laboratories spent from three to six months at the factory. They developed new test facilities and production techniques and tools, conducted many experiments with production lots to investigate causes for rejects, and made changes in design to improve yields. By mid-1942, production had reached a rate of 24,000 units per day, and by 1946, a total of 13 million units of various types had been shipped to the military. During this same period, large quantities of the high-frequency clamped-type AT- and BT-cut crystal units were produced at Western Electric's facilities in Kearny and Clifton, New Jersey.

Of the many developments during the early effort to increase production for the military, two were particularly important because they led to increased stability. First, the cause of rapid initial aging in high-frequency units was found to be due principally to a small disoriented layer of crystalline material left on the surface of the quartz plate by the grinding process; the layer could be easily removed by etching with hydrofluoric acid. Completing the final frequency adjustment by etching resulted in crystal units of very low aging and increased stability.

The second development was the investigation by I. E. Fair of instabilities in low-frequency face-shear resonators used in the FM transmitter-receiver. He observed that these instabilities were caused by changes in the wire mounting system.[15] The energy absorbed from the crystal resonator through the soldered wire support set up a flexure wave in the wire supports. This energy could be absorbed or reflected back to the resonator by clamping or soldering the wire to a rigid support at an even or odd quarter-wave flexure length. The use of small solder weights or supports, when controlled at even quarter-wavelengths, materially reduced rejects, simplified the frequency adjustment operation, and gave improved stability with time and temperature.

As part of its commitment to offer the military as many procurement options as possible, Bell Laboratories published advances in crystal technology in the *Bell System Technical Journal* as they occurred.[16] Early publication made key innovations available to potential suppliers of the new technology.

## IV. ULTRASONIC DELAY LINES

Another class of devices based on piezoelectric materials is that of ultrasonic delay lines. Their initial applications were for military radar systems. Before the 1940s, such systems used transmission lines for time calibration and other delay applications. During the 1940s, military radar development began to need longer and more accurate delays for time calibration and stationary target cancellation. Long sections of transmission lines (coaxial cable, twisted pair, etc.) were thus needed, since their propagation delay is only $10^{-9}$ second per foot (sec/ft). In addition, they were lossy and unstable with changes in temperature and configuration. An ultrasonic delay line with a propagation delay of about $10^{-4}$ sec/ft has an intrinsic size advantage for these functions.

In a delay line, an input transducer converts electrical signals into acoustical signals that travel through the delay medium at the speed of sound. [Fig. 6-8] An output transducer converts the acoustical signals back into electrical signals. Since much of the design and fabrication technology for ultrasonic delay lines was similar to that for crystal resonators, Bell Laboratories was a natural resource for this new area of development. Delay devices were also thought to be generally useful in the kinds of circuits used in the telephone system. This promise of potential application contributed to the Bell System interest in delay line development.

The course of the resulting development was dictated by the material properties needed for the delay line medium and for the transducers. Many applications were digital, requiring a delay line medium having low loss at high frequencies and wideband transducers.

Low loss at high frequencies and low cost could be achieved with only a few existing delay media. The best of these materials seemed to be fused

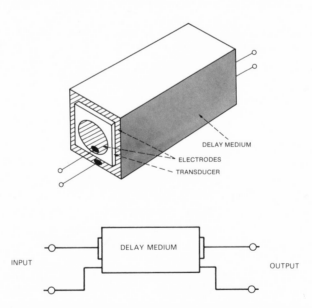

Fig. 6-8.   Delay line schematic. The input transducer converts electrical signals into acoustical signals that travel through the delay medium at the speed of sound and are converted back to electrical signals by the output transducer. [Smits and Sittig, *Ultrasonics* **7** (1969): 168.]

silica or glasses, which had been developed for use in optical lenses and prisms. Fortunately, some of the optical material needs for homogeneity and cost and some of the fabrication techniques were similar to those for the ultrasonic delay line.

The ultrasonic transducer needs to be a bandpass device with a wide bandwidth. Early theoretical and experimental work showed that piezo-electric coupling, a parameter of the piezoelectric transducer material, is of secondary importance for a crystal resonator, but of primary importance for a delay line transducer. This parameter controls the compromise between transducer loss and bandwidth. Unfortunately, piezoelectric coupling is small for the best resonator material, quartz. Piezoelectric ceramics, such as barium titanate, on the other hand, have very high piezoelectric coupling. These materials must be polarized with a high electric field, and early work was concentrated on thickness-polarized material. Transducers made with this material generate dilatational waves. Multiple-reflection delay lines with dilatational transducers generate unwanted modes because of strong mode conversion at the reflecting surfaces. Since a properly polarized shear wave reflects with no mode conversion, considerable effort was directed at the development of shear mode transducers. This development included techniques for fabricating transverse polarized ceramic transducers and techniques for converting dilatational motion into shear motion.

Early delay line designs emphasized large delays per unit volume. Complicated polygon structures were designed in which the acoustic path was folded to traverse the delay medium many times. The first such design was by D. L. Arenberg of the Massachusetts Institute of Technology in 1950.[17] In a design by H. J. McSkimin in 1954, spurious signals were suppressed by absorbing materials placed on those surfaces at which reflection from the direct path should not occur.[18]

From the late 1940s to the mid-1960s, many delay lines were designed by McSkimin,[19,20] M. D. Fagen,[21] J. E. May, Jr.,[22] and A. H. Meitzler.[23] These lines had center frequencies in the range of 10 to 60 MHz, delays from a few microseconds to a millisecond, and bandwidths up to 40 percent of the center frequency. Delay media were mostly fused quartz or glass, and transducers were quartz or barium titanate ceramic.

Later developments included the invention of the depletion layer transducer by D. L. White in 1962,[24] in which the depletion layer in a GaAs p-n junction functioned as an ultrasonic transducer. Because the depletion layer can be very thin, it extended the frequency of operation to the gigahertz range. Another technique for achieving this very high-frequency performance was to apply thin-film transducers of zinc oxide or cadmium sulfide to the storage medium by evaporation, sputtering, or pyrolysis. Development of these techniques was carried out by N. F. Foster,[25] D. Beecham,[26] and D. L. Denberg[27] between 1965 and 1971.

In the 1950s, work on radar technology led to the realization that a dispersive delay line could be used to improve the range and signal-to-noise ratio of target-tracking radars. At Bell Laboratories, May suggested that a delay line using the torsional, flexural, and longitudinal modes in a wire could have a large delay dispersion.[28] He was able to demonstrate several hundred microseconds of delay change for a 300-kHz bandwidth centered around 2 MHz. None of the other techniques being considered could match the performance of this dispersive delay line. A problem with this kind of delay line, however, was the large number of other modes that were very difficult to suppress. T. R. Meeker continued these developments by using the first longitudinal mode in a strip.[29] This structure had adequate performance and was used in the Nike Zeus acquisition radar tests at White Sands, New Mexico and at Kwajalein in the South Pacific in the late 1950s. (For more information on this defense work, see a companion volume in this series subtitled *National Service in War and Peace (1925-1975)*, Chapter 7, section 1.6.3.)

M. R. Parker, Jr. and R. S. Duncan[30] proposed a diffraction delay line, in which the dispersion was produced by the transmission properties of the radiation field of a grating transducer with a nonuniform grating spacing. One useful structure was a wedge with a solid transducer on one side and a grating transducer on the other,[31] as shown in Fig. 6-9. A second structure had two grating transducers perpendicular to each other.[32] One design of this perpendicular delay line was produced by Western Electric for the missile site radar used in a system called Safeguard. This delay line had a 10-MHz bandwidth at 30 MHz, with 20 microseconds ($\mu$s) of delay change. A 75-MHz version of this delay line, with a 30-MHz bandwidth and 30 $\mu$s of delay change, was developed but never manufactured.

J. H. Rowen,[33] E. K. Sittig,[34] G. A. Coquin and H. F. Tiersten,[35] Meitzler,[36] and H. Seidel and White[37] proposed a surface acoustic wave structure for this function in the second half of the 1960s. This structure was a planar device using an interdigitated transducer on a quartz substrate. Although this device was never used in Safeguard, it was the forerunner of surface acoustic wave devices used extensively in telecommunications equipment of the mid-1970s.

The large delay and low loss of the aluminum-strip delay lines developed for the Nike Zeus acquisition radar suggested that this type of delay line could be used in a sequential digital memory. In the late 1950s, Fair and Meitzler[38] designed a nondispersive aluminum-strip delay line that was used in a high-speed (3 megabit/sec), high-capacity (1000 bit/channel) memory for a message store and forward system (see a companion volume in this series subtitled *Switching Technology (1925-1975)*, Chapter 9, section 6.3). The delay line was manufactured by Bliley Electric Company, Anderson Laboratories, and other companies. This delay line had the lowest cost per bit of any device available at that time. Later, however, the devices were replaced by integrated circuit (IC) shift registers.

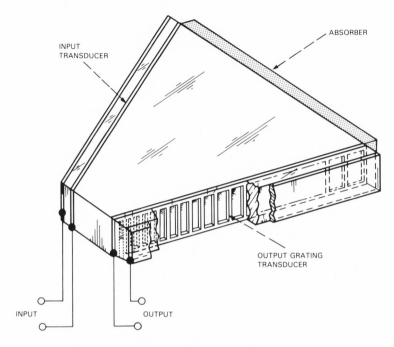

Fig. 6-9. Dispersive delay line, which uses a nonuniform grating output transducer. Because of interference effects, high-frequency signals are detected at the
output transducer with a shorter delay than the lower-frequency signals.

## V. PLATED ELECTRODE UNITS

A major change in quartz crystal development came in 1942 with the
formation of a group at the Murray Hill, New Jersey site to develop a
high-frequency crystal unit of the thickness-shear type by using plated
electrodes and applying ideas learned earlier in the development of crystal
units for telephone carrier systems.[39] While the previous clamped-typed
crystal unit required an etching of the crystal plate surface for the final
frequency adjustment, the plated unit could be adjusted to the desired
frequency by the addition of electrode material while the unit was controlling an oscillator. This approach substantially increased the production
yield of units by simplifying the structure. The method of final frequency
adjustment is shown in Fig. 6-10. Most of the high-frequency crystal units
used in military systems were replaced by the plated unit. Several million
were produced by Western Electric at the Kearny, Clifton, and Hawthorne
facilities.

Several important ideas emerged from the plated-unit development. It
was evident that the plated-unit device could be used to study in greater

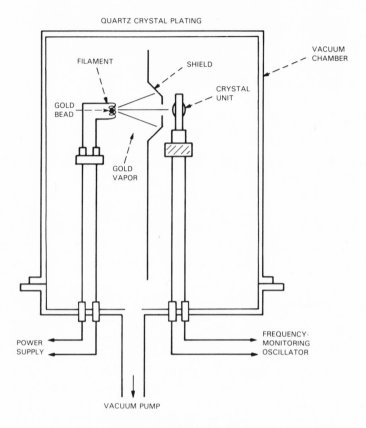

Fig. 6-10. Frequency adjustment for the plated unit: the evaporator method. Because the frequency can be monitored during the evaporation process, precise adjustment of the resonator frequency is greatly simplified. [Adapted from Sykes, *Proc. IRE* **36** (1948): 6.]

detail unwanted response spectra, plating materials, effects of power dissipation, effects of dislocations, and imperfections in the quartz material, as well as causes for frequency aging over time due to atmospheric contamination and mounting systems.

By 1944, a new department had been formed to work on frequency control and selection. It combined the existing groups on radio application development, telephone apparatus development, and fundamental development. At this time, the military needed greater quantities of better crystal units to meet new and more stringent requirements for aging and frequency control. In addition, the Bell System needed increased carrier system production to meet the wartime increase in toll telephone traffic.

Researchers in the new department undertook a complete revision of crystal unit development and design. They also developed a new series

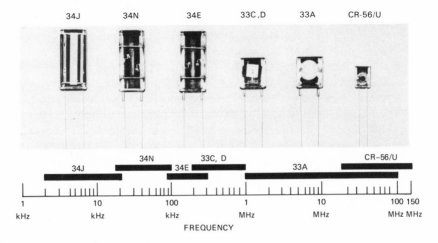

Fig. 6-11. Solder-sealed crystal units (above), showing frequency ranges covered by each unit (below).

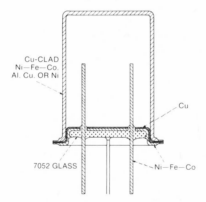

Fig. 6-12. Cross section of the cold-welded quartz crystal enclosure. This enclosure replaced the solder-sealed enclosure of the units shown in Fig. 6-11.

of units that were equally applicable to frequency control and use in filters. Figure 6-11 shows the result of this effort. All types of resonators from the flexure, extensional, face, and thickness-shear types were employed in solder-sealed metal containers using common piece parts and identical processing, where practical. Identical types were also used for both network and frequency control applications, with only the quartz plate dimensions modified to meet particular impedance requirements. Shortly after the completion of the standardization program, the military accepted the metal enclosure designs and called for their use in all new communications equipment. Most of the CR-type military crystal unit codes of the mid-1970s were derived from this original series.

The enclosures shown in Fig. 6-11 were later replaced by a series of cold-welded circular types following some of the technology developed for transistors.[40] Figure 6-12 shows the cross section of such an enclosure, and the frequency range covered by each type of crystal unit appears in Fig. 6-13. The fact that the enclosure base and cover could be hydrogen fired prior to use, plus the fact that the unit could be vacuum sealed following final adjustment before being exposed to a contaminating atmosphere, solved most of the problems associated with long-term frequency drift.

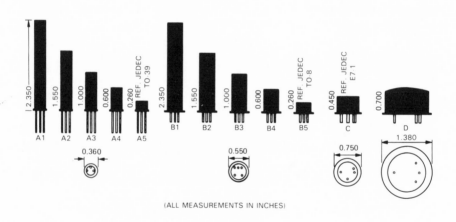

(ALL MEASUREMENTS IN INCHES)

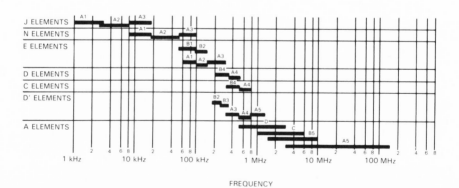

FREQUENCY

Fig. 6-13.   Standardized series of quartz crystal units in cold-welded enclosures (above) and frequency range covered by each enclosure type (below).

## VI. PRECISION FREQUENCY CONTROL

During the development of the high-frequency plated crystal unit, and particularly following the introduction of hermetically sealed enclosures, it appeared that further investigation should result in a unit of higher stability and precision than the 100-kHz GT-plate frequency standard developed earlier by Mason.[41] Experiments by A. W. Warner, H. E. Bommel, and Mason[42,43] on highly contoured, fifth-overtone thickness-shear modes at 2.5 and 5 MHz yielded units of very high Q and stability, with significantly reduced aging.[44] Figure 6-14 shows an early model of one of these crystal units. The mounting and surface loss of this unit were reduced to such a degree that it could be used to measure the intrinsic loss of the material. Figure 6-15 shows the most probable loss in natural quartz as a function of frequency, with measurements on selected fifth-overtone resonators. Resonators of this type were used for a number of years to evaluate the quality of cultured quartz (see section VIII), particularly for Q values of 1 to 2 million or more at 5 MHz. The infrared technique, a much simpler and faster method, could be used for lower Q values.[45]

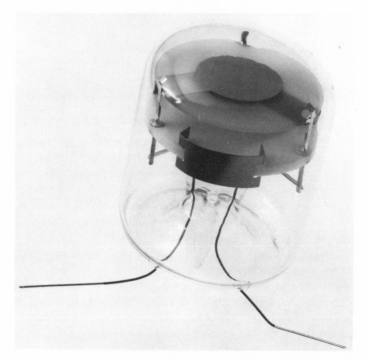

Fig. 6-14. High-precision 2.5-MHz contoured crystal unit. Sealing the unit into an evacuated glass envelope improved frequency stability.

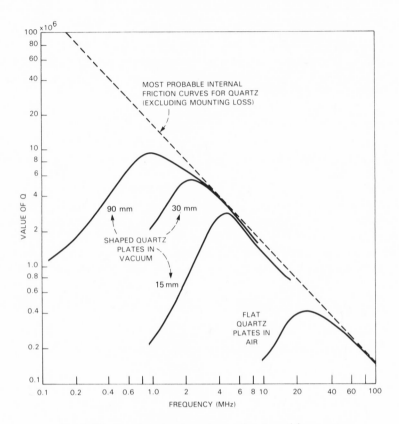

Fig. 6-15.   Loss in natural quartz as a function of frequency.

As the development of the high-frequency precision unit brought lower aging and increased stability, it became necessary to develop better oscillator circuitry and precise temperature control in order to measure the resulting improvements.[46] Subsequent advances in the crystal unit came about with thermocompression-bonded, strain-free mounting ribbons, final adjustment in an oil-free ion-pumped system, and vacuum sealing in cold-welded enclosures that were hydrogen fired prior to use. Further improvements were the use of high-gain active elements to give very low crystal current and double-oven proportional temperature control. The resulting precision oscillators were essentially drift free, and frequency changes were of only a long-term nature. Even with the development of atomic and molecular frequency sources, crystal-controlled oscillators still play an important role in frequency standards. Figure 6-16 shows a comparison of the performance of crystal-controlled atomic and molecular frequency sources as a function of measurement period. For periods less than one minute, the crystal-controlled oscillator remains superior.

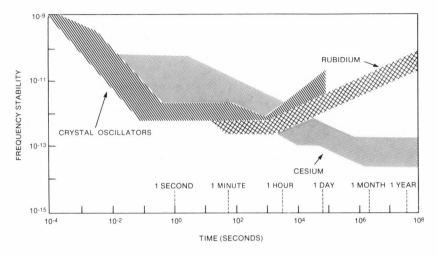

Fig. 6-16. Frequency stability of various precision sources for short and long periods. [Kartaschoff and Barnes, *Proc. IEEE* **60** (1972): 498.]

## VII. MATHEMATICAL MODELING OF MODE SPECTRA

With the plated high-frequency crystal unit, detailed resonance spectra could be studied by the simple addition of mass to the electrode by the evaporation process; subsequently, sufficient experimental data was obtained to set up a mathematical model of this complex problem. R. D. Mindlin[47] of Columbia University studied the data in the early 1950s and set up boundary conditions that gave the first satisfactory mathematical solution to the coupled high-frequency thickness-shear and flexure modes. Figure 6-17 shows a comparison of theory and experimental data over a wide range of dimensional ratios. A further extension to the understanding of the mode spectra problem was made in the early 1960s by W. J. Spencer,[48] who applied the technique of X-ray topography to the study of strain patterns in quartz plates resulting from resonances and imperfections. Figure 6-18 shows the many resonances near that of the principal high-frequency thickness-shear mode at 3200 kHz. The photographs of the strain patterns for the modes were obtained by driving the crystal plate by an oscillator at the particular frequency. The patterns, due to the strain induced by the mounting system, illustrate the sensitivity of the method.

Following many experiments and X-ray topographs, Mindlin and Spencer[49] were able to formulate a theory to account for these more complex modes in the mid-1960s. Other workers extended the theory to include the effect of energy trapping.[50,51] The application of this theory and the

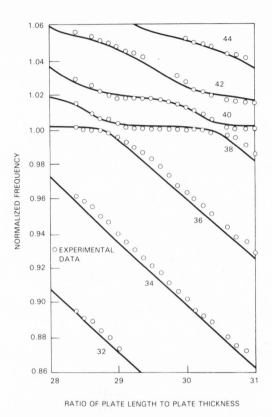

Fig. 6-17.   Comparison of theory and experimental data
for resonant frequencies of the even modes of vibration
of an AT-cut quartz plate; length varied from 25 to
21.8 millimeters, with mode number as parameter.

design techniques developed in the 30 years following World War II made
it possible to obtain a high-frequency crystal unit with effectively a single
resonance, an achievement thought nearly impossible as late as 1965.

During the same period that studies were conducted on the high-frequency mode spectra, the same kind of data was accumulated on the low-
frequency extensional, flexure, and face-shear modes. While the practical
problems associated with the application of crystal units to wave filters
were not as severe as in the high-frequency case, certain empirical rules
had to be established for dimensional ratios to reduce rejects with unwanted
responses. Even though a physical model representing the low-frequency
modes had been postulated correctly and empirical formulas developed
for their use, it was not until Spencer's work on the study of mode shapes
using X-ray diffraction techniques in the early 1960s that a mathematical

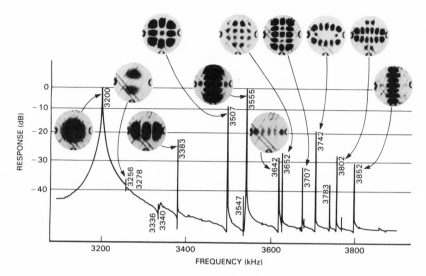

Fig. 6-18.   Strain patterns associated with resonances near the principal thickness-shear mode of an AT-cut quartz plate at 3200 kHz, shown by X-ray topographs. [Gerber and Sykes, *Proc. IEEE* **54** (1966): 107.]

solution could be found.[52] Closed-form solutions of coupled extensional, flexural, and width-shear modes were obtained for thin rectangular plates with free edges by P. C. Y. Lee of Princeton University.[53] This analysis agreed closely with experimental data, so after a search of more than 30 years, a firm mathematical foundation was laid for the design and application of piezoelectric crystal units.

## VIII. CULTURED QUARTZ GROWTH

During World War II, the import of natural quartz from Brazil was threatened because of submarine attacks, and it became apparent that the supply of natural quartz would not be adequate for the expected expansion of telephone carrier systems. In particular, substitutes were needed for the large quartz crystal plates used in channel filters (60-108 kHz). As discussed in a companion volume subtitled *Physical Sciences (1925-1980)*, pp. 529-530 and 576, alternate piezoelectric materials were developed in response to these needs. These included ammonium dihydrogen phosphate (ADP)[54] and ethylene diamine tartrate (EDT).[55] Crystal units made with EDT for carrier telephone channel filters performed acceptably.[56,57] Since this material was water soluble and soft compared to quartz, entirely different processing techniques had to be developed for crystal units used in the channel and 64-kHz pilot filters. The completed units, however, were interchangeable with those they replaced.

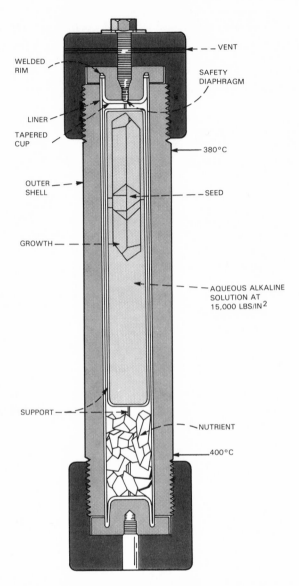

Fig. 6-19.    Cross section of an autoclave used in the growth of quartz crystals.

While this work was in progress, experiments were initiated to grow quartz by the hydrothermal process reported initially by R. Nacken in Germany.[58] Techniques for the successful growth of cultured quartz were developed between 1948 and 1958, eliminating the need for alternative materials. (For a general discussion of crystal growth technology, see *Physical Sciences (1925-1980)*, Chapter 19, section 1.1.)

At high pressures (about 1000 atmospheres) and temperatures (about 450 degrees C), quartz is soluble in water. Figure 6-19 shows the hydrothermal process, in which quartz is grown in an autoclave, where the nutrient (small natural quartz crystals), placed in the lower portion of the autoclave, goes into solution and is redeposited on the seeds held in the top portion. Large-size crystals can be formed in this manner.

Early experiments yielded material with high acoustic losses, even though the crystals looked clear optically. J. C. King studied this problem by measuring the Q of thickness-shear-type resonators over a wide temperature range.[59] He found that a major relaxation peak existed at a temperature of 50 kelvins (K) and that a general background loss peaked at 20 K. The peak at 50 K was found to be due to a sodium impurity. This peak could be removed by electrolytically sweeping out the sodium or by replacing it with lithium during the growth process of cultured quartz.[60] By 1968, the commercial growth of cultured quartz had progressed to a point that all quartz devices, except those requiring the highest frequency stability, were made with the cultured material.[61,62]

## IX. MONOLITHIC CRYSTAL FILTERS

The history of high-frequency crystal unit development for use in filter networks paralleled that of oscillators. The unwanted and anharmonic modes just above the thickness-shear resonance produce unwanted transmission bands that make the design and use of filters more difficult than oscillators; in an oscillator, these modes can be avoided because their impedance is usually higher than the main shear mode.

The first application of thickness-shear mode filter units was for the 2064- and 3096-kHz pilot filters for the L1 coaxial system, which were developed as early as 1935, as mentioned in section II. Figure 6-20 shows the crystal plate and electrode arrangement of these filters. While the dual electrodes on one side of the plate produced symmetric as well as antisymmetric anharmonics, the unbalanced network avoided the use of transformers and was satisfactory for transmission.

At the time, the hybrid transformer filter held greater promise,[63] and developments of high-frequency band filters followed this course for the next 30 years. They were still in use in the mid-1970s in single-sideband radio communication equipment. Figure 6-21 shows the schematic of the hybrid transformer filter with its electrical equivalent lattice. The advantage of the hybrid coil approach was that it gave a wider band, and by using several sections in series, the unwanted mode problem could be solved by displacing these modes in succeeding sections.

Fig. 6-20. The 92F crystal unit used in the 3096-kHz pilot filter for the L1 coaxial-cable carrier system. Dual electrodes produced symmetric and antisymmetric anharmonics; no transformers were needed.

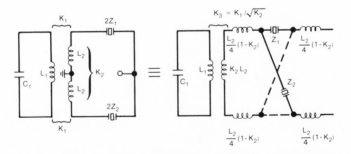

Fig. 6-21. Circuit schematic of the hybrid transformer filter and electrically equivalent lattice. The hybrid coil approach offered a wider band and allowed displacement of unwanted modes into succeeding series sections.

In 1965, upon examining resonance data on resonator pairs, Sykes and W. D. Beaver noted that the separation of the symmetric and antisymmetric modes of electrode pairs was controlled by the size, mass, and spatial separation, and not by plate dimension alone. They indicated that it would be possible to produce a monolithic wave filter of controlled bandwidth with such resonator pairs.[64,65] The improved theoretical understanding of the unwanted mode spectra and the introduction of the trapped energy concept discussed in section VII helped in understanding these observations. Further theoretical work by Spencer on wave propagation and by W. L. Smith on network synthesis led to the multiresonant monolithic filter.[66] Figure 6-22 shows an X-ray topograph of the modes in a six-resonator monolithic filter at the critical frequencies of the filter structure. In Fig. 6-18, we saw similar modes, but they depend on a single electrode and the crystal plate dimension. In other words, the anharmonic modes of Fig. 6-

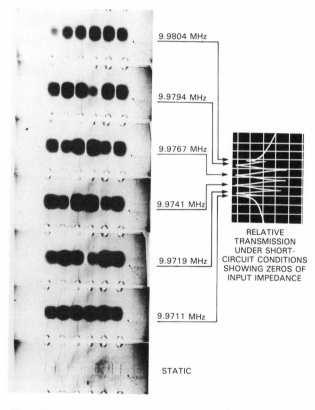

9.9804 MHz

9.9794 MHz

9.9767 MHz

9.9741 MHz

9.9719 MHz

RELATIVE
TRANSMISSION
UNDER SHORT-
CIRCUIT CONDITIONS
SHOWING ZEROS OF
INPUT IMPEDANCE

9.9711 MHz

STATIC

Fig. 6-22. An X-ray topograph of the strain pattern in a six-resonator monolithic filter at the frequencies corresponding to the zeros of input impedance.

18 are all controlled by the plate dimension alone, whereas those shown in Fig. 6-22 are the anharmonics of the electrode array, which can be controlled and placed at the critical frequencies of the filter. It was, therefore, a strange turn of fate that gave us control of these modes that for nearly 30 years researchers from many institutions had tried to eliminate.

This work on multielectrode devices culminated in the development of the eight-resonator monolithic crystal filter shown in Fig. 6-1(b), which was introduced into manufacture in 1972 for use in the A6 channel bank.[67,68] The eight-resonator monolithic crystal filter was still being manufactured in 1984, when this book was being prepared for publication.

## X. CONCLUDING REMARKS

This history covers developments of piezoelectric devices through 1975. Since then, ICs have reached levels of complexity that permit complex filter functions to be realized in an IC. But this realization is not possible without the precise frequency reference of quartz resonators. Thus, the complex filters described in this history are more and more being replaced by ICs supported by simple quartz resonators. But because ICs can perform complex functions very economically, they have also opened up broad new markets for quartz resonators, such as the ubiquitous quartz-controlled watches and digitally tuned AM/FM receivers of the 1980s.

## REFERENCES

1. A. M. Nicolson, U.S. Patent No. 1,495,429; filed April 10, 1918; issued May 29, 1924.
2. A. M. Nicolson, U.S. Patent No. 2,212,845; filed April 13, 1923; issued August 27, 1940.
3. W. G. Cady, "The Piezoelectric Resonator," *Proc. IRE* **10** (April 1922), pp. 83-114.
4. G. W. Pierce, "Piezoelectric Crystal Resonators and Crystal Oscillators Applied to the Precision Calibration of Wavemeters," *Proc. Amer. Acad. Arts Sci.* **59** (1923), pp. 81-106.
5. G. A. Campbell, U.S. Patent No. 1,227,113; filed July 15, 1915; issued May 22, 1917.
   G. A. Campbell, U.S. Patent No. 1,227,114; filed June 5, 1916; issued May 22, 1917.
6. G. A. Campbell, "Physical Theory of the Electric Wave Filter," *Bell Syst. Tech. J.* **1** (November 1922), pp. 1-32.
7. O. J. Zobel, "Theory and Design of Uniform and Composite Electric Wave Filters," *Bell Syst. Tech. J.* **2** (January 1923), pp. 1-46.
8. K. S. Johnson, *Transmission Circuits for Telephonic Communication* (New York: D. Van Nostrand Co., 1927).
9. E. H. Colpitts and O. B. Blackwell, "Carrier Current Telephony and Telegraphy," *Trans. Amer. Inst. Elec. Eng.* (1921), pp. 205-300.
10. L. Espenschied, U.S. Patent No. 1,795,204; filed January 3, 1927; issued March 3, 1931.
11. W. P. Mason, "Electrical Wave Filters Employing Quartz Crystals as Elements," *Bell Syst. Tech. J.* **13** (July 1934), pp. 405-452.
12. C. E. Lane, "Crystal Channel Filters for the Cable Carrier System," *Bell Syst. Tech. J.* **17** (January 1938), pp. 125-136.
13. W. P. Mason and R. A. Sykes, "Electrical Wave Filters Employing Crystals with Normal and Divided Electrodes," *Bell Syst. Tech. J.* **19** (April 1940), pp. 221-248.
14. R. A. Sykes, "Modes of Motion in Quartz Crystals, the Effects of Coupling and Methods of Design," *Bell Syst. Tech. J.* **23** (January 1944), pp. 52-96.
15. R. A. Sykes, "Principles of Mounting Quartz Plates," *Bell Syst. Tech. J.* **23** (April 1944), pp. 178-189.

16. See *Bell Syst. Tech. J.* 22 (July and October 1943) and 23 (April and July 1944). These publications were later summarized, with new material added: R. A. Heising, *Quartz Crystals for Electrical Circuits* (New York: D. Van Nostrand Co., 1946).

17. D. L. Arenberg, U.S. Patent Nos. 2,505,515 and 2,512,130; filed April 2, 1946; issued April 25, 1950.

18. H. J. McSkimin, "Wedge Geometry for Ultrasonic Delay Lines," *J. Acoust. Soc. Amer.* 26 (January 1954), p. 146.

19. H. J. McSkimin, U.S. Patent No. 2,505,364; filed March 9, 1946; issued April 25, 1950.

20. H. J. McSkimin, "Transducer Design for Ultrasonic Delay Lines," *J. Acoust. Soc. Amer.* 27 (March 1955), pp. 302-309.

21. M. D. Fagen, "Performance of Ultrasonic Vitreous Silica Delay Lines," *Proc. Nat. Electron. Conf.* 7, Chicago (October 23, 1951), pp. 380-389.

22. J. E. May, "Characteristics of Ultrasonic Delay Lines Using Quartz and Barium Titanate Ceramic Transducers," *Proc. Nat. Electron. Conf.* 9, Chicago (September 1953), pp. 264-277.

23. A. H. Meitzler, "Methods of Measuring Electrical Characteristics of Ultrasonic Delay Lines," *IRE Trans. Ultrason. Eng.* **PGUE-6** (December 1957), pp. 1-16.

24. D. L. White, "The Depletion Layer Transducer," *IRE Trans. Ultrason. Eng.* **UE-9** (July 1962), pp. 21-27.

25. N. F. Foster, "Cadmium Sulphide Evaporated-Layer Transducers," *Proc. IEEE* 53 (October 1965), pp. 1400-1405.

26. D. Beecham, "Ultrasonic Transducers for Frequencies Above 50MHz," *Ultrasonics* 5 (January 1967), pp. 19-27.

27. D. L. Denburg, "Wide-Bandwidth High-Coupling Sputtered ZnO Transducers on Sapphire," *IEEE Trans. Son. Ultrason.* **SU-18** (January 1971), pp. 31-35.

28. J. E. May, Jr., "Wire-Type Dispersive Ultrasonic Delay Lines," *IRE Trans. Ultrason. Eng.* **UE-7** (June 1960), pp. 44-53.

29. T. R. Meeker, "Dispersive Ultrasonic Delay Lines Using the First Longitudinal Mode in a Strip," *IRE Trans. Ultrason. Eng.* **UE-7** (June 1960), pp. 53-58.

30. M. R. Parker, Jr. and R. S. Duncan, U.S. Patent No. 3,522,557; filed July 19, 1963; issued August 4, 1970.

31. M. R. Parker, Jr., U.S. Patent No. 3,387,233; filed June 11, 1964; issued June 4, 1968.

32. M. R. Parker, Jr., U.S. Patent No. 3,562,676; filed December 18, 1967; issued February 9, 1971.

33. J. H. Rowen, U.S. Patent No. 3,289,114; filed December 24, 1963; issued November 29, 1966.

34. E. K. Sittig, "Dispersive Ultrasonic Delay Lines Using Multielement Arrays," *Proc. 5th Int. Congress Acoust.* **1A**, Liege, Belgium (September 7-14, 1965), Paper D-22.

35. G. A. Coquin and H. F. Tiersten, "An Analysis of the Excitation and Detection of Piezoelectric Surface Waves in Quartz by Means of Surface Electrodes," *J. Acous. Soc. Amer.* 41 (April 1967), pp. 921-939.

36. A. H. Meitzler and H. F. Tiersten, U.S. Patent No. 3,409,848; filed October 30, 1967; issued November 5, 1968.

37. H. Seidel and D. L. White, U.S. Patent No. 3,406,358; filed October 30, 1967; issued October 15, 1968.

38. A. H. Meitzler, "Ultrasonic Delay Lines Using Shear Modes in Strips," *IRE Trans. Ultrason. Eng.* **UE-7** (June 1960), pp. 35-43.

39. R. A. Sykes, "High-Frequency Plated Quartz Crystal Units," *Proc. IRE* 36 (January 1948), pp. 4-7.

40. R. J. Byrne and R. L. Reynolds, "Design and Performance of a New Series of Cold Welded Crystal Unit Enclosures," *Proc. 18th Ann. Frequency Contr. Symp.*, Atlantic City, New Jersey (May 4-6, 1964), pp. 166-180.

41. W. P. Mason, "New Quartz Crystal Plate, Designated the GT, Which Produces a Very

Constant Frequency Over a Wide Temperature Range," *Proc. IRE* **28** (May 1940), pp. 220-223.

42. A. W. Warner, "Frequency Aging of High Frequency Plated Crystal Units," *Proc. IRE* **43** (July 1955), pp. 790-792.

43. H. E. Bommel, W. P. Mason, and A. W. Warner, "Dislocations, Relaxations, and Anelasticity of Crystal Quartz," *Phys. Rev.* **102** (April 1956), pp. 64-71.

44. A. W. Warner, "Design and Performance of Ultraprecise 2.5-mc Quartz Crystal Units," *Bell Syst. Tech. J.* **39** (September 1960), pp. 1193-1217.

45. A. M. Dodd and D. B. Fraser, "The 3000-3900 cm$^{-1}$ Absorption Bands and Anelasticity in Crystalline $\alpha$-Quartz," *J. Phys. Chem. Solids* **26** (April 1965), pp. 673-686.

46. R. A. Sykes, W. L. Smith, and W. J. Spencer, "Performance of Precision Quartz-Crystal Controlled Frequency Generators," *IRE Trans. Instrum.* **I-11** (December 1962), pp. 243-247.

47. R. D. Mindlin, "Thickness-Shear and Flexural Vibrations of Crystal Plates," *J. Appl. Phys.* **22** (March 1951), pp. 316-323.

48. W. J. Spencer, "Modes in Circular AT-Cut Quartz Plates," *IEEE Trans. Son. Ultrason.* **SU-12** (March 1965), pp. 1-5.

49. R. D. Mindlin and W. J. Spencer, "Anharmonic, Thickness-Twist Overtones of Thickness-Shear and Flexural Vibrations of Rectangular, AT-Cut Quartz Plates," *J. Acoust. Soc. Amer.* **42** (December 1967), pp. 1268-1277.

50. W. S. Mortley, "Energy Trapping," *Marconi Rev.* **30** (Second Quarter 1967), pp. 53-67.

51. W. Shockley, D. R. Curran, and D. J. Koneval, "Trapped Energy Modes in Quartz Filter Crystals," *J. Acoust. Soc. Amer.* **41** (April 1967), pp. 981-993.

52. P. C. Y. Lee and W. J. Spencer, "Shear-Flexure-Twist Vibrations in Rectangular AT-Cut Quartz Plates with Partial Electrodes," *J. Acoust. Soc. Amer.* **45** (March 1969), pp. 637-645.

53. P. C. Y. Lee, "Extensional, Flexural and Width-Shear Vibrations of Thin Rectangular Crystal Plates," *Proc. 25th Ann. Frequency Contr. Symp.*, Atlantic City, New Jersey (April 26-28, 1971), pp. 63-69.

54. A. C. Walker, "Piezoelectric Crystal Culture," *Bell Lab. Rec.* **25** (October 1947), pp. 357-362.

55. A. C. Walker and G. T. Kohman, "Growing Crystals of Ethylene Diamine Tartrate," *Trans. Amer. Inst. Elec. Eng.* **67** (1948), pp. 565-570.

56. J. P. Griffin and E. S. Pennell, "Design and Performance of Ethylene Diamine Tartrate Crystal Units," *Trans. Amer. Inst. Elec. Eng.* **67** (1948), pp. 557-561.

57. E. S. Willis, "Crystal Filters Using Ethylene Diamine Tartrate in Place of Quartz," *Trans. Amer. Inst. Elec. Eng.* **67** (1948), pp. 552-556.

58. G. E. Guellich, J. White, and C. B. Sawyer, "Report on Questioning of Prof. Pahl, Prof. Nacken, Prof. Spangenberg, Prof. Joos, Prof. Gunther, Dr. Chytrek, et al.: Artificial Quartz Crystals," Report C-65, U.S. Dept. of Commerce, Off. Pub. Bd., Report PD 006498 (1945).

59. J. C. King, "The Anelasticity of Natural and Synthetic Quartz at Low Temperatures," *Bell Syst. Tech. J.* **38** (March 1959), pp. 573-602.

60. J. C. King, A. A. Ballman, and R. A. Laudise, "Improvement of the Mechanical Q of Quartz by the Addition of Impurities to the Growth Solution," *J. Phys. Chem. Solids* **23** (July 1962), pp. 1019-1021.

61. A. A. Ballman, R. A. Laudise, and D. W. Rudd, "Synthetic Quartz with a Mechanical Q Equivalent to Natural Quartz," *Appl. Phys. Lett.* **8** (January 15, 1966), pp. 53-54.

62. A. A. Ballman, R. A. Laudise, and D. W. Rudd, U.S. Patent No. 3,356,463; filed September 24, 1965; issued December 5, 1967.

63. W. P. Mason, *Electromechanical Transducers and Wave Filters* (New York: D. Van Nostrand Co., 1942), p. 262.

64. W. D. Beaver and R. A. Sykes, U.S. Patent No. 3,564,463; filed June 17, 1966; issued February 16, 1971.

65. R. A. Sykes and W. D. Beaver, "High Frequency Monolithic Crystal Filters with Possible Application to Single Frequency and Single Side Band Use," *Proc. 20th Ann. Frequency Contr. Symp.*, Atlantic City, New Jersey (April 19-21, 1966), pp. 288-308.

66. R. A. Sykes, W. L. Smith, and W. J. Spencer, "Monolithic Crystal Filters," *IEEE Int. Conv. Rec., Pt. 11-Speech* **15** (March 1967), pp. 78-93.

67. S. H. Olster, I. R. Oak, G. T. Pearman, R. C. Rennick, and T. R. Meeker, "A6 Monolithic Crystal Filter Design for Manufacture and Device Quality," *Proc. 29th Ann. Frequency Contr. Symp.*, Atlantic City, New Jersey (May 28-30, 1975), pp. 105-112.

68. H. F. Cawley, J. D. Jennings, J. I. Pelc, P. R. Perri, F. E. Snell, and A. J. Miller, "Manufacture of Monolithic Crystal Filters for A6 Channel Bank," *Proc. 29th Ann. Frequency Contr. Symp.*, Atlantic City, New Jersey (May 28-30, 1975), pp. 113-119.

# Chapter 7

# Relays and Switches

*As the Bell System moved from manual to automatic switching, relay tech-nology played a key role in logic, memory, and switching functions previously performed by human operators. Research and development efforts focused on reliable and inexpensive relays and switches to handle the demands of the rapidly growing telephone system. The number of contacts per relay was in-creased to permit more complex functions, while the final assembly was simplified through the use of premolded subassemblies. Sealed enclosures improved the quality of relay contacts. As semiconductor devices started to perform the logic functions, there was a demand for miniature relays compatible with the physical design of semiconductor electronics. In other developments, crossbar switches replaced panel and step-by-step switches, and later the ferreed switch was designed to provide the switching path in electronic switching systems.*

## I. INTRODUCTION

In 1925, the Bell System was serving about 12 million telephones. Manual switching was still dominant, although automatic switching was being introduced. Two types of automatic systems had been developed in which the talking path was established by special switches: step-by-step and panel switching systems. In addition to these switches, general-purpose relays were required to perform various peripheral logic functions.

The development of these systems is described in the companion volume *The Early Years (1875-1925)*. Another companion volume, *Switching Tech-nology (1925-1975)*, covers the system developments after 1925, while this chapter covers the relays and talking path switches for these systems.

As of 1925, Western Electric was manufacturing about 3 million relays per year.[1] The E- and R-type relays were the general-purpose relays of the period; they used punched parts to permit economical large-scale man-ufacture. Contact springs were stud activated and single contacts were used. The relay magnet was capable of activating a maximum of 12 contact springs—4 transfer (break-make) contact sets, for example.[2,3]

---

Principal author: S. J. Elliott

During the next five decades, the number of Bell System telephones grew to more than 100 million; mechanized switching almost completely displaced manual switching, and the demand for relays increased dramatically. By 1970, Western Electric was manufacturing more than 80 million relays per year.

In addition to increasing the need for relays, the evolution and expansion of mechanized switching also led to more stringent demands on relay performance. As relays and other electromechanical devices took over logic and memory functions formerly performed by human operators, the average number of relay operations per call also increased dramatically. For example, a manual line (in which an operator selected the desired lines of both calling and called customer, rang the station, checked for busy signals, and kept track of charges) needed 2.5 relays per line. The mechanized step-by-step line of the mid-1950s, on the other hand, required seven relays. The No. 5 crossbar system of the same period needed 30 times the number of relays performing over 12 times the number of relay operations per call.

The increasingly large capital investment required for complex automatic switching machines dictated that the useful service life of such machines be increased also. Consequently, where a relay life of 10 million operations had been adequate in the early manual switching systems, some later relays had to operate 100 million times in step-by-step and panel systems, and as many as 1 billion times in crossbar and electronic switching systems.

At the same time, the switching networks used in the talking path were undergoing dramatic changes. The early step-by-step and panel switches were replaced by crossbar switches, which were incorporated into switching systems with common control units. Hermetically sealed ferreed contacts became the basic building blocks in the switching networks for electronic switching systems.

## II. EXPOSED-CONTACT RELAYS

During the first half of the 1930s, development started on crossbar switching systems using a matrix switch in the talking path (see section IV) and a common control unit for establishing connections. Because the control unit is shared by many users, it is heavily used while executing complicated logic. Requirements for a new family of general-purpose relays to meet these needs led to the development of the U-type relays. [Fig. 7-1] Compared with the earlier E- and R-type relays, the U-type relays provided up to twice as many contacts per relay, faster operation, generally more reliable contacts, and longer life. The U-type relay continued to lean heavily on punched parts, including flat, stud-actuated contact springs. But twin contacts were provided, and one spring of each pair was bifurcated to provide a degree of independence between the two contacts—a form

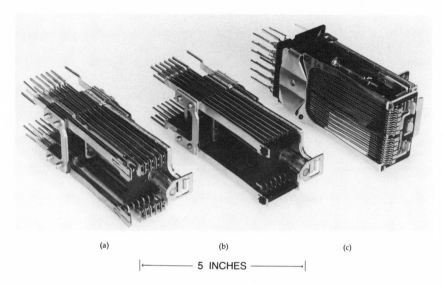

|← ———— 5 INCHES ———— →|

Fig. 7-1.   The three types of general-purpose relays developed during the period from 1925 to 1952. (a) The U type was the first Bell System relay to employ twin contacts and bifurcated contact springs. (b) The UB type was the first to use card-release actuation, and the AF type (c) was the first of the wire spring relays.

of redundancy to increase contact reliability.[4] Manufacture of U-type relays started in 1936.

A decade later, in a refinement of the U-type relay design, stud actuation was replaced by card-release actuation in the UB-type relay. Card actuation was an important step in improving reliability by substantially decreasing the tendency of wear to reduce contact forces and by reducing the tendency of contacts to lock mechanically when "pip-and-crater" erosion occurred. [Fig. 7-2] The design also permitted a longer slot in the bifurcated contact springs, thereby increasing the independence of the twin contacts and further improving their reliability.[5] [Fig. 7-3]

The AF-type wire spring relays, the next generation of general-purpose relays, also used card-release actuation and continued the trend set by the U-type relay of providing completely independent twin contacts, more contacts per relay where required, either low battery drain or faster operation, greater contact reliability, and longer mechanical life. [Fig. 7-4] In addition, the wire spring design permitted substantial reductions in the amount of hand labor required to assemble and adjust a relay, an important factor in holding down the cost of telephone equipment during the rapid rise of hourly wages following World War II. For example, where the assembly of a U-type relay required the handling of many individual contact springs, insulators, bushings, and screws, the contact springs and insulators for an AF-type relay were provided as molded subassemblies,

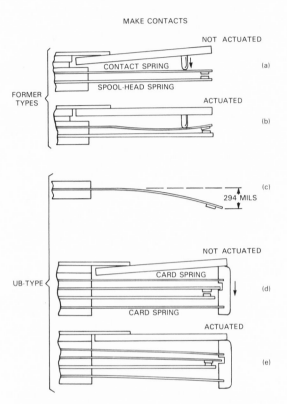

MAKE CONTACTS

Fig. 7-2.  Actuation of contact springs. (a) and (b) illustrate stud actuation of a normally open contact in a U-type relay. Without applied pressure, the spring carrying the moving contacts is positioned away from the fixed contact. A stud, moved by the armature, presses against the contact spring a short distance behind the contact to close the contacts when the relay is operated. In this case, the desired contact force of 20 to 30 grams is developed by a spring deflection of about 13 mils. (c), (d), and (e) illustrate the card-release actuation used in the UB-type relay. Here, without an external force, the moving contact spring is positioned against the fixed contact. To produce the desired contact force, the unconstrained spring has a deflection of about 294 mils. When the relay is not actuated, the card holds the movable contact away from the fixed contact. When the relay is actuated, the armature presses the card toward the fixed contact, allowing the contacts to close. The card no longer touches the contact spring when the relay is fully operated.

so that the assembly operator had to handle only two or three wire blocks; a spring clamp took the place of the screws. Moreover, a considerable amount of individual spring bending was done to obtain the desired contact

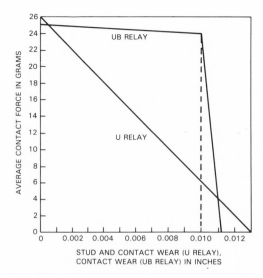

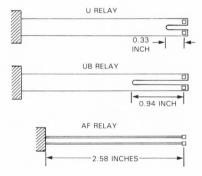

Fig. 7-3. Stud and contact wear for the U-type relay and contact wear for the UB-type relay. An important characteristic of the UB-type relay's card-release actuation is that wear has very little effect on contact force until the wear reaches the stage where the card fails to separate from the movable contact spring when the contact is closed. By contrast, contact force in the stud-actuated U-type relay drops off fairly rapidly as the contacts and the stud wear.

Fig. 7-4. Evolution of twin-contact springs from the slightly bifurcated flat spring used in the U-type relay to the completely independent twin wire springs of the AF-type relay.

forces and contact gaging in U-type relays. In the wire spring relays, on the other hand, the combination of low-stiffness movable springs and card-release actuation avoided the need for any contact force adjustment, and a simple mass adjustment was sufficient to meet the gaging requirements.[6,7,8,9] Regular production of wire spring relays started in 1952, reaching a peak level of over 40 million in 1970.

By the late 1950s, interest shifted from electromechanical switching to electronic switching. Semiconductor devices were beginning to take over the logic functions that were performed by relays, and there was little support for development of still another generation of general-purpose relays.

Despite the inroads of semiconductor devices for logic functions, however, circuit designers continued to specify relays for many switching functions, particularly where very high speed was not required, where a very high ratio of open-circuit impedance to closed-circuit impedance was needed, where electrical isolation from other circuits was important, and where it clearly would cost less to use relays than to use semiconductor devices.[8] Circuits became mixtures of comparatively large relays with much smaller semiconductor devices, resistors, and capacitors. It was awkward to implement such circuits on the increasingly popular printed wiring boards, for the protruding relays prevented close spacing of circuit packs in equipment units.

The response to this problem was the development of smaller relays that could be readily mounted on printed wiring boards: the MA and MB family of relays. [Fig. 7-5] The MA-type relay was about one-tenth the volume of an AF-type wire spring relay and was available with up to four transfer contact sets. The slightly larger MB-type relay was available with up to six transfer contact sets.[10]

The MA- and MB-type relays were developed on an urgent schedule as part of a major effort of the Bell System to reduce the size and increase the ruggedness of equipment installed on customer premises in key telephone systems. The relays had to be manufacturable in large numbers at low cost and without excessive preparation time or expense. Flat spring designs were chosen to avoid the expense and long manufacturing preparation time associated with the molded wire blocks for wire spring relays. A number of the desirable features of the wire spring relay design were utilized, however, including card-release actuation, low-stiffness pretensioned springs, twin contacts, and coplanar spring groups to simplify manufacture. Some of the compromises that were made to minimize manufacturing cost limited the mechanical life of the MA- and MB-type relays to about ten million operations. Although this fell far short of the life of AF-type wire spring relays, it was adequate for the key telephone system application. Manufacture of MA- and MB-type relays started in 1962.

Another line of smaller relays, the miniature wire spring (BF through BM types), was developed to meet the more stringent performance re-

(a)                           (b)                           (c)

|—————— 2 INCHES ——————|

Fig. 7-5.   Three types of miniature general-purpose relays: the MA type (a) and the MB type (b) use flat contact springs; the BF type (c) uses wire springs.

quirements of central office applications. This design utilized wire spring technology and a more powerful magnet than the MB-type relay. The BF-type relay was equipped with up to six transfer contact sets, and its mechanical life was about 200 million operations.[11] Manufacture started in 1965.

In response to recurring needs for relays having even lower profiles than the MA, MB, and miniature wire spring types, the LM- and LP-type relays were developed. They projected no more than 5/8 inch above the printed wiring board on which they were mounted and provided up to eight transfer contact sets; manufacture started in 1975.

Still lower in profile was the LR relay family, which had up to four transfer contact sets. [Fig. 7-6] LR-type relays used flat, bifurcated springs stamped from spinodal copper alloy stock, and could be assembled either by hand or automatically. In response to the changing demands for relays brought on by the rise of electronic switching machines, a life of 10 million operations was considered satisfactory, but frame temperatures of the new solid-state systems required relay operation up to 80 degrees C, a much warmer environment than that seen by the AF series of frame-mounted relays. As with all relay families designed after the AF-type relay, the LR-type relay was intended only for mounting on printed wiring boards. Manufacture began in 1979.

The descriptions in the preceding paragraphs have focused on the general-purpose relays that were developed in Bell Laboratories (with considerable collaboration by Western Electric) during the period from 1925

|— 1 INCH —|

Fig. 7-6. A typical LR-type relay intended for printed wiring board mounting in solid-state systems.

to 1980. In addition to the general-purpose designs, however, a number of variations were developed to meet specialized needs. Some of these are noted in the next few paragraphs.

In the U-type relay family, one variation was the UA-type relay, which had an enlarged pole face to increase its sensitivity. Another variation was the Y-type relay, equipped with a copper or aluminum sleeve over its core to obtain slow release.[12] Another variation of the U-type relay family, the previously mentioned UB type, was important because of the very large number of operations encountered in an accounting center application where added reliability was required.

In the original wire spring relay family, the AF-type relay was the general-purpose type. Although the design provided space for as many as 12 transfer contact sets, the AF relay magnet was not strong enough to handle a full complement of contacts. Six transfer contact sets plus six make contact pairs or the equivalent was the limit. Where more contacts were required, a longer, heavier armature was used to provide a more powerful electromagnet. The AJ series, with a more powerful magnet, could be equipped with 12 transfer contact sets, or 24 make or break pairs.

Sensitive relays and marginal relays were also included in the AJ series. AG-type relays were equipped with copper or aluminum sleeves to obtain a slow release.[6,7,13] AL-type relays utilized remanent magnetic material to produce magnetic latching.

The wire spring relay family also included a dual-armature design. This variation provided two relays in a single unit mounted in the same space as the AF-type relay. Each half could be equipped with up to five transfer contact sets. Nonlatching dual-armature relays were coded in the AK series; magnetic latching was provided in the AM series.[14,15]

In the miniature wire spring relay family, the BF-type relay was the general-purpose type, equipped with up to six transfer contact sets. The BG type provided slow release; the BL and BM types provided magnetic latching with 6 and 12 transfer contact sets, respectively; and the BJ type provided additional contacts—up to 12 transfer contact sets in a nonlatching version.[11]

The need for a new connection arose with the development of the wire spring relay, because the standard method of applying connections to the flexible and closely spaced wires was very expensive and not satisfactory. A better and cheaper connection was important, because in 1953 about 50 million relay connections were scheduled for the wire spring relay alone. This prompted the development of a tool by H. A. Miloche[16] that could wrap a few turns of wire around the terminals of a relay. Initially, these connections were then soldered, but later tests showed that for specific terminal shapes, solder was not needed and wrapped connections were satisfactory by themselves.[17]

Solderless wrapped wire connections permit automatic tape-controlled wiring, which has profound implications on the wiring cost of complex systems. After the Gardner-Denver Company developed a commercial automatic wiring machine, the approach was heavily used by Western Electric and others.

Another exposed-contact relay type was the so-called multicontact relay that was designed for use in crossbar systems to close momentarily large numbers of circuit paths between various parts of the switching equipment. (An earlier multicontact relay was developed for the coordinate switching system and later used in the panel decoder connector.) The first version of the multicontact relay was developed in parallel with the U-type relay and the early crossbar switch, and was strongly influenced by both. It was a flat spring relay that provided up to 60 make contact pairs. It employed two electromagnets, each of which actuated half the contacts;[18] it was succeeded by a wire spring design in the 1950s that was equipped with 30 make contact pairs and a single electromagnet. Two of these relays, mounted on a single bracket, replaced the earlier 60-contact, flat spring type.[19,20]

In addition, a number of other relay types were designed for special purposes. These included polarized relays, frequency-sensitive relays (vibrating reed selectors), and the miniature, highly reliable, latching relay that was used first in the Telstar satellite and later in submarine cable systems.[21]

## III. SEALED CONTACTS

The contacts of the relays described so far were exposed to the atmosphere, limiting the contact reliability. In the 1930s, development started on relay contacts hermetically sealed inside a glass envelope. A. C. Keller, who was associated with electromechanical device development for many decades and later was director of switching systems development, has described the emergence of these contacts in the following words:

> "The wire spring relay and the relays using glass-enclosed contacts are related through similar and interesting origins. About 1935, Dr. Oliver E. Buckley, who later became President of Bell Laboratories, wanted to stimulate fundamental development and invention that would eventually lead to a new and improved general-purpose relay. To do this, he spoke to H. C. Harrison and W. B. Ellwood about the problem. Both of these men, creative and inventive members of the Laboratories, were soon on their way to carry out the broad and general instructions Dr. Buckley gave them. However, they did not go in the same direction even though their instructions were presumably the same.

> "Henry Harrison visualized, as the best solution, a 'code-card' operated relay using wire springs to carry the contacts. The design permitted a high degree of mechanization in manufacture and assembly, which in turn realized the desired low costs. This work eventually led to the family of wire spring relays now in large-scale production. . . .

> "In contrast to Harrison's approach, Walter Ellwood saw the ultimate in reliability in a glass-enclosed contact containing the cleanest and best atmosphere, completely free of outside influences such as dust and dirt. He also visualized low manufacturing costs and product uniformity through a high degree of mechanization. His work also led to a family of relays, in this case with glass-enclosed contacts. These are also now being produced and used in large quantities. . . ."[22]

Two general types of glass-enclosed sealed contacts were developed at Bell Laboratories. One was the dry-reed type,[23] the other was the mercury-wetted type.[24]

Dry-reed sealed contacts consist of two magnetic reeds sealed in a glass tube. [Fig. 7-7] The mating contact surfaces are plated with a precious metal such as gold, or with a combination such as gold and silver; this is sometimes diffused in a controlled atmosphere to achieve low and stable contact resistance and to avoid sticking. Once the reeds are hermetically sealed in the glass envelope, they are no longer vulnerable to such external influences as dust and hostile atmospheres. Careful control of manufacturing processes is required, however, to assure internal cleanliness and reliable performance.[8]

In the early designs of dry-reed contacts, the reeds were formed from soft magnetic materials. In a magnetic field of sufficient strength, the magnetic attraction between the two reeds causes the contact to close. The

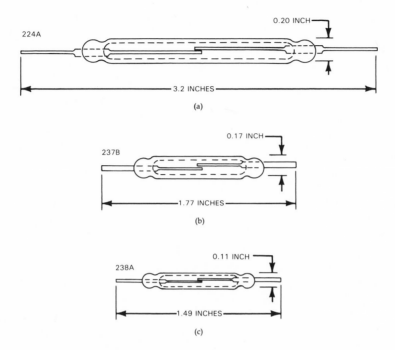

Fig. 7-7. Three widely used types of dry-reed sealed contacts. (a) The 224A contact was the first dry-reed sealed contact to be used widely in telephone applications. (b) The 237B contact was used extensively in ferreed switches, particularly in early 1ESS* electronic switching equipment. (c) The 238A contact is the remanent-reed contact used in remreed switch packages, particularly for the later 1ESS electronic switching equipment networks.

contact remains closed as long as the magnetic field persists. When the magnetic field is removed (or nullified by an opposing field), the contact opens.

In the early applications of the dry-reed contact, sealed contacts were mounted in a solenoidal winding to produce a relay with normally open contacts. Where normally closed contacts were required, a small permanent magnet was provided to hold the contacts closed, and the contacts would be opened by passing current through the winding in the direction to oppose the permanent magnet.[25,26]

Manufacture of relays with dry-reed contacts for military applications began in 1944.[27] [Fig. 7-8] Although there had been earlier exploratory use of such contacts in Bell System applications, large-scale telephone uses came only after the war, in a combined assembler-computer for automatic

* Trademark of AT&T Technologies, Inc.

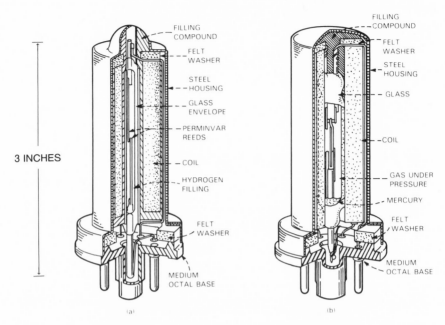

Fig. 7-8. Sealed-contact relays as manufactured for military applications during World War II: (a) relay employing a dry-reed sealed contact; (b) relay employing a mercury-wetted sealed contact.

message accounting, for which two types of sealed-contact relays were developed. One was the 289 digit register relay, which actually was an assembly of five relays in a single package. Each of the five relays consisted of an operating coil, a current-limiting resistor, and a single 224A dry-reed contact. The 289 digit register relay was also used as a digit register in the No. 5 crossbar system. The second type was the 290 connector relay, which contained twelve 224A sealed contacts mounted side by side within a single flat operating coil.[28]

The most extensive Bell System use of dry-reed contacts, however, was in switching networks for electronic switching systems. Invention of the ferreed in the late 1950s provided a switching network for ESS electronic switches, which had metallic contacts, required no holding power, could be switched with speeds compatible with the rest of the system, and were low in cost.[29] The original ferreed concept combined a magnetically hard ferrite member with the magnetically soft dry-reed contact and an exciting winding to provide a magnetically latched metallic contact. A current pulse of only a few microseconds in the exciting winding could magnetize the ferrite member. [Fig. 7-9] The remanent magnetic field of the ferrite then was sufficient to operate the associated dry-reed contact, even though it might take several hundred microseconds more for the contact to close.

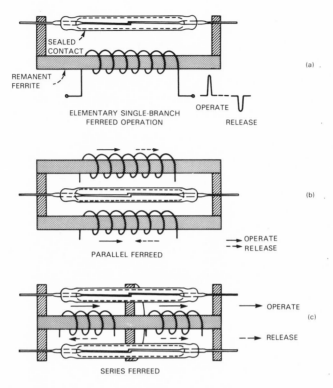

Fig. 7-9. Three types of ferreeds as originally conceived. The remanent ferrite material, being virtually nonconductive, can be magnetized by very short current pulses in an energizing coil. It was envisioned that a current pulse of only a few microseconds duration in the single-branch ferreed (a) would magnetize the ferrite member. The remanent magnetic field of the ferrite would then operate the associated dry-reed contact, even though it might take several hundred microseconds more for the contact to close. A current pulse of the right magnitude in the reverse direction would demagnetize the ferrite and permit the contact to open. The parallel ferreed (b) has two ferrite members. When they are magnetized in the same direction, the contact closes. When they are magnetized in opposite directions, the contact opens. In the series ferreed (c), the contacts close when the two halves of the ferrite member are magnetized in the same direction. The contacts open when the two halves are magnetized in opposite directions.

A major attraction of the original concept was this ability to control comparatively slow-acting metallic contacts at electronic speeds.[30]

Although the early ferreed used a cobalt ferrite as the remanent material, later designs used remendur, a cobalt-iron alloy, instead. Because remendur is an electrically conducting material, it cannot be switched as fast as a cobalt ferrite. However, it avoids difficulties with the ferrite, most notably, the strong temperature dependence of its magnetic properties and low flux density, which led to structures of large cross section and poor efficiency. Furthermore, as more thought was given to the ferreed as a system component, it was found that the originally postulated microsecond specifications for actuation of the ferreed could be avoided, which also led to easier driving requirements.[31] So, the somewhat slower remendur version was adopted.

The 237B dry-reed sealed contact, smaller than the earlier 224A contact, was developed for the ferreed, as shown in Fig. 7-7. An even smaller contact, the 237A type, had been developed for military applications, but its characteristics were not completely suitable for electronic switching networks.[8,32] Manufacture of 237B contacts started in 1962 and reached a peak production level of about 130 million in 1973.

It had been recognized in the early stages of the ferreed development that the efficiency and sensitivity of the device could be improved by making the contact reeds themselves out of remanent magnetic material instead of providing separate remanent members.[8,30] Because of materials processing problems, however, this did not appear to be an economical approach for large-scale manufacture. Later development work surmounted the processing problems, permitted the development of the 238A remanent-reed (remreed) sealed contact shown in Fig. 7-7, and led to dramatic reductions in the size of electronic switching networks.[32] Manufacture of 238A contacts began in 1972. Approximately 50 million were manufactured in 1973, and the annual production level reached almost 380 million in 1980.

Development of mercury-wetted sealed contacts occurred in parallel with the dry-reed contact development. Compared with solid metallic contacts, liquid-mercury contacts were attractive, because they will not stick by locking or welding and they do not chatter. Most of the mercury-contact relays and switches that were available commercially in the 1930s, however, were designed to move fairly large quantities of mercury by gravitational acceleration. Consequently, they tended to be too slow for widespread use in telephone circuits.[24]

The Bell Laboratories design approach was to wet solid metallic contacts with a film of mercury and to maintain the mercury film by means of a capillary connection to a mercury reservoir below the contacts. This minimized the amount of mercury that had to be put in motion for operation and permitted the moving contacts to be carried by a light armature capable of being moved at high speed. In the early designs, the armature capillary path was produced by welding two vertical wires to the armature. The two wires were touching each other and the groove between them formed the capillary path.[24] In later designs, capillary paths were produced by rolling grooves into the armature surface.[33]

The original mercury-wetted sealed contact was a simple transfer contact set. The stationary front contact was mounted on a magnetic member, the back contact on a nonmagnetic member. [Fig. 7-10] In the unoperated state, the movable armature contact was held against the back contact by spring force. The contact was operated by a magnetic field that would cause the magnetic armature to be attracted toward the front contact.

Manufacture of mercury-wetted contacts and of relays utilizing them began in 1944 to fill a military need.[27] In this first large-scale application,

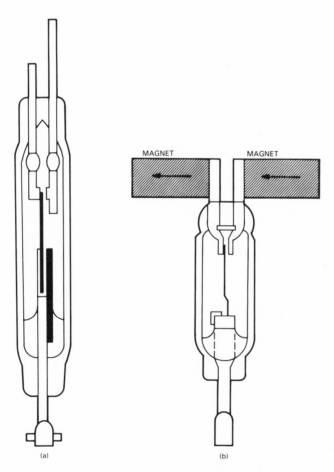

Fig. 7-10.    Two types of mercury-wetted sealed contacts: (a) the early neutral type, in which the armature contact was tensioned against the nonmagnetic back contact and was attracted toward the magnet support of the front contact in the presence of a magnetic field; (b) the later polar type, showing external permanent magnets adjusted to provide the desired operate and release sensitivities.

mercury-contact relays were used to control the direction and rotational speed of motors in servomechanisms.[24]

The principal telephone applications of mercury-wetted contacts came after World War II. These applications used the 218-type sealed contact, which was functionally similar to the simple transfer contact set described previously, but which contained two front contacts, two back contacts, and a common armature contact.

Two types of relays were manufactured at first. The 275-type relay was a neutral relay containing a 218-type sealed contact and an operating winding enclosed in a steel electron-tube housing equipped with an octal tube base. The 276-type relay included, in addition, two small permanent magnets. These magnets were adjusted to cause the relay to operate or release on specific current values. The available adjustments included magnetic latching.[26]

A later design of mercury-wetted contact, the 222 type, utilized a symmetrical polar structure with a simple, very light magnetic reed armature.

|←—————— 1 INCH ——————→|

Fig. 7-11. Examples of the miniature mercury relay family. The 345-type and 356-type relays (marked with arrows) require fixed mounting orientation. The 354-type and 358-type relays can be mounted in any position.

Combined with appropriate permanent magnets and winding, it provided greater sensitivity and faster operation than could be obtained with the 218-type contact.[33] It was used extensively in the 223-type coaxial switch for switching high-frequency circuits, but it was soon replaced for other applications by the similar but shorter 226-type sealed contact.[8] The 222- and 226-type contacts were both simple transfer contact sets.

Because the mercury-wetted contacts contained more parts than the dry-reed contacts and required more manufacturing steps, they were more costly than dry-reed contacts. Nevertheless, their ability to switch fairly high voltages and currents rapidly, with negligible chatter, for a billion or more operations led to significant use of such contacts in the Bell System. Western Electric manufactured nearly 13 million of them in 1981.

Later mercury-wetted contact relays reduced size and cost while providing greater versatility in application. [Fig. 7-11] In the 345-type relays, which had one make contact pair, the height above the printed wiring mounting board was reduced to 0.27 inch. Other versions of this structure included the 352-type relay, which had a permanent-magnet-biased break contact pair; the 353-type relay, which had two make contact pairs; and the 356-type relay, which had a transfer contact set. Through allowance of a reduced contact capability and use of only the amount of mercury that would cling firmly to the wetted reed, a series of position-insensitive relays was developed. These included the 355 (one make contact pair), 354 (one break contact pair), and 358 (two make contact pairs). Production of these reduced-size relays began in 1976 with the 345A relay and five years later was at an annual rate of over 5 million units.

## IV. SWITCHES

In 1925, Bell System automatic switching systems utilized two types of switches: the step-by-step switch and the panel switch. In both, contacting brushes were required to travel rather long distances (inches), sliding over stationary terminals during part of the travel, to establish the desired connections. Consequently, the operating times of these switches tended to be long compared with the operating times of relays, where movable contacts were required to travel no more than a small fraction of an inch. Moreover, the switch-driving mechanisms were fairly complex, and the contacts themselves were usually made of base metals, because then known precious metals, which might have provided very much quieter talking circuits, wore out rapidly under the sliding mode of operation.

During the period from 1925 to 1975, three new types of switches were developed and used extensively in Bell System telephone switching networks. The crossbar switch, using exposed relay-type contacts, came first. The ferreed switch, using dry-reed sealed contacts with magnetically soft reeds, was next. The remreed switch, using dry-reed sealed contacts with magnetically hard (remanent) reeds, was last.

The crossbar switch was invented in 1913 by J. N. Reynolds of Western Electric.[34] It did not look attractive, however, for the types of switching systems that were being considered at that time. Its development was not pursued by the Bell System until some years later, after the Swedish telephone administration had contributed improvements to the original concept.

The crossbar switch, as developed in the 1930s, was essentially a rectangular array of relay-type contacts disposed in a pattern of horizontal rows and vertical columns.[35] The early switches had ten horizontal rows and either 10 or 20 vertical columns. [Fig. 7-12] Each intersection of a row and a column was called a crosspoint; each switch contained either 100 or 200 crosspoints. A cluster of normally open contacts was provided at each crosspoint—usually three contact pairs (for the tip, ring, and sleeve circuits), but up to six pairs appear in later designs.

To actuate the contacts, the switch was equipped with a selecting electromagnet for each horizontal row and a holding electromagnet for each vertical column. Mechanical linkages between magnets and crosspoint contacts were provided by horizontal selecting bars and vertical holding bars—and the resultant pattern of crossed bars led to the name "crossbar switch."

Any particular cluster of crosspoint contacts could be closed by first energizing the selecting magnet for that horizontal row and then energizing

(a)
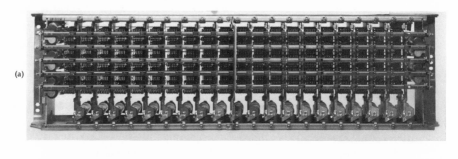

(b)
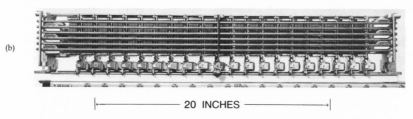

|←————————————— 20 INCHES —————————————→|

Fig. 7-12.   Crossbar switches. (a) The largest of the crossbar switches manufactured before 1969. It contained 200 crosspoints, with six normally open contacts at each crosspoint. (b) The largest of the later designs, which also contained 240 crosspoints, also had six normally open contacts at each crosspoint, but had significantly smaller physical dimensions.

the holding magnet for that vertical column. The selecting magnet could be disconnected after the contacts had closed, and they would remain closed as long as the holding magnet remained energized. The selecting magnet then was free to be used again, if required, for selecting another crosspoint in a different vertical column.

The contacts and flat contact springs used in the early crossbar switches were very similar to those used in U- and Y-type relays, which were developed concurrently with the crossbar switch. Twin contacts of palladium, with bifurcated moving springs, were provided for contact reliability. All the corresponding stationary contacts in each vertical column were connected together, and all the corresponding movable contacts in each horizontal row frequently were connected also, although horizontal connecting was optional.

Regular manufacture of crossbar switches began in 1936. Numerous design changes were introduced during the next 30 years to reduce the cost of manufacture, to improve performance, and to provide a magnetic latching version,[36] but the general configuration remained very similar to the original design. From time to time, various major design changes were explored, particularly with the idea of reducing size and increasing operating speed. But most of these proposals were abandoned, because at the time they did not offer advantages that would justify the costs of development and introduction.

In the 1960s, however, two circumstances began to make small crossbar switches more attractive: (1) a number of crossbar central offices were beginning to run short of trunking capacity, but had no space left in which to add more equipment, and (2) inflation was driving the cost of new buildings upward at a distressing rate. So a new and much smaller family of crossbar switches was developed to permit expansion of trunking capacity without requiring additional space. Wire spring relay technology was utilized in the new design, and provision was made for 12 horizontal levels, where required, in place of 10 horizontals.[37] Regular production of the small crossbar switch began in 1969.

Production of crossbar switches (large and small types combined) reached a peak of 634,000 switches in 1971 but subsequently declined as the telephone operating companies turned increasingly to the then evolving electronic switching systems.

When the development of the 1ESS switching equipment began in the 1950s, it was widely expected that electronic crosspoints (that is, electron tubes or semiconductor devices) would be used in the switching networks. As development progressed, however, it became clear that all-electronic networks would carry a number of handicaps: the need for different and perhaps more costly telephone sets, new protection problems, and limitations in transmission properties.[30]

On the other hand, existing electromechanical switches operated far too slowly to be compatible with the goal of controlling an entire central office switching network (serving up to perhaps 65,000 lines) with a single, very fast, electronic central processor. The solution to the 1ESS electronic switch network problem was the ferreed sealed contact, described in section III, because ferreed contacts provided a metallic transmission path that could be switched at high speed.

The ferreed switches designed for the 1ESS switching equipment were rectangular arrays of ferreed contacts. [Fig. 7-13] The most widely used configuration was an 8-by-8 array in which each crosspoint contained two magnetically soft dry-reed sealed contacts and two magnetically hard re-mendur plates, all surrounded by two pairs of energizing coils. One pair of coils was connected in series with the corresponding coils of the other crosspoints in the same horizontal row. The second pair of coils was connected in series with the corresponding coils of the other crosspoints in the same vertical column. The arrangement was such that simultaneous current flow in a horizontal string of coils and in a vertical string closed the contacts at a common crosspoint and caused the contacts at all other crosspoints in that row and column either to open or to remain open. Short, coincident current pulses in the horizontal and vertical strings of coils were sufficient to magnetize the remendur plates at the common

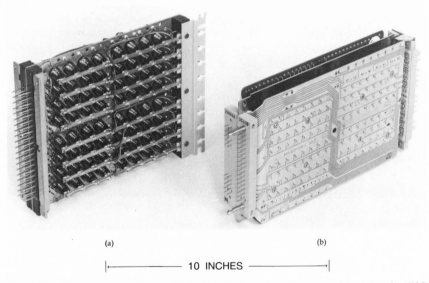

(a)                                              (b)

|——————— 10 INCHES ———————|

Fig. 7-13. Two remreed switches. (a) An 8-by-8 ferreed switch as used in the early 1ESS electronic switch. (b) A dual 8-by-8 remreed switch package as used in the later 1ESS electronic switches. This package was equivalent to two ferreed switches and, in addition, contained a number of diodes and transistors that served to steer operating current pulses through the desired crosspoint coils.

crosspoint, and the remanent magnetic field of those plates was sufficient to operate the associated sealed contacts and to hold them closed thereafter.

Manufacture of ferreed switches began in 1962. By 1970, more than one million ferreed switches of various types were being manufactured annually, most of them for central offices with 1ESS switching equipment.

An initial impediment to the 1ESS electronic switch was the cost of a new switching office as compared with a crossbar or step-by-step office. Much effort was made to reduce the cost of 1ESS switching equipment and also to reduce its size so that it could be housed in smaller, less costly buildings. This effort led to development of the remreed sealed contact, described in section III, and the remreed switching network.[38]

In the remreed switch, the magnetically hard (remanent) material required for magnetic latching was in the reeds of the sealed contacts themselves instead of in external plates. This design had several important advantages. Elimination of the external plates permitted some reduction in the size of a crosspoint. Bringing the remanent material right up to the working gap of the reed contact improved the magnetic circuit, permitting remreed crosspoints to be operated by lower coil currents. This in turn reduced the magnetic interference between crosspoints, permitting closer spacing of crosspoints. The better magnetic coupling also permitted the use of small, low-cost semiconductor devices (instead of relay contacts) for steering operating current pulses through the desired crosspoint coils. In fact, the steering diodes and transistors were small enough to be mounted right in the remreed switch packages. Many of the interconnections between these pulse-steering components and the crosspoint coils were provided by flexible printed wiring that was part of the remreed switch package, eliminating much of the external cabling formerly required between ferreed switch packages and external steering relays. These were some of the factors that enabled circuit and equipment designers to make dramatic reductions in the size of 1ESS switching equipment networks in the 1970s.[32,38]

Manufacture of remreed switch packages began in 1972. Approximately 100,000 were manufactured in 1973, and the production level was almost 1.5 million in 1981.

It is worth noting that the crossbar, ferreed, and remreed switches were designed primarily for use in voice-frequency circuits. Where switching of much higher frequencies was necessary in carrier systems and radio systems, a number of coaxial relays and switches were designed for those purposes.

The simplest was a coaxial-cable relay designed by Ellwood for the L1 carrier system and the TD-2 radio system. It consisted of a single dry-reed sealed contact in the central conductor of a coaxial structure. The capacitance across the open gap of the dry-reed contact, of course, limited the isolation loss obtained at any particular frequency.

A more complex design, utilizing four mercury-wetted contacts, was used in the TD-2 radio system and in the L3 and L4 carrier systems. It provided suitable isolation loss and other desired transmission characteristics at frequencies up to 80 megahertz (MHz), but it was a rather expensive piece of apparatus.

A less costly coaxial switching element, called a cable switch, was invented in the 1960s.[39] Two or more dry-reed sealed contacts were connected in series in the central conductor of the coaxial structure, and, very importantly, an appreciable length of coaxial transmission line was provided between adjacent contacts. Even a short section of transmission line between the adjacent contacts greatly enhanced the increase in isolation loss obtained by adding series contacts. High-frequency switches based on the cable-switch principle were first manufactured in 1967 for a trial of a broadband restoration system. Those switches, with three series contacts in each cable switch element, provided at least 95 dB of isolation loss at frequencies up to 100 MHz. [Fig. 7-14]

|←——————— 10 INCHES ———————→|

Fig. 7-14.   An 8-by-8 cable switch matrix as used for broadband restoration systems.

## V. CONCLUDING REMARKS

The beginning of the time period covered by this volume is characterized by the introduction of "artificial intelligence" into the telephone plant— at that time based entirely on relay technology. Relays provided the logic and memory function previously performed by operators, and they established the connections between subscribers, replacing patch cords. The end of the time period is dominated by the pervasive expansion of solid-state logic, which long since had replaced relay logic in all new telephone switching systems.

Even though there have been many attempts at replacing relays in the switching function as well (see Chapters 1 and 2), relays continue to play a key role in this area. Their very low ON resistance, combined with a very high OFF resistance and a tolerance to high-voltage surges that appear on telephone lines, make them difficult to replace. Even the speed advantage of solid-state switching devices is of little consequence in these applications as long as analog signals are being switched. The low, stable resistances of metallic contacts in electromechanical relays do not introduce additional noise or distortion into analog signals as solid-state devices normally do.

The move towards solid state led to the expectation that electromechanical relay demand would decrease significantly. This, however, has not materialized, and there are many special uses of electromechanical relays in otherwise solid-state systems. Even at the time of publication of this volume, as time-division switching starts to spread and solid-state switching is becoming a significant factor, large numbers of electromechanical relays are still being produced at three facilities of AT&T Technologies (Columbus, Kansas City, and New River Valley*) and by other leading producers. Miniaturization of relays makes them quite compatible with semiconductor components, so one can expect that electromechanical relays will continue to thrive in a world otherwise dominated by semiconductors.

## REFERENCES

1. R. Mueller, "Relays in the Bell System: Facts and Figures," *Bell Lab. Rec.* **35** (June 1957), pp. 227-230.
2. A. C. Keller, "The DDD Relay," *Bell Lab. Rec.* **38** (September 1960), pp. 322-327.
3. S. P. Shackleton and H. W. Purcell, "Relays in the Bell System," *Bell Syst. Tech. J.* **3** (January 1924), pp. 1-42.
   E. D. Mead, "Saving by Swaging," *Bell Lab. Rec.* **3** (November 1926), pp. 83-86.
4. H. N. Wagar, "The U-Type Relay," *Bell Lab. Rec.* **16** (May 1938), pp. 300-304.
5. H. M. Knapp, "The UB Relay," *Bell Lab. Rec.* **27** (October 1949), pp. 355-358.

---

* In Columbus, Ohio; Lee's Summit, Missouri; and Fairlawn, Virginia.

6. A. C. Keller, "A New General Purpose Relay for Telephone Switching Systems," *Bell Syst. Tech. J.* **31** (November 1952), pp. 1023-1067.

7. H. M. Knapp, "New Wire Spring General Purpose Relay," *Bell Lab. Rec.* **31** (November 1953), pp. 417-425.

8. A. C. Keller, "Recent Developments in Bell System Relays—Particularly Sealed Contact and Miniature Relays," *Bell Syst. Tech. J.* **43** (January 1964), pp. 15-44.

9. R. L. Peek, Jr. and H. N. Wagar, *Switching Relay Design* (Princeton, New Jersey: D. Van Nostrand Co., 1955).

10. W. W. Werring, "Miniature Relays for Key Telephone Systems," *Bell Lab. Rec.* **40** (December 1962), pp. 414-417.

11. C. B. Brown, "A Series of Miniature Wire Spring Relays," *Bell Syst. Tech. J.* **46** (January 1967), pp. 117-147.

12. F. A. Zupa, "The Y-Type Relay," *Bell Lab. Rec.* **16** (May 1938), pp. 310-314.

13. O. C. Worley, "Slow Release Wire-Spring Relay," *Bell Lab. Rec.* **32** (September 1954), pp. 351-355.

14. T. H. Guettich, "The 'Two-in-One' Wire-Spring Relay," *Bell Lab. Rec.* **36** (December 1958), pp. 458-460.

15. T. G. Grau and A. K. Spiegler, "Magnetic Latching Wire Spring Relays," *Bell Lab. Rec.* **42** (April 1964), pp. 139-140.

16. H. A. Miloche, "Mechanically Wrapped Connections," *Bell Lab. Rec.* **29** (July 1951), pp. 307-311.

17. J. W. McRae, "Solderless Wrapped Connections: Introduction," *Bell Syst. Tech. J.* **32** (May 1953), pp. 523-524.

    R. F. Mallina, "Solderless Wrapped Connections: Part I—Structure and Tools," *Bell Syst. Tech. J.* **32** (May 1953), pp. 525-555.

    W. P. Mason and T. F. Osmer, "Solderless Wrapped Connections: Part II—Necessary Conditions for Obtaining a Permanent Connection," *Bell Syst. Tech. J.* **32** (May 1953), pp. 557-590.

    R. H. Van Horn, "Solderless Wrapped Connections: Part III—Evaluation and Performance Tests," *Bell Syst. Tech. J.* **32** (May 1953), pp. 591-610.

    W. P. Mason and O. L. Anderson, "Stress Systems in the Solderless Wrapped Connection and Their Permanence," *Bell Syst. Tech. J.* **33** (September 1954), pp. 1093-1110.

    S. J. Elliott, "Evaluation of Solderless Wrapped Connections for Central Office Use," *Bell Syst. Tech. J.* **38** (July 1959), pp. 1033-1059.

18. B. Freile, "The Multi-Contact Relay," *Bell Lab. Rec.* **17** (May 1939), pp. 301-302.

19. R. Stearns, "A Wire-Spring Multicontact Relay," *Bell Lab. Rec.* **35** (April 1957), pp. 131-134.

20. I. S. Rafuse, "A New Multicontact Relay for Telephone Switching Systems," *Bell Syst. Tech. J.* **33** (September 1954), pp. 1111-1132.

21. C. Schneider, "A Miniature Latch-In Relay for the Telstar Satellite," *Bell Lab. Rec.* **41** (June 1963), pp. 241-243.

22. See reference 2, p. 323.

23. W. B. Ellwood, "Glass-Enclosed Reed Relay," *Elec. Eng.* **66** (November 1947), pp. 1104-1106.

24. J. T. L. Brown and C. E. Pollard, "Mercury Contact Relays," *Elec. Eng.* **66** (November 1947), pp. 1106-1109.

25. O. M. Hovgaard and G. E. Perreault, "Development of Reed Switches and Relays," *Bell Syst. Tech. J.* **34** (March 1955), pp. 309-332.

26. P. Husta and G. E. Perreault, "Magnetic Latching Relays Using Glass-Sealed Contacts," *Bell Syst. Tech. J.* **39** (November 1960), pp. 1553-1571.

27. C. G. McCormick, "Glass-Sealed Switches and Relays," *Bell Lab. Rec.* **25** (September 1947), pp. 342-345.

28. M. D. Fagen, ed., *Impact: A Compilation of Bell System Innovations in Science and Engineering Which Have Helped Create New Industries and New Products* (Bell Laboratories, 1971), p. 99.

29. A. Feiner, C. A. Lovell, T. N. Lowry, and P. G. Ridinger, "The Ferreed—A New Switching Device," *Bell Syst. Tech. J.* **39** (January 1960), pp. 1-30.

30. A. Feiner, "The Ferreed," *Bell Syst. Tech. J.* **43** (January 1964), pp. 1-14.

31. I. S. Rafuse, "New Miniaturized Dry Reed Glass Sealed Contacts," *Bell Lab. Rec.* **42** (March 1964), pp. 99-100.

32. E. G. Walsh and G. Haugk, "The Development and Application of Remanent Reed Contacts in Electronic Switching Systems," *Int. Switching Symp. Rec.*, Cambridge, Massachusetts (June 6-9, 1972), pp. 343-347.

33. J. T. L. Brown and C. E. Pollard, "Balanced Polar Mercury Contact Relay," *Bell Syst. Tech. J.* **32** (November 1953), pp. 1393-1411.

34. J. N. Reynolds, U. S. Patent No. 1,131,734; filed May 10, 1913; issued March 16, 1915.

35. J. N. Reynolds, "The Crossbar Switch," *Bell Lab. Rec.* **15** (July 1937), pp. 338-343.

36. F. A. Zupa, "Magnetic-Latching Crossbar Switches," *Bell Lab. Rec.* **38** (December 1960), pp. 457-462.

37. R. P. Holtfreter, "A Switch to Smaller Switches," *Bell Lab. Rec.* **48** (February 1970), pp. 46-50.

38. R. J. Gashler, W. A. Liss, and P. W. Renaut, "The Remreed Network: A Smaller, More Reliable Switch," *Bell Lab. Rec.* **51** (July/August 1973), pp. 202-207.

39. M. B. Purvis and R. W. Kordos, "Reed-Contact Switch Series for the I. F. Band," *Bell Syst. Tech. J.* **49** (February 1970), pp. 229-254.

# Chapter 8

# Capacitors and Resistors

*Impregnated paper capacitors played a major role as general-purpose capacitors for most of the time period covered by this history. Various developments resulted in smaller size, leakproof seals, improved moisture resistance, and techniques for predicting and measuring life expectancy. A second material, mica, was used in channel filter capacitors because of its high stability and low loss. Organic and synthetic film capacitors played large roles in later designs, as did units based on metalized paper. With the advent of transistor circuitry, tantalum solid electrolytic capacitors came into prominence. Among resistors, the basic wire-wound type was popular for over 50 years. As demand for resistors increased, a search for new and better materials was initiated, and deposited carbon resistors were introduced during World War II. Finally, tantalum nitride resistors, with superior stability, precision, and a low temperature coefficient of resistance, became the dominant resistors.*

## I. CAPACITORS

### 1.1 Introduction

Since the days of the Leyden jar and early wireless telegraphy, the capacitor has played an important role in electronics. It is an energy storage device analogous to a mechanical spring. In its simplest form, the storage medium (the dielectric) completely fills the gap between a pair of parallel metallic plates. Energy is stored in the displacement of charges under the action of an electric field induced by an applied voltage difference. In combination with an inductor (the electrical analog of mass), resonant energy storage systems may be constructed, and from combinations of these circuit elements, various filtering functions may be devised. Other applications stem from the nature of the electrical impedance of the ideal capacitor. Since its impedance varies inversely with frequency and capacitance value, dc isolation between circuits in a signal path may be

---

Principal author: C. A. Goddard

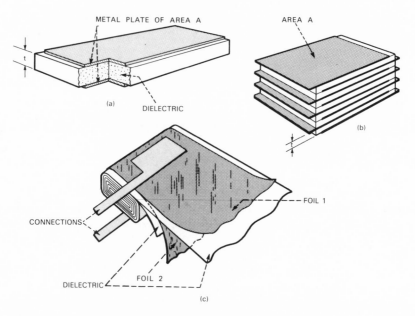

METAL PLATE OF AREA A                    AREA A

(a)          DIELECTRIC                                                    (b)

CONNECTIONS

DIELECTRIC          FOIL 2                    FOIL 1

(c)

Fig. 8-1.    Capacitor structures. (a) The basic structure has a capacity $C = \epsilon A/t$, where $\epsilon$ is the dielectric constant of the dielectric material having a thickness $t$, and $A$ is the area of the capacitor. (b) Stacking n layers increases the capacitance n-fold. Given electrodes of negligible thickness, the capacitor volume is $V = ntA$, giving a capacitance density of $C/V = \epsilon/t^2$. Stacking is done for mica and ceramic capacitors. (c) With flexible dielectrics, such as paper, Mylar*, or polystyrene, the wound structure is preferred.

obtained. Separately, this characteristic permits low-impedance bypassing of signal currents or power-supply ripple currents to ground.

For power-supply filtering and low-frequency bypassing, large capacitance values are required. In contrast with this, lower values of capacitance but very tight tolerances and long-term stability are necessary for frequency-shaping networks. In some electromechanical switching circuits, the time constant of resistor-capacitor networks plays an important role. These circuits operate to reduce the clicks and pops in signals that the opening and closing of contacts may induce, and they serve the additional function of minimizing the spark erosion of relay contacts. Finally, for high-frequency and broadband transmission applications, the additional constraints of small size and low self-inductance are essential.

Several capacitor structures are shown in Fig. 8-1. The capacitance value is proportional to the dielectric constant and the surface area of the dielectric, and inversely proportional to its thickness. Since nature thus prescribes a very thin structure of large area, a primary motivation throughout the

---

* Trademark of E.I. DuPont de Nemours Co., Inc.

history of capacitor development was a search for materials having physical properties that permit stacking, winding, or other fabrication techniques to attain a three-dimensional assembly of minimum volume. Capacitance density (capacitance per unit volume), however, is proportional to the dielectric constant and inversely proportional to the square of the thickness of the dielectric (plus the two metallic electrodes). The general direction of capacitor developments is discernible from these observations.

The major challenge was to find stable, thin dielectric materials of high dielectric constants and high dielectric strength. The physical properties of both the dielectric and the metal electrodes played an important role in structural designs and in the economics of manufacture. The history of capacitor developments in the Bell System was, of course, closely linked to the changing nature of the telephone business. As a passive circuit element, the capacitor complements the inductor and the resistor to provide the array of functions needed in the telephone system.

## 1.2 Paper Capacitors

In 1927, P. R. Coursey[1] compiled the dielectric properties of over 125 liquid and solid materials. At that early date, the total demand for capacitors was dominated by the telephone system, and only a few of Coursey's materials were in use. Rag stock paper derived from linen and impregnated with paraffin or other waxes provided the dielectric for a major portion of the early demand; mica and anodic aluminum oxide took care of the balance of Bell System needs.

For impregnated paper capacitors, the metal electrodes were generally thin sheets of lead-tin, and the units were wound as described in Fig. 8-1 to obtain high volumetric efficiency. In 1929, Western Electric manufactured about 7 million capacitors of more than 50 types based primarily on this structure. (In contrast, the manufacture of capacitors of all types by Western Electric was on the order of 200 million in 1974.) With manufacture at that level and growing rapidly, there was a continuing search for new or improved materials and processes and for innovative manufacturing methods.

In permeating the structure of paper capacitors, the impregnant fills the voids and serves to solidify or cement the structure to enhance its physical and electrical properties. Paraffin, with occasional wax additives, was the principal impregnant until the late 1920s. Its principal deficiencies were its low dielectric constant and its propensity to leak or flow from the structure at elevated temperatures. In 1926, a mineral wax (No. 14 compound) was introduced to manufacture, followed shortly by an improved chlorinated napthalene (Halowax*) in 1928. Increasing demand for re-

---

* Trademark of Koppers Co., Inc.

duction in size played an important role in new designs of paper capacitors, and Halowax offered an attractive solution, since its dielectric constant of about 5.5 permitted a 50-percent increase in volume efficiency. Initially, the presence of less highly chlorinated phases (monochlor) in Halowax severely limited the life of the product, because it tended to react with the asphaltic compounds used as a structural sealant to keep out moisture. An early effort in the research area was therefore devoted to the purification of this material.

In 1929, improvements in the quality of linen rag stock paper permitted a reduction in dielectric thickness from 0.0005 inch to 0.0003 inch, and improvements in manufacturing processes to permit capacitor fabrication with this thinner paper were introduced by Western Electric.[2] These advances were complemented by facilities for pressing wound capacitors into flat elliptical forms before and after impregnation to compact and stabilize the structure. A primary incentive for these changes was the need for capacitors of smaller size for use in station sets and in central office equipment. Pressing after impregnation was intended to maintain an excess of wax at the end of the unit as a moisture barrier; this was difficult to achieve because of shrinkage of the paper during cooling. A solution developed in 1935 was the use of a lead can that was compressed along with the capacitor unit and subsequently filled under vacuum with a soft mineral wax.[3] The lead can design is shown in Fig. 8-2.

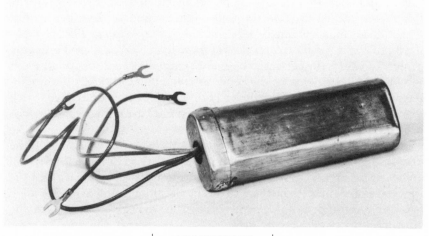

|—— 5 CENTIMETERS ——|

Fig. 8-2.    The lead can capacitor, introduced in 1935 as a solution to the problem of providing a moisture barrier at the end of wax-impregnated paper devices. The can was compressed along with the capacitor unit and filled under vacuum with a soft mineral wax.

At about this same time, wood pulp (Kraft*) paper of suitable quality and thickness was introduced for larger capacitance values. These developments, together with the introduction of 0.00025-inch aluminum foil for the electrodes, contributed to size reductions culminating in the first "combined" station set, in which all components were housed in the base of the telephone.

Because of variability in the final size of the pressed lead capacitor container, the conversion of this structure to stud mounting for central office use was complex and not economical. Aluminum cans had been considered for this use, but cost factors and the poor adherence of sealing compounds to aluminum prevented the metal's early introduction. The continuing decrease in the cost of aluminum, however, and the development of the soft mineral wax referred to above combined to make this structure a feasible venture. A manufacturing trial was completed in 1939. The results were economically and esthetically attractive, but quantity manufacture was delayed until 1946 because of critical material shortages during the war years.

Paper capacitors for high-voltage applications were manufactured by Western Electric in relatively small quantities beginning in 1925. Early in this period, it was observed that very small amounts of moisture were a cause of short life—especially when continuous operation at high voltage and relatively high temperatures was required. Early impregnants were mixtures of petrolatum and paraffin that behaved reasonably well but tended to soften and flow from the containers at high temperatures.[4] A compression-type rubber seal was introduced in 1931, but moisture sealed within the container continued to plague the product. A demand for smaller designs and the development of a rubber seal for leakproof containers rekindled an earlier interest in liquid impregnants for high-voltage use. Among materials studied, polychlorinated biphenyl (PCB) had particularly attractive dielectric properties. Its use in ac capacitors was common with outside manufacturers, but little was known of its dc properties. Following an intensive study at Bell Laboratories, a series of capacitor designs was introduced, reflecting a 30-percent reduction in volume compared to capacitors using mineral wax. The low viscosity and solvent characteristics of PCB reopened the problem of leakage in the field and prompted the development of plastic molded terminals that could be soldered to the metal container. [Fig. 8-3]

A few years earlier, a rash of field failures of small central office capacitors had prompted an extensive program of life and reliability studies at Bell Laboratories. Samples were drawn at random from Western Electric's monthly production to establish a broad data base for product control and

---

* Trademark of Tuttle Press Co.

|—— 5 CENTIMETERS ——|

Fig. 8-3. The 289A capacitor. A 30-percent reduction in volume of high-voltage capacitors was achieved by using polychlorinated biphenyl (PCB) as the liquid impregnant. Because the impregnant is liquid, plastic molded terminals soldered to the metal container were introduced to avoid leaks.

improvement. The techniques were extended to the evaluation of PCB as an impregnant for high-voltage capacitors with very interesting results. Although PCB generally showed excellent chemical stability at high temperatures under moderate electric stress, its use in high-voltage dc capacitors was initially disappointing because of rapid deterioration. A series of careful studies by D. A. McLean, G. T. Kohman, L. A. Wooten, L. Egerton, and M. Brotherton defined the chemical and electrochemical reactions and led to the development of additive stabilizers.[5,6] Most important among these was anthraquinone, which was added to PCB in high-voltage capacitor manufacture in 1939. This stabilizer was added to Halowax for low-voltage capacitors a year later. Process information for this significant means of extending capacitor life was made available, royalty free, to the U.S. government for use in military equipment during the war years.

A final contribution to paper capacitor quality was made in 1942 when it was found that controlled oxidation of the capacitor paper was highly beneficial to long capacitor life.[7] This work stemmed from earlier cooperative programs with the paper manufacturers, which were continued through 1944, bringing changes in paper processing with significant improvements in the quality of this basic material.

The years 1935 to 1945 were marked in particular by the development of techniques for the accelerated life test evaluation of capacitor materials and structures. The success of programs cited above was critically dependent on statistical methods and the determination of acceleration factors to forecast life expectancy from short-term tests.[8,9]

### 1.3 Mica Capacitors

Mica is a natural mineral occurring in small blocks that require skillful splitting into sheets by hand. It is of interest as a dielectric for capacitors because high-quality mica has excellent dielectric properties—primarily very low intrinsic loss and a high degree of stability. However, it frequently contains undesirable mineral and vegetable impurities that detract from its quality. Since mica is a relatively stiff and brittle material, volumetric efficiency is obtained by stacking alternate layers of mica and electrode material in the manner described in Fig. 8-1. A finished unit is shown in Fig. 8-4. Because of the amount of hand labor required, mica capacitors are considerably more expensive than paper capacitors.

Starting in 1925, mica played a small but important role in capacitor manufacture. Its primary use was in capacitors for channel filters in carrier systems, where low dielectric loss and a high degree of stability were important. Since the capacitance values for channel filters were nominally quite low, only a few sheets were required in a stack. Finally, the precise values needed for this application could be obtained by selecting the final sheet in a stack or by selecting the size of the last metal electrode in the stack. Designs for precision capacitors introduced in 1930 used copper-foil electrodes to provide compliance with the somewhat irregular mica surfaces, while avoiding the plastic flow associated with lead-tin foil. Temperature compensation was attained by the use of hardened-steel clamping plates with brass screws to provide uniform pressure over a moderate temperature range.

Even with a compliant electrode material and clamping, air voids are not completely preventable. For this reason, improved processes were desirable in which the electrode material is in direct contact with the mica. The first significant change toward process improvements occurred in 1930 and 1931. Cathodic sputtering of silver and platinum by ion bombardment

|←——— 4 CENTIMETERS ———→|

Fig. 8-4. A mica capacitor. As early as 1925, mica was used because of its low intrinsic loss and high stability.

in a high-vacuum system was successfully performed at that time. Although this process was to form the basis for the tantalum thin-film technology (see Chapter 9 in this volume), the high sheet resistance of these thin films added too much loss to mica capacitor characteristics to be attractive. Chemical deposition and electroplating were tried as an alternative means of depositing metallic films on mica, but these were abandoned. Finally, a commercial silver paste used for decorating china and bonded to the surface by heat was found to offer an attractive solution. This technique was first used at Bell Laboratories on fused quartz for very low capacitance values. In 1937, the method was extended to Pyrex* glass as a low-cost replacement for fused quartz. Fired silver electrodes on mica were introduced about the same time, but the general use of silvered mica in production did not occur until 1940. Metalization by this technique eliminated the air voids so that the need for clamping the structure under pressure was reduced, permitting plastic-molded mica capacitors to be introduced.

### 1.4 Ceramic Capacitors

The dielectric properties of ceramics were studied by researchers at Bell Laboratories as early as the second half of the 1920s. Although ceramic played an important role in electron tubes and other Western Electric

---

* Trademark of Corning Glass Works.

products, capacitor development was not pursued at that time. Ceramic capacitors with rather wide tolerances became available from outside manufacturers in the mid-1930s. Although these capacitors competed with mica units in their performance characteristics, the requirements for precision in channel filter use were not easily duplicated except by selectively pairing capacitor elements. For these reasons, Bell System needs were limited, and all needs were filled by outside purchases. Only during World War II did Western Electric produce a small number of high-voltage titanium dioxide ceramic capacitors for military applications.

## 1.5 Substitutes for Mica Capacitors

The high demand for small, radio-frequency bypass capacitors and the restricted supply of mica created a critical supply shortage during World War II. This situation was alleviated by the development of a series of small, molded, oil-impregnated paper capacitors. These capacitors used a relatively viscous cable oil found to be superior under voltage stress to the commonly used lighter oils. This development was timely, since it preceded the growth in demand for pulse-forming networks required in radar systems. Even though the capacitance values were small in pulse-forming networks, meeting high-voltage requirements would have consumed large amounts of mica. The methods of reliability evaluation developed earlier were applied to these oil-impregnated capacitors to determine that they could withstand the high voltages and high pulse currents of the intended application. Design information and the results of Bell System tests were shared with other manufacturers of pulse network capacitors to effect a significant reduction in demand for quality mica during the war.

## 1.6 Organic Film Capacitors

It was recognized as early as the 1920s that a homogeneous dielectric film should be superior to impregnated cellulose fiber (paper), in that such films should permit a more uniform thickness control with fewer defects than reprocessed "natural" materials. However, the results of these early studies were frustrating for many years, and synthetic organic dielectric films did not play an important role in Bell System components until rather late in the time span covered by this history.

In 1921, Western Electric engineers fabricated experimental capacitors in which the dielectric was an enamel of the type used as an insulation coating on wire and cable. When films of cellulose acetate about 0.0005-inch thick became available in 1928, these films were evaluated together with metal foil coated with cellulose acetate. The dielectric strength was found to be inadequate or, at best, to be no better than wax-impregnated paper. Furthermore, a difficulty with generating dielectric films on metal foil from the liquid was that surface tension created thinning of the film at corners and edges. This markedly reduced the breakdown strength, so this effort was abandoned.

Among precast films, cellophane was investigated extensively in 1931 and 1932, but its dielectric properties were not attractive in spite of cooperative work by the manufacturer to improve the material.

Polystyrene was available in film form at about the same time, and it initially looked very attractive because of its low ac loss. However, the thin films available at that time were too brittle and too stiff for winding but not stiff enough for stacking in the manner used in mica capacitor structures. A limited amount of polystyrene film was used for a special capacitor in a military computer in 1941. Here a special heat-treatment process was devised to provide a very stable capacitor with low dielectric loss.

Polystyrene was used sparingly in Bell System applications until the late 1950s, when an improved flexible film from a source in Germany appeared on the market. Foil polystyrene capacitors thermally processed to heat-shrink the body and to fuse the ends of the wound unit were then possible, and by the 1970s, a moderately large number of these devices were used in the Bell System. They had excellent low-loss characteristics approaching that of mica, and their temperature coefficient of capacitance was a good match for a particular ferrite compound used in the fabrication of inductors. Since the units could be fabricated to rather close tolerances and the stability was quite good, there was a moderate demand for this product.

The major breakthrough in the field of organic dielectric films was the development by DuPont of polyethylene terephthalate (Mylar), which was introduced in products for Bell System use in 1952. Since the material has moderately low dielectric loss and high physical strength, it had a profound impact on capacitor designs of the following 20 years. Its high mechanical strength in combination with the lower voltage requirements of transistorized circuits permitted the design of capacitors using films as thin as 0.0001 inch.

Bell Laboratories played an important role in the development of Mylar capacitors, since DuPont's interest in Mylar was spurred primarily by the growing market for transparent, high-strength plastic packaging materials. (Mylar's desirable dielectric properties were brought to light as part of a continuing search by Bell Laboratories for new capacitor materials. This search involved regular contacts by McLean[10] and others with the major chemical firms following World War II.) Interactions between DuPont and Bell Laboratories were particularly gratifying in the postwar years as they related to optimization of the physical properties of Mylar as a capacitor dielectric.

A later contender in the organic film arena was polypropylene, a low-cost material with low-loss characteristics approaching those of polystyrene. Endowed with high dielectric strength and a marked insensitivity to moisture, the film was first used with foil electrodes in capacitors for Bell System station equipment in 1971.

## 1.7 Metalized Film Capacitors

In England, G. F. Mansbridge[11] recognized as early as 1900 that self-healing in the vicinity of a dielectric defect is possible if thin metallic films are used in place of foil electrodes. Localized heating caused by a short-circuit current at a defect in the dielectric vaporizes the thin metal layer to isolate the defect region from the capacitor proper. Some modest effort toward attaining thin-film coatings on paper started in Bell Laboratories in the 1930s with the work of Kohman, who pursued these developments following the advent of vacuum evaporation techniques for depositing aluminum and other metals on quartz crystals for frequency control. Models of metalized paper capacitors were fabricated, and the process was evaluated for manufacture. However, the low vapor pressure of aluminum limited the deposition rate, and it was found that vaporized aluminum tended to penetrate the porous structure of capacitor paper, resulting in low breakdown voltage. For these reasons, the process did not appear sufficiently attractive to consider manufacture.

At the end of World War II, an investigating team from the U.S. Department of Commerce found that the Robert Bosch Company in Germany had solved the two problems cited above. They had developed a means of coating capacitor paper with a very thin film of lacquer to fill the porous structure. The dielectric strength was enhanced, and the possibility of corrosion of the thin metal film under humidity and potential was reduced. Further, they used zinc instead of aluminum for the vaporant because of its high vapor pressure and consequent high rate of evaporation. Pure zinc does not adhere well to dielectrics, but the Germans had discovered that the dielectric could be sensitized by adding a trace of silver to the vaporant charge to obtain bright adherent coatings of zinc with good conductivity. The Department of Commerce contracted with Western Electric to evaluate the lacquering and metalizing machinery and to demonstrate this new production technique to the American electronics industry. Supplied with a coater and an evaporator from Germany and for a one-dollar fee, Bell Laboratories fulfilled the contract.[12] Capacitors made from metalized paper were introduced into Western Electric manufacture about 1949.

Development of capacitors with metalized electrodes on Mylar was started by Bell Laboratories and Western Electric in 1951. This type of capacitor was produced for military applications beginning in 1955 and

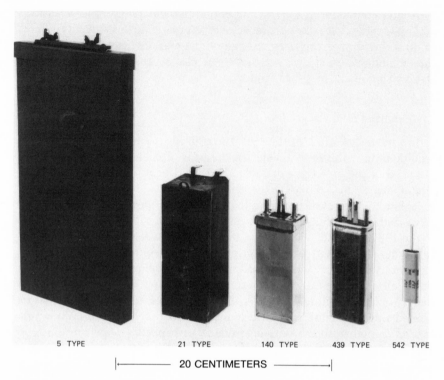

5 TYPE          21 TYPE          140 TYPE          439 TYPE     542 TYPE

|———————— 20 CENTIMETERS ————————|

Fig. 8-5.   Reduction in capacitor size. Paper (and later, Mylar) metalized capacitors showed a significant size reduction over the years. Shown are 2-microfarad, 200-volt capacitors. The 5 type was in use until 1905, when it was replaced by the 21 type. The 140 type was introduced in 1929, continuing a trend of size reductions through thinner paper and thinner foil. Replacement of the soldered can of the 140 type with an extruded aluminum can achieved an even more compact structure, the 439 type. Metalized paper and metalized Mylar capacitors continued the trend. The 542-type Mylar capacitor, besides having a thinner dielectric and thinner electrodes, requires no metal can because it is less sensitive to moisture than metalized paper capacitors.

for Bell System use starting in 1956. These capacitors were manufactured by Western Electric in rapidly increasing quantities, reaching an annual rate of 1 million in 1959 and almost 10 million in 1960. Usage continued in very large quantities after 1960. Their numbers were exceeded only by outside suppliers' ceramic types, which were particularly attractive for wide-tolerance applications in transistor circuits. Figure 8-5 shows the size reduction attained since 1905, a trend in which metalized Mylar represents the last step.

### 1.8 Electrolytic Capacitors

The electrolytic capacitor was discovered in the late 1800s. These capacitors excel in providing large blocks of capacitance at low cost. Their disadvantage is that they are polarized and have a relatively high series resistance. However, these shortcomings are inconsequential in applications at low frequency and in filtering the output of power supplies and motor generators for charging telephone central office batteries.

The four elements of an electrolytic capacitor are shown in Fig. 8-6. The first electrode material was selected from one of the metals such as aluminum or tantalum, whose surface may be converted to its oxide by anodization techniques. The amorphous oxide so generated forms the capacitor dielectric. Extremely thin and uniform oxides grown by this method show excellent dielectric strength, and it is this characteristic that contributes to the high-volume efficiency of these components. The function of the second electrode is performed by the electrolyte, but its more important function is to provide a supply of oxygen by electrochemical processes to heal faults occurring in the normally continuous oxide film.

This healing of the dielectric occurs only when the metal electrode is the anode and the electrolyte is the cathode of the capacitor structure. With the polarity reversed, breakdown occurs readily and the capacitor forms a short circuit. Thus, the structure behaves like a rectifier. Until the mid-1950s, this rectification mechanism was not understood correctly. It was believed that the oxide was an electronic conductor and the metal cathode was a source of electrons. Conduction would then occur with the metal as the cathode. The electrolyte was seen as a source of ions and not as a source of electrons, so conduction should not occur with the electrolyte as the cathode. Since at the time the term "valve" was used

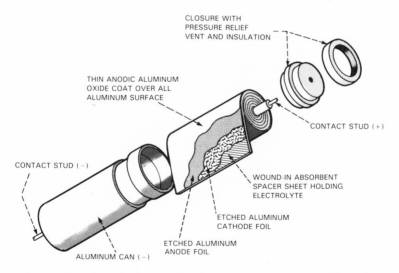

Fig. 8-6.   Schematic of an electrolytic capacitor. Electrolytic capacitors provide large blocks of capacitance at low cost but are restricted to low-frequency applications, such as power-supply filtering.

for a rectifier, metals that could be oxidized by anodization techniques were referred to as "valve metals."

Prior to 1930, comparatively few electrolytic capacitors were used in the Bell System plant except for large, liquid-filled glass container types used for filtering the output of dc generators, because the quality of so-called dry electrolytes (using a paste instead of a liquid), developed by other manufacturers, had generally been found marginal for Bell System capacitor use. However, the plans for a self-contained speech network for station set use created a potential demand that warranted an intensive development starting in 1932. Since the miniaturization of paper capacitors discussed in section 1.2 offered a solution, this effort was short-lived. Further, since Western Electric annual purchases of aluminum electrolytic capacitors were somewhat less than $25,000 until about 1938, the introduction of such a product to manufacture was not warranted. The results of these developments were used to assist suppliers in improving their product and in preparing specifications and acceptance documents.

Tantalum was one of the valve metals studied for electrolytic capacitors. The attractive high dielectric constant of its oxide (approximately 26) was offset in part by the high cost of the metal. Experimental capacitors were made in 1935 using cathodically sputtered tantalum films on an aluminum base—a forerunner of the thin-film work of the 1950s. By 1935, the Fansteel Metallurgical Corporation was marketing porous tantalum electrolytic rectifier and capacitor elements. In a program of collaboration with Fansteel, Bell Laboratories worked to improve the structure and purity of tantalum porous anodes and to continue developments of capacitor products. Laboratory models were used in development circuits in 1941, but little additional work was done until after the war. By 1948, it appeared that there would be a significant Bell System demand for wet tantalum capacitors, but it was decided to purchase this product from outside vendors.

The aqueous electrolyte detracted from the other excellent features of electrolytic capacitors, as electrolytes might leak, dry out, or congeal at low temperatures. In the late 1940s, H. E. Haring and R. L. Taylor of the chemical research area conceived a tantalum solid electrolytic capacitor that uses a completely solid nonaqueous medium.[13,14] They proposed a structure involving a semiconducting oxide as a part of the counterelectrode. It provides adequate conductivity for normal capacitor use but limits the current at defects in the tantalum oxide dielectric, while supplying oxygen required for healing the defects. After studying the properties of manganese dioxide and lead dioxide, they soon concentrated on the former because of its simplicity in application and control.

In 1950, a concerted development effort was launched under McLean to engineer the process for manufacture. A porous tantalum anode was chosen over tantalum foil because of its large ratio of surface area to volume. Development of a method of penetrating this porous mass was a key contribution to the success of the new capacitor. After the anodic oxide was formed, the porous body was saturated with a manganous nitrate solution that was then heated to decomposition, leaving a residue of manganese dioxide. This process was repeated several times to produce essentially complete filling of the pores in the tantalum anode. A coating of colloidal graphite and solder then completed the structure, which for most Bell System applications was protected by a molded plastic body. [Fig. 8-7]

The tantalum solid electrolytic capacitor was a natural complement to the transistor. The low impedance level of modern transistor circuits demanded a proportionate increase in capacitance, but the operating voltages were reduced. With voltage ratings of 35 volts or less, very thin dielectric oxides (precisely controlled by the oxide formation voltage) could be used, and with a dielectric constant of 26, the volumetric efficiency was very high.

With the increasing use of solid-state logic in the Bell System, the demand for tantalum solid electrolytic capacitors increased dramatically. By the mid-1970s, tens of millions per year were manufactured. The development of batch processing methods of fabrication resulted in very low cost. Improvements resulting from major developments in the 1960s included the

|← 3 CENTIMETERS →|

Fig. 8-7. Tantalum solid electrolytic capacitors molded in a plastic body. A porous tantalum anode is oxidized to form a dielectric and is filled with a semiconducting oxide of manganese, which becomes the cathode and supplies oxygen needed to heal any defects in the dielectric. Colloidal graphite and solder form the contact to the cathode.

attainment of higher purity in tantalum anode processing and a better understanding of geometric effects on ac properties and the role of manganese dioxide in the self-healing properties of these capacitors.

## II. RESISTORS

### 2.1 Introduction

In the early years covered by this history, resistors were used primarily as current-limiting devices for battery charging or in networks for the distribution of power to switching and other central office facilities. Stability and long life at power levels of a few watts were primary requirements. Except for a small demand in the input circuits of vacuum tubes and in radio applications, resistor values were generally below 10,000 ohms ($\Omega$), and the parasitic effects of inductance and capacitance associated with wire-wound resistors were not detrimental to circuit performance. Only as switching systems were further developed and as radio and carrier systems came into being did these parameters assume increasing importance. Quite ingenious structures were devised to meet the new needs.

### 2.2 Wire-Wound Resistors

Flat-type, wire-wound resistors using nichrome wire on phenolized cards were popular prior to 1926, and the rack-mounted 18- and 19-type resistors shown in Fig. 8-8 were already used in fairly large numbers. Indeed, these particular codes were still in manufacture at a level of about 10 million per year as of 1977. Over the years, many design improvements were made. The product of the mid-1970s was assembled on a nickel-silver core with a phenolized asbestos wrapping to provide good insulation and stability at relatively high temperatures. Terminals for solderless wrapped connections were held precisely in place by phenolic molding techniques developed in the 1930s (see Chapter 7, section II). Low-tolerance, higher-power resistors took the form of wire wound on ceramic cores with a vitreous enamel cover coat.

For power levels below 1 watt (W), molded carbon composition resistors were initially obtained from the Allen Bradley Company, following interactions with Western Electric engineers shortly after 1900. The International Resistor Company version of a similar unit followed a number of years later. For applications requiring precise values and long-term stability, wire-wound types with variations in winding techniques were used. The inductance in high-resistance codes was minimized by bifilar and reverse winding or other schemes.

Perhaps the most significant Bell Laboratories resistor development prior to World War II was a series of wire-wound resistor designs based on the

|← 5 CENTIMETERS →|

Fig. 8-8. Flat-type resistors using nichrome wire wound around phenolized cards, which were in use prior to 1926. Improved models of the same basic design were still in use over 50 years later.

use of mandrelated wire. In these structures, nichrome wire with a diameter of 0.002 inch or less was wound on a small-diameter cotton core at 150 turns per inch to obtain a resistance of about 800 Ω per linear inch of core. When impregnated with Bakelite* varnish, this structure could be wound on a molded phenolic core with integral axial terminating leads. Under the code numbers 106 and 107, these designs were used in precision circuits and in carrier telephone applications where the need for low inductance and low distributed capacitance was particularly important. [Fig. 8-9] Precision (with tolerances in the range of 0.1 percent to 1 percent) and low noise were among the characteristics particularly desired.

A variation of the 106 and 107 types saw extensive service in World War II. Precision resistors in large quantities were required for computers, gun directors, and other control systems where reliability and resistance

---

* Trademark of Union Carbide Corp.

|⟵ 3 CENTIMETERS ⟶|

Fig. 8-9.   Mandrelated–wire-wound resistors, which use a cotton core wrapped with nichrome wire at 150 turns per inch, that in turn is wound on a molded phenolic core. They provide high precision, low noise, low inductance, and low distributed capacitance.

values in the range from 50 kilohms to 1 megohm were needed. One such design was the cartridge type. Many layers of mandrelated wire were wound on a bobbin in the fashion common in relay manufacture and housed in a brass shell with terminals exiting through a phenolic plate at the end. These resistors were wound directly to a 1-percent tolerance and then padded by a special 106-type resistor to obtain resistance ratios that tracked to 100 parts per million (ppm).[15] The units were designed for use from the tropics to the arctic, and it was reported that they performed particularly well in the African campaign.

### 2.3 Deposited Carbon Resistors

Because of an earlier interest in carbon as a resistor element, several researchers at Bell Laboratories had been exploring the pyrolytic cracking of hydrocarbons for depositing carbon films on ceramic bodies. Resistors of this type had been produced earlier in Europe by methods that were more art than science. Since the demand for resistance wire was starting to outstrip capacity, work was accelerated in the early years of World War II. Film resistors had been of interest earlier because of apparent resistance change in wires due to skin effect at high frequencies. Further, except for alloys such as nichrome, the temperature coefficients for metal films were known to be high.

Processes and facilities for the pyrolytic deposition of carbon films from methane gas were developed by R. O. Grisdale, A. C. Pfister, and W. van

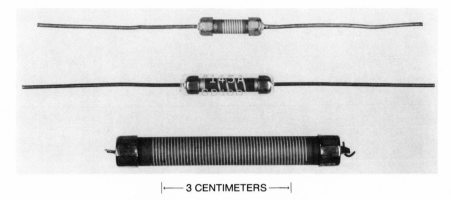

|—— 3 CENTIMETERS ——|

Fig. 8-10. Carbon film resistors, made by grinding helical grooves through the film by automated machinery developed by Western Electric engineers.

Roosbroeck.[16,17] Instabilities that were observed in carbon films deposited on Lenox* china rods were traced to the migration of free alkali metals in the presence of dc fields. M. D. Rigterink, a research ceramist, developed a particular alkaline earth porcelain, called the R-3 composition, that was free from polarization of the type observed.[18] Resistance adjustment included film thickness control based on both the controlled mixture of methane in nitrogen and on time and temperature considerations. Automated machinery was developed by engineers at Western Electric for grinding helical grooves through the carbon film to obtain precision adjustment to the desired value. [Fig. 8-10]

Deposited carbon resistors were used extensively in aircraft radar systems and in computers for the M8 gun director. For applications requiring the greatest stability, resistors were individually sealed in glass and back-filled with helium to provide an inert environment.

Following World War II, several miniature versions of the deposited carbon resistor saw extensive use in the Bell System. Characterized by low noise, excellent high-frequency performance, tight tolerances, and long life, these resistors enjoyed popularity from about 1946 to 1970.

### 2.4 Tantalum Film Resistors

The shortcoming of deposited carbon resistors is their relatively high temperature coefficient. In the search for alternative material systems, sputtered tantalum films were found to have many desirable features. This material system was being investigated as part of the thin-film technology discussed in Chapter 9 of this volume. In the course of this work, while

---

* Trademark of Lenox, Inc.

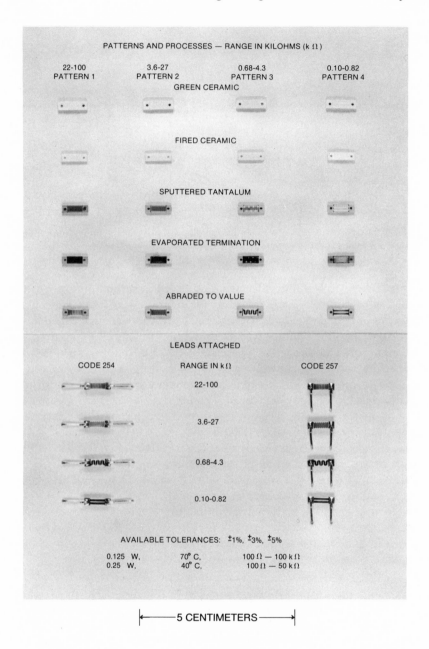

PATTERNS AND PROCESSES — RANGE IN KILOHMS (k Ω )

| 22-100 PATTERN 1 | 3.6-27 PATTERN 2 | 0.68-4.3 PATTERN 3 | 0.10-0.82 PATTERN 4 |

GREEN CERAMIC

FIRED CERAMIC

SPUTTERED TANTALUM

EVAPORATED TERMINATION

ABRADED TO VALUE

LEADS ATTACHED

| CODE 254 | RANGE IN k Ω | CODE 257 |
|---|---|---|
| | 22-100 | |
| | 3.6-27 | |
| | 0.68-4.3 | |
| | 0.10-0.82 | |

AVAILABLE TOLERANCES:  ±1%, ±3%, ±5%

| 0.125 W, | 70° C, | 100 Ω — 100 k Ω |
| 0.25 W, | 40° C, | 100 Ω — 50 k Ω |

|←——— 5 CENTIMETERS ———→|

Fig. 8-11.   Planar resistor structures, based on tantalum nitride. Developed in the late 1940s, they offered a low temperature coefficient and long-term drift stability. A serpentine channel embossed in the active face during pressing is first covered with the tantalum nitride film. After processing, the structure is passed through a grinder to cut away the hills, leaving the film only in the channel.

exploring the effect of admitting small amounts of reactive gases to the vacuum sputtering chamber in 1962, D. Gerstenberg found that near-stoichiometric tantalum nitride films of composition $Ta_2N$ showed particularly attractive properties as a resistor material.[19,20] The temperature coefficient of resistivity for these films ($-100$ ppm per degree C) was about one-fourth the value for deposited carbon film. Further, the long-term drift stability of $Ta_2N$ films was markedly better than that of deposited carbon, and it was further enhanced by thermal processing in air at elevated temperatures. A number of Bell Laboratories and Western Electric engineers collaborated in devising a technique for fabricating and trimming a planar resistor structure based on $Ta_2N$ films.[21] [Fig. 8-11]

The concept started with a molded R-3 porcelain body having a serpentine channel embossed in the active face during the pressing operation. The surface structure was preserved through firing, with the result that the sputtered $Ta_2N$ film conformed to the hills and valleys of the surface topography. After the unit was thermally processed, it was passed beneath the wheel of a surface grinder to cut away the hills, leaving the metal film only in the serpentine valley. Further, the width of the serpentine stripe was a function of the depth of the grinding cut. It was a relatively simple matter to use the measured resistance value as a feedback control of the depth of surface grinding to obtain precision resistance trimming. Indeed, the precision attainable was such that it was not economic to continue the two initial codes of 1- and 3-percent trim tolerance, and only the tighter tolerance was produced as of the mid-1970s. These resistors found broad applications in Bell System equipment; by 1975, just under 100 million resistors were fabricated annually by Western Electric at its Winston-Salem, North Carolina plant at a cost competitive with commercial resistor codes of lesser quality.

## III. CONCLUDING REMARKS

In this resume of developments in support of discrete capacitors and resistors, a few fundamentals stand out in historical perspective. Each phase of the development programs was dictated by unique needs of the Bell System at successive points in time. A common thread is the search for improved stability, long life, and often, tightly controlled parameters. For example, precision capacitors were required for carrier system channel filters long before the industry at large was faced with appreciable demand for tight-tolerance units. Space considerations dictated by the station set and by the economic pressure to reduce floor space for central office equipment played an important and continuing role in the drive to miniaturization. Finally, the advent of the transistor and modern microelectronics placed extreme pressure on the attainment of very small size. In part, this was simply to obtain a compatible reduction in system size. More importantly, however, systems with increasing transmission capacity required ever-increasing bandwidth and thus excellence in performance at higher and higher frequencies. This need led in due time to the popularity of tantalum solid electrolytics and the ceramic capacitor of outside suppliers. Similarly, $Ta_2N$ film resistors play a key role in all modern electronic systems.

Of course, the potential for cost and size reductions is limited in any discrete component system. In silicon technology, this interest in size reduction led to the development of the integrated circuit technology, as discussed in Chapter 2. The same driving forces spawned the tantalum thin-film technology reviewed in Chapter 9 of this volume, which has as its foundation many developments described in the present chapter.

## REFERENCES

1. P. R. Coursey, *Electrical Condensers* (London: Sir Isaac Pitman and Sons, 1927).
2. C. A. Purdy, "Systematizing Manufacture," *J. Western Soc. Eng.* **37** (August 1932), pp. 181-188.
3. F. J. Given, U.S. Patent No. 2,181,695; filed July 29, 1936; issued November 28, 1939.
4. E. T. Hoch, U.S. Patent No. 1,780,917; filed May 29, 1926; issued November 11, 1930.
5. D. A. McLean and G. T. Kohman, "The Influence of Moisture upon the D.C. Conductivity of Impregnated Paper," *J. Franklin Inst.* **226** (August 1938), pp. 203-220.
   D. A. McLean and L. A. Wooten, "Cation Exchange in Cellulosic Materials," *Ind. Eng. Chem.* **31** (September 1939), pp. 1138-1143.
   D. A. McLean, L. Egerton, G. T. Kohman, and M. Brotherton, "Paper Dielectrics Containing Chlorinated Impregnants," *Ind. Eng. Chem.* **34** (January 1942), pp. 101-109.
6. D. A. McLean, U.S. Patent No. 2,259,978; filed August 31, 1938; issued October 21, 1941.
   L. Egerton, U.S. Patent No. 2,287,421; filed November 14, 1940; issued June 23, 1942.
   D. A. McLean, U.S. Patent No. 2,339,091; filed November 14, 1940; issued January 11, 1944.
   L. Egerton, U.S. Patent No. 2,391,685; filed November 14, 1940; issued December 25, 1945.
   D. A. McLean, U.S. Patent No. 2,391,686; filed August 31, 1938; issued December 25, 1945.
   D. A. McLean, U.S. Patent No. 2,391,687; filed January 9, 1941; issued December 25, 1945.
   D. A. McLean, U.S. Patent No. 2,391,688; filed August 31, 1938; issued December 25, 1945.
   L. Egerton, U.S. Patent No. 2,391,689; filed November 14, 1940; issued December 25, 1945.
7. D. A. McLean, "Paper Capacitors Containing Chlorinated Impregnants: Benefits of Controlled Oxidation of the Paper," *Ind. Eng. Chem.* **39** (November 1947), pp. 1457-1461.
8. H. A. Sauer and D. A. McLean, "Direct Voltage Performance Test for Capacitor Paper," *Proc. IRE* **37** (August 1949), pp. 927-931.
9. J. R. Weeks, "Capacitor Life Testing," *Bell Lab. Rec.* **24** (August 1946), pp. 296-299.
10. H. A. Sauer, D. A. McLean, and L. Egerton, "Stabilization of Dielectrics Operating under Direct Current Potential," *Ind. Eng. Chem.* **44** (January 1952), pp. 135-140.
11. G. F. Mansbridge, British Patent No. 19,451; issued 1900.

12. D. A. McLean, "Metallized Paper for Capacitors," *Proc. IRE* **38** (September 1950), pp. 1010-1015.

    J. R. Weeks, "Metallized Paper Capacitors," *Proc. IRE* **38** (September 1950), pp. 1015-1018.

    G. P. McGraw, "The Manufacture of Metallized Mylar Capacitors," *W. Elec. Eng.* **1** (July 1957), pp. 16-22.

13. H. E. Haring, "The Mechanism of Electrolytic Rectification," *J. Electrochem. Soc.* **99** (January 1952), pp. 30-37.

14. R. L. Taylor and H. E. Haring, "A Metal-Semiconductor Capacitor," *J. Electrochem. Soc.* **103** (November 1956), pp. 611-613.

15. E. C. Hagemann, "Precision Resistance Networks for Computer Circuits," *Bell Lab. Rec.* **24** (December 1946), pp. 445-449.

16. R. O. Grisdale, A. C. Pfister, and W. van Roosbroeck, "Pyrolytic Film Resistors: Carbon and Borocarbon," *Bell Syst. Tech. J.* **30** (April 1951), pp. 271-314.

17. A. C. Pfister, "Precision Carbon Resistors," *Bell Lab. Rec.* **26** (October 1948), pp. 401-406.

18. M. D. Rigterink, "Ceramics for Electrical Applications," *Bell Lab. Rec.* **25** (December 1947), pp. 464-468.

19. D. Gerstenberg, U.S. Patent No. 3,242,006; filed November 1, 1961; issued March 22, 1966.

20. D. Gerstenberg and E. H. Mayer, "Properties of Tantalum Sputtered Films," *Proc. Electron. Compon. Conf.* **57**, Washington, D.C. (May 8-10, 1962), pp. 57-61.

21. W. van Roosbroeck, "High-Frequency Deposited Carbon Resistors," *Bell Lab. Rec.* **26** (October 1948), pp. 407-410.

# Chapter 9

# Thin-Film Circuits

*Thin-film technology grew out of work on discrete devices. Sputtered tantalum films were found to be suitable resistor material, and anodically oxidized tantalum was demonstrated to be a good dielectric for capacitors. Thin films deposited on glass and later on ceramic surfaces, patterned by photolithographic techniques, led to integrated resistor and capacitor circuits compatible with the evolving semiconductor device technology. Initially used with discrete transistors, thin-film technology was soon adapted to beam-lead silicon integrated circuits by the inclusion of an interconnection and crossover technology compatible with thermocompression-bonded silicon devices. The resulting hybrid circuit technology became a major means for Bell Laboratories to provide unique circuit functions across a broad spectrum of applications, including precision filters for transmission systems, Touch-Tone generators for telephone sets, and the family of 1A circuit packs that formed the basis of an advanced processor for switching systems.*

## I. INTRODUCTION

Deposited carbon resistors were introduced as cost-effective alternatives to wire-wound resistors in the late 1940s, as described in Chapter 8 of this volume. Since carbon films have a relatively high temperature coefficient of resistance, on the order of 400 parts per million per degree C (ppm/degree C), alternatives to carbon film resistors were sought. At other research laboratories, a search was started in the 1950s for alternative materials that could be deposited in thin-film form. At Bell Laboratories, boro-carbon films were investigated but found to be highly unstable. Therefore, an investigation of a broad spectrum of metals and alloys was started in the mid-1950s by a group that was also responsible for the solid tantalum electrolytic capacitor development. Consequently, it was recognized early in the work that tantalum is suitable as a material for both resistors and capacitors. It is from this effort that thin-film technology developed, a technology that was soon recognized to hold great potential

---

Principal authors: R. W. Berry and D. Gerstenberg

for networks of capacitors and resistors, and which later provided an effective means for interconnecting beam-lead integrated circuits (ICs).

## II. EARLY DEVELOPMENTS LEADING TO RESISTOR CIRCUITS

The traditional method for depositing metal films involves evaporation techniques. It is suitable for most nonrefractory metals, but the deposition of alloys with a well-defined composition is quite difficult. Sputtering was considered as an alternative in 1956 by H. Basseches, because it provides a method suitable for the deposition of alloys and metals, including refractory metals. After assembling the necessary equipment, Basseches started a systematic investigation of the properties of thin films of various metals and alloys.[1]

At the same time, R. W. Berry started to investigate compounds such as tantalum silicide as a resistor material. Using Basseches's sputtering equipment, he deposited tantalum films on quartz substrates for later conversion to the silicide. When he realized that these films were optically smooth and had no inclusions (as was characteristic of mechanically polished bulk tantalum), he decided to use such films for some fundamental measurements of the dielectric properties of anodic tantalum oxide, the dielectric used in solid tantalum electrolytic capacitors. As mentioned in Chapter 8, these capacitors were increasingly important, but at the time, the physics of the structure was not adequately understood.

Films thick enough to be anodized were sputtered on microscope slides and anodized to 100 volts (V). In order to have a well-defined area of oxide film to measure, a gold electrode was evaporated onto the anodized tantalum, and the capacitance of the structure was measured, using a small-signal ac capacitance bridge. In accordance with the theory of rectification of an electrolytic cell involving a "valve" metal (see Chapter 8), the test structure using gold as the cathode was expected to be essentially a short circuit to dc. In reality, the structure could withstand voltages close to the anodization voltage. This not only changed the understanding of the solid tantalum electrolytic structure but established the feasibility of what were then called tantalum printed capacitors.[2]

While these studies were in progress, samples of tantalum films for resistor applications had been set aside and were remeasured after sitting in the laboratory for about six months. They were found not to have changed in resistance value. This result was better than had been found for any of the other metals and alloys that Basseches had evaluated; accordingly, he decided to begin a much more intensive study of tantalum films for resistor applications.[3,4] Initially, tantalum films with a sheet resistance of 10 to 100 ohms per square ($\Omega/\square$) with a temperature coefficient of $-200$ ppm/degree C were produced, and methods for adjusting the resistance and protecting the film by anodic oxidation were developed by Basseches, P. L. McGeough, and D. A. McLean.[5]

With these findings, it was realized that the tantalum thin-film system permitted the fabrication of both resistors and capacitors on the same substrate. Printed capacitor and resistor networks, later known as thin-film ICs, appeared feasible, and further development efforts were directed at these goals under the leadership of McLean [Fig. 9-1], who recognized the potential of the technology and who subsequently made significant contributions to its development.

For producing patterns in the sputtered films, the photolithography techniques under development for transistors and ICs were adopted (see Chapters 1 and 2 of this volume). The photoresist process initially developed was not compatible with the mixture of hydrofluoric and nitric acids required to etch tantalum. Thus, a patterning technique involving a sacrificial copper film was developed, and in 1958 a demonstration circuit was made with 27 resistors and 9 capacitors on a single ceramic substrate; holes were drilled into the ceramic for transistor leads. The area required for all the components was less than two inches square, showing the potential of

Fig. 9-1. D. A. McLean, a pioneer in tantalum thin-film technology.

the technology for combining several thin-film resistors, capacitors, and transistors on a ceramic substrate of a size considerably smaller than an equivalent discrete component circuit.[6] Shortly thereafter, a photoresist process was developed with which tantalum could be etched directly. This permitted significantly improved pattern definition and simplified the process.

But merely being able to build both capacitors and resistors on the same substrate is insufficient to allow for the fabrication of complete circuits. A means of interconnecting those components as well as providing external electrical connections to the substrate is needed. The first material system developed for these purposes was a combination of evaporated nichrome (to provide adhesion to the substrate) and gold (to provide high conductance and joinability). While this metal system was adequate for internal connections, the gold dissolved too quickly in the solder used to make external connections. Accordingly, an alternative system was developed, consisting of nichrome followed by a relatively thick layer of copper, which was then overcoated by a thin layer of palladium to ensure solderability.[7] This NiCr-Cu-Pd metalization system was used for all circuits fabricated before the introduction of beam-lead silicon devices.

In work directed at improving the understanding of the properties of tantalum films, D. Gerstenberg and C. J. Calbick found that the structural and electrical properties of tantalum films were strongly affected by elements such as nitrogen, carbon, and oxygen, which are present in most vacuum stations in the form of background gases.[8] This led to the development by Gerstenberg and E. H. Mayer of tantalum nitride ($Ta_2N$), which, as mentioned in Chapter 8, became the major resistive film used for discrete resistors.[9]

A later development in resistor films was tantalum oxynitride,[10] which could be made to have a temperature coefficient of resistance equal in magnitude, but opposite in sign, to the temperature coefficient of capacitance of tantalum film capacitors, thus permitting temperature compensation in resistor-capacitor (RC) networks. Another later addition to the family of tantalum-based film was a tantalum-aluminum alloy developed by C. A. Steidel in 1969,[11,12] a very stable film that can be made with about twice the sheet resistance of the nitride.

The first circuit chosen for a feasibility evaluation of the thin-film technology was a modulator pad used in A5 channel modems containing ten resistors and two capacitors. [Fig. 9-2] In a joint effort involving Bell Laboratories and the Western Electric Allentown Works in Allentown, Pennsylvania, 500 modulator pads were fabricated. The circuits were installed in A5 channel modems by the Western Electric Merrimack Valley Works in North Andover, Massachusetts and were successfully placed into service in many locations throughout the country in 1962.

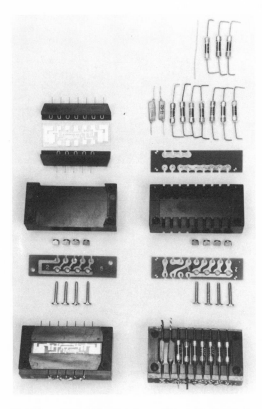

Fig. 9-2. A5 modulator pad assembly—thin film (left) and standard (right). The first field trial of circuits with tantalum film resistors took place in 1962.

Although this experiment showed the viability of thin-film RC networks in a systems environment, the yield of the capacitors was not adequate to commit RC circuits to production. Accordingly, at that time, only circuits with resistor networks were made available for commercial applications. The first of these was a resistor-transistor logic circuit for the 101ESS* electronic switch private branch exchange (PBX) system. The design had to be electrically and mechanically equivalent to an existing design based on discrete components. Resistor and conductor patterns were generated on a chemically stable glass substrate[7] of high quality [Fig. 9-3], which, with its associated transistor, was mounted and interconnected by welding the circuit terminations to leads molded within a phenolic cradle. The

---

* Trademark of AT&T Technologies, Inc.

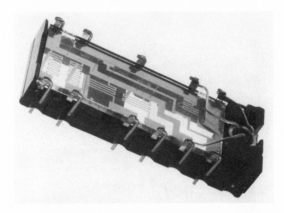

Fig. 9-3. Transistor-resistor logic gate for the 101ESS electronic switch. A thin-film resistor network on a glass substrate is connected to leads molded in a phenolic cradle. The transistor (right side) is connected to the same leads.

leads of the assembled gate were inserted into via holes in a printed wiring board for soldering. Manufacture of this circuit began at Western Electric in Allentown in 1962.

These circuits established the feasibility of thin-film technology in a systems environment. Accordingly, systems designers became confident in the technology, and further designs were not constrained by the requirement of mechanical and electrical interchangeability with a discrete component design.

Because glass is mechanically weak and a poor heat conductor, alumina ceramic was introduced as a substrate material. Improvements in ceramic technology permitted the fabrication of flat, thin, glazed substrates with numerous holes positioned as desired, permitting direct mounting of several transistors and thus the implementation of several logic gates in a single circuit. In addition, conductor paths on both sides could be connected through the holes to provide more flexibility for interconnections between gates. Because substrates as large as 4 by 5 inches were available, as many as 12 circuits could be processed on one piece of ceramic; they were separated by diamond sawing after all thin-film processing was completed. Transistors, leads, and through-connect eyelets were then inserted into the proper holes in the substrate and wave-soldered to the NiCr-Cu-Pd metalization. Since the leads were directly mounted on the ceramic, no plastic carrier was needed, as in the case of the A5 and 101ESS electronic switch designs. The resistor networks for 101ESS switching equipment were redesigned on alumina ceramic utilizing these advantages and went into production in 1964. [Fig. 9-4]

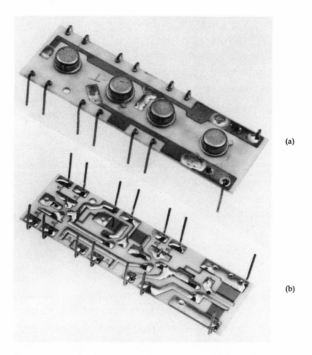

Fig. 9-4.   Multiple transistor-resistor logic gates implemented
on a ceramic substrate: (a) top view; (b) bottom view. Because
holes can be drilled through the ceramic, transistors and leads
can be mounted directly to the ceramic, and both sides of the
ceramic can be used for circuitry.

## III. CAPACITORS AND THERMOCOMPRESSION BONDING

A major development effort was launched in 1962 to improve capacitor yields and to establish long-term reliability. This effort lasted for several
years and resulted in a better understanding of the parameters affecting
the quality of tantalum film capacitors. F. Vratny, B. H. Vromen, and
A. J. Harendza-Harinxma found that capacitor yield and life test performance depended on the crystalline structure and mechanical properties
of the bottom tantalum electrode.[13]

It was found that in thin-film form, tantalum can exist in a crystallographic structure different from that of bulk tantalum. Structural analysis
showed this structure, named $\beta$-Ta, to be either tetragonal, as first reported
by M. H. Read and C. Altman in 1965,[14] or to have a hexagonal subcell
that was transformed irreversibly to body-centered cubic (bcc) tantalum,
the structure of pure bulk tantalum, upon heating in a vacuum to 700–
750 degrees C. However, there was no adequate understanding of the
detailed role of such variables as sputtering parameters, composition of
sputtering gas, and substrate surface chemistry, that initiated and controlled

the formation of $\beta$-Ta. Only in 1968 and 1969 did the work by L. G. Feinstein and R. D. Huttemann lead to a better understanding of the conditions that cause the formation of the $\beta$-Ta phase.[15] Their work showed that $\beta$-Ta was an allotrope of tantalum formed in the nucleation stage. The presence of gaseous impurities was not necessary for its formation, but surface oxygen or hydroxyl groups played a major role in the nucleation of $\beta$-Ta.

At the time, dielectrics formed by oxidation of $\beta$-Ta produced thin-film capacitors superior to those made with dielectrics prepared from sputtered bcc Ta. This improvement resulted in higher initial yields (based on leakage current measurements) and a much smaller number of failures during accelerated life testing. The striking difference between the performance of dielectrics prepared from bcc and $\beta$-Ta appeared to be due primarily to the high incidence of mechanical faults in the bcc Ta, which resulted in weak spots of tantalum oxide formed on it.[13]

Concurrently, Vromen and J. Klerer studied the effect of anodization voltage on the behavior of leakage current, reverse and forward breakdown voltage, capacitance, and dielectric loss.[16] Their work revealed that tantalum film capacitors formed on $\beta$-Ta improved markedly in yield and in failure rates, as shown by tests over a wide range of voltage and temperature stress conditions, when the anodization voltage was increased to over 200 V from the 130 V used in the early work. The effect of primary film composition on capacitor quality was studied by Gerstenberg.[17] The results of this study demonstrated that addition of small amounts of either carbon or nitrogen to tantalum was desirable not only for consistently meeting the 50-V dc leakage current criterion used on 230-V tantalum film capacitors, but also for optimizing capacitor reliability.

The observation that capacitors with pure gold counterelectrodes exhibited a sensitivity of capacitance to changes in humidity led in 1966 to the use of the NiCr-Au interconnection metal system for counterelectrodes. The addition of the thin layer of NiCr along with scrupulous pre-evaporation cleaning essentially eliminated the sorption of moisture at the top electrode-oxide interface; it also reduced the changes in capacitance observed in cycling such devices between 0 and 87 percent relative humidity from 3 percent with gold counterelectrodes to less than 0.2 percent with NiCr-Au electrodes.[18]

During the period from 1963 to 1966, it was also shown that $Ta_2N$ could be anodized and used as the basis of a thin-film capacitor.[19] However, the dielectric constant was found to be lower than that of anodic oxide films formed on $\beta$-Ta or bcc Ta containing less than 5 atomic percent nitrogen. In addition, the dielectric constant of the anodic film changed as the material was anodized, thus making control of capacitance difficult.

The usefulness of a layer of oxygen-deficient manganese oxide ($MnO_2$)

as part of the counterelectrode to make a thin-film version of the solid tantalum electrolytic capacitor was demonstrated in 1966 by McLean and F. E. Rosztoczy.[20] The self-healing property of $MnO_2$ permitted thinner dielectrics, and hence higher capacitance density, as well as fabrication on unglazed ceramic substrates. While the breakdown voltage of such capacitors was markedly higher than that of tantalum film capacitors without $MnO_2$, their properties, like their bulk counterpart, were also found to be more frequency dependent. At the same time, a duplex capacitor structure consisting of evaporated silicon monoxide (SiO) on tantalum pentoxide ($Ta_2O_5$) as the capacitor dielectric was developed by H. N. Keller, C. T. Kemmerer, and C. L. Naegele for thin-film capacitor applications where the high capacitance density of the $Ta_2O_5$ might be a disadvantage.[21] The capacitance in this structure depended almost entirely on the thickness of the SiO layer, because of its low dielectric constant (6, compared to 26 for $Ta_2O_5$).

Thus tantalum film capacitors covered a range from one microfarad ($\mu f$), when $MnO_2$ was part of the counterelectrode structure, to a few picofarads for capacitors with the SiO-$Ta_2O_5$ duplex structure.

The first use of thin-film RC networks for precision RC filter elements was in a joint effort between the data set and components development groups, in the design of the 402B data set.[22,23] In this application requiring precision filters, the resistors and capacitors were fabricated on glazed ceramic. [Fig. 9-5] The glaze was protected by a tantalum oxide film during etching of the tantalum by a hydrofluoric-nitric acid mixture. Although this data set did not go into manufacture as designed, it proved the feasibility of the concept.

Extensive studies of the methods available for attaching silicon devices and external leads to thin-film substrates were carried out by a group in Allentown under the direction of J. E. Clark. This work showed thermocompression bonding to be the most suitable technique for bonding beam-lead silicon devices and gold-plated copper leads to thin-film substrates.[24] This method brought the advantages of batch processing to the bonding operation (see Chapter 2). Tools for simultaneously bonding large numbers of beam-lead devices were developed to realize this potential and were widely used in manufacture during the 1970s.

For thermocompression bonding to be compatible with thin-film circuits, a variant of the metalization used for silicon devices was developed for hybrid circuits. The metalization in use for beam-lead ICs was titanium (for adhesion), followed by platinum (as a diffusion barrier), followed by gold (for conduction as well as protection). Platinum, however, was not compatible with the processing for hybrid circuits. A detailed study by J. S. Fisher and P. M. Hall in 1967-68 showed that palladium could be substituted as a very near equivalent.[25] Unfortunately, the use of titanium

Fig. 9-5.    Tantalum component side of a thin-film module. In this first application of resistor-capacitor (RC) networks for use as precision filters, the resistors and capacitors were on glazed ceramic. The glaze was protected by a tantalum oxide film.

as the adhesion layer reduced the yield of thin-film capacitors drastically, and for circuits containing capacitors, nichrome had to be used in place of titanium.

In addition, the glaze used to smooth the surface of the ceramic so that resistors and capacitors could be fabricated was not compatible with ther-mocompression bonding. While the answer appeared to be unglazed surfaces, the surface morphology and surface chemistry of unglazed ceramic were variable and deleterious to the component's properties and film adhesion. Considerable ceramic substrate developmental effort was expended from the late 1960s through the early 1970s to solve the problems of adhesion of Ti-Pd-Au films,[26] resistor variability, and interconnection integrity. A fine-grained 99-percent alumina substrate was developed by D. J. Shanefield and R. E. Mistler,[27] using a tape cast process that solved the adhesion problem and resistor variability. R. C. Sundahl and E. J. Sedora[28] made a detailed quantitative analysis of the relationship between the distribution of various surface defects (pits and burrs) and interconnection performance.

An alumina ceramic pilot line was established at Allentown in 1970 to develop these high-quality substrates. Through experience gained at this facility and information exchanged with both alumina raw material and alumina substrate suppliers, the substrate quality and consistency available from the major substrate suppliers improved to the point that film adhesion and resistor reproducibility were excellent; capacitors, however, still required a glazed surface.

To protect the circuits after completion, the active side has to be coated with a suitable encapsulant to provide protection against condensed moisture, particulate matter, damage during assembly, and—for some silicon devices—light. Two critical requirements for the encapsulant are a low curing temperature, to prevent changes in the values of precisely adjusted tantalum film resistors and capacitors, and the capability to fill completely the volume under silicon chips and other components. Earlier work by M. White in 1966 demonstrated that Dow Corning room-temperature vulcanizing silicone rubber (RTV-3145) exhibits excellent performance as a moisture protection coating for silicon ICs (SICs).[29] Subsequent work by Basseches, J. H. Heiss and A. Pfahnl, reported by Basseches,[30] and later work by Heiss, D. G. Jaffe and R. G. Mancke, reported by Jaffe,[31] was directed toward the evaluation and further development of various Dow Corning silicone dispersions for use as encapsulants for hybrid ICs, including silicon devices and appliquéd components. Major findings of the work were that a dispersion of RTV silicone rubber in xylene could readily be applied to circuits by a single-step, flow-coat technique, and that no masking of external leads to prevent silicone creep was required. The silicone rubber, after curing, gave good underchip coverage as well as adequate coverage of discrete components. The maximum temperature during the curing cycle did not exceed 120 degrees C. Low leakage currents between closely spaced conductors aged under bias and high-humidity conditions demonstrated its excellent moisture-protection capability.

The first application of thin-film RC networks to go into manufacture was a tone generator for the Touch-Tone Trim Line* telephone.[32] It was also the first application to take advantage of the compatibility of thin-film passive devices and beam-lead SICs (see Chapter 2). The circuit consisted of two switchable oscillators that generated a series of tones in prescribed pairs, one pair for each dialed digit. A total of seven tones was required. The feedback loop in each amplifier included a notch filter to control the frequency of oscillation. The pair of tones was coupled to the output by buffer amplifier stages in the SIC. The passive network that controlled the Touch-Tone frequencies was made of tantalum thin-film resistors deposited on one substrate that was connected by gold leads to another substrate containing the thin-film capacitors. The transistors, power supply diodes, and amplifier resistors were integrated into a beam-lead sealed-junction SIC. The SIC was attached to the resistor substrate by thermocompression bonding. [Fig. 9-6]

---

* Registered Trademark of AT&T.

Fig. 9-6.   Early version of a hybrid IC for the Touch-Tone dial system. In this first application of precision RC circuits, the capacitors were formed on one glass substrate; resistors, beam-lead ICs, and other discrete components were on a second glass substrate. The substrates were stacked and interconnected by gold leads; another set of leads connected the entire assembly to a printed circuit. The gold wire leads were later replaced by tape leads.

Tuning the notch circuit to the desired oscillating frequency required the development of an economic adjustment procedure. By taking advantage of the fact that each frequency is a function of a resistance-capacitance product of the circuit, the frequency-controlling tantalum resistors were designed 5 percent below the required value. By the anodic conversion of some of the tantalum to oxide, the resistance was increased until the desired RC product was achieved.[32] The RC tone generator, a unique product of the combination of thin-film technology and SIC technology, provided cost, size, and weight reductions, as well as improvements in reliability over the more conventional inductor-capacitor design. Production of these circuits started in 1969.

Reduction in size and weight and improved circuit performance were major motivations for applying thin-film circuits to transmission equipment such as the T1 carrier repeater, the repeatered line for the L5 transmission system, and the D3 channel bank. In the case of the T1 repeaters, small size was the main consideration that led to their redesign in 1965. The thin-film version carried most of its circuitry on a single ceramic substrate containing tantalum film components, silicon devices, and appliquéd capacitors. This circuit required a system of conducting metals that not only permitted thermocompression bonding of the beam-lead silicon devices but that was also compatible with the solder attachment of the appliquéd capacitors. When solder is used on the titanium-palladium-gold conductor system, solder will dissolve the gold and creep from the joint area into some of the adjacent thin-film conductor area. In a joint effort between engineers at Western Electric in Allentown and in Merrimack Valley, a new system was developed in which rhodium and gold layers were selectively plated over the area to be soldered. The gold layer was etched to expose a rhodium dam that oxidized during the resistor stabilization process and acted as a barrier to the solder creep.[33]

The hybrid IC version of a dual T1 repeater, shown in Fig. 9-7, occupied only one-third the volume of the earlier repeater, and in its apparatus case occupied about one-half the wall space of its predecessor. In addition, it

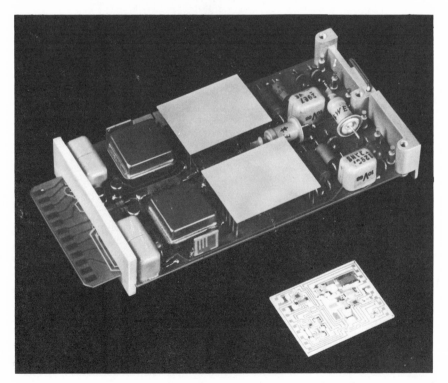

Fig. 9-7.  Hybrid IC version of the T1 repeater. White areas are identical circuits, mounted component-side down; the face of one such circuit is shown in front. The hybrid version occupied only one-third the volume of the earlier repeater.

simplified installation by the operating companies, because it included automatic line impedance adjustment and thus eliminated the need for special networks as well as the engineering associated with their proper selection.

Work on active RC filters for the D3 channel bank was initiated in the late 1960s. The active RC networks that served as antialiasing and reconstruction filters were made on a single ceramic substrate that initially was glazed everywhere except in the bonding areas; later, the glaze was retained only in the capacitor areas. [Fig. 9-8] As in the case of the Touch-Tone application, the D3 filters were adjusted by anodization. The resistor and capacitor values were measured before operational amplifier chips were added. From the measured values of the capacitors, a computer program calculated the desired resistor values, and the resistors were then adjusted in an automated, computer-controlled test set. After the beam-lead operational amplifiers were attached to the substrates, the filter performance

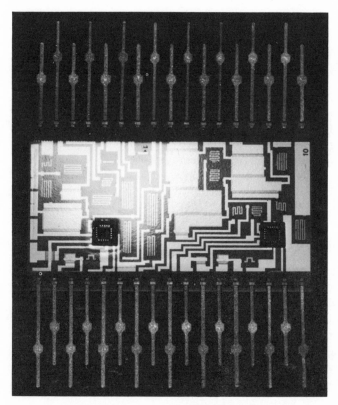

Fig. 9-8.   The RC active filter for the D3 channel bank. In this circuit, capacitors, resistors, and ICs were contained on a single piece of ceramic to which leads were directly bonded.

was measured at two frequencies. If the filter did not perform adequately, a second test set adjusted the feedback resistors until the filter characteristics were within the required limits. The work on the D3 filters serves as a good example of the way in which computer aids provided powerful new methods for designing, producing, and testing thin-film circuits. Production of these circuits began in 1971.

In 1973, Bell Laboratories introduced the universal "building block" concept to analog filters with the standard active resonator (STAR) filter.[34] Instead of requiring the filters to be designed individually, the STAR circuits are tailored to a wide range of precision filter applications by trimming the resistors during the final steps of circuit production. STAR circuits are general-purpose, economical circuit building blocks that perform bandpass, low-pass, notch, high-pass, and other filter functions. When the STAR technology was first developed, it consisted of $\beta$-Ta film capacitors anodized to 190 V and high-sheet-resistance tantalum oxynitride resistors. Subsequently, it was found that bcc Ta containing 10 to 20 atomic percent of nitrogen improved the long-term aging characteristics and the usable frequency range of tantalum film capacitors,[35] and that the anodically formed dielectric could withstand higher temperatures than its counterpart formed on $\beta$-Ta. Furthermore, films of bcc Ta sputtered under these conditions exhibit a much lower incidence of mechanical faults than the early bcc Ta films. The reduced sensitivity to heat permitted the implementation of a simple, more economical fabrication process of STAR and other high-precision RC filter circuits.[36] Thus, in the mid-1970s, bcc Ta replaced $\beta$-Ta as the preferred film material. In addition, the application of the emerging laser-machining technology to resistor trimming provided significant advantages to the implementation of the STAR concept.[37] Various types of STAR filters are used in more than 25 Bell System products. Among these are data sets, Touch-Tone signaling receivers (key telephone, PBX, and central office), DIMENSION* PBX, and R2 signaling transceivers.

## IV. INTERCONNECTION WITH CROSSOVERS

As more and more silicon devices were used on single thin-film substrates, the interconnection density increased and a need arose for a structure that would allow thin-film conductors to cross each other. This growth in complexity of the interconnections was precipitated by the development of semiconductor memories and other circuits for 1ESS and 4ESS switching equipment in the late 1960s and early 1970s. Several of these circuits also required a metalized ground plane on the back of the substrate with metalized via holes connecting it to the metalization on the front.

A crossover technology was developed so that very large circuits could

---

* Registered Trademark of AT&T.

be realized with up to 52 SICs, requiring long conductors and connections to metalized backplanes by via holes. A process, originally conceived by M. P. Lepselter for SICs,[38] was adapted by Basseches, Pfahnl, and W. Worobey in 1968-69 to thin-film circuits.[39,40]

In its final form, as used in the 1970s with selectively plated Ti-Pd-Au bottom conductors, the technology can briefly be described as follows. First the conductor pattern is generated by coating the ceramic substrate with Ti-Pd, applying the photoresist, delineating the conductors, and selectively plating the gold. The photoresist is stripped, and the palladium and gold surfaces are carefully cleaned. Now, to form the free-standing conductor bridges, a 25-micrometer-thick copper spacing layer is plated over the entire area and coated with a chromate conversion coating for improved resist adherence. A positive photoresist layer is applied, the pillar pattern exposed and developed, and the pillar holes etched into the copper. A second exposure is used to delineate the beam pattern in the resist. Both pillars and beams are then plated in a gold-plating bath at 65 degrees C. After stripping the photoresist, the copper, palladium, and titanium layers are etched, with the gold acting as an etching mask.

With the reliability of encapsulated beam crossovers established by extensive life testing,[41] crossover circuit production began in 1972. The circuits had from 200 to 3000 crossovers, with the largest circuits having a size of 3.25 by 4.00 inches.

With such large numbers of crossovers on a circuit, individual crossover preparation and assembly yield requirements were quite stringent. For example, an average defect density of one defect in 10,000 crossovers results in a circuit yield of only 70 percent for 3000 crossovers per circuit. Fortunately, the repair of defective crossovers was quite simple and was applied extensively in production. It was done by lifting a shorted crossover with a fine tool or replacing an open crossover with a stitch-bonded wire. In complex wiring schemes, however, it proved far more difficult to locate defective crossovers (as well as other faulty components) than to repair them. Therefore, much effort was spent on the development of automatic fault-location equipment.[42] Placing a supplementary insulating layer of organic or inorganic composition over the bottom conductor under the beam helped greatly to reduce the number of shorts that could develop during preparation and assembly of the hybrid circuits.[43,44]

The steady increase in the cost of precious metals, such as palladium and gold, in the early 1970s led to the development of lower-cost, but equally reliable, alternatives to the Ti-Pd-Au conductor system. J. H. Thomas, J. M. Morabito, and N. G. Lesh demonstrated satisfactory reliability for a Ti-Cu-Ni-Au system compatible with the processing steps used for the fabrication of film circuits.[45] In this system, copper provides the conductivity and only a thin gold layer is included as surface protection.

Two other approaches to crossover and crossunder structures were studied in the early 1970s. One developed at the Western Electric Engineering Research Center is batch bonding of separately fabricated crossover arrays to conductor substrates. Such circuits were introduced into manufacture with Ti-Cu-Ni-Au conductor metalization.[46] The second approach, developed by Western Electric in Merrimack Valley, made use of both thick- and thin-film materials.[47] First, thick-film gold crossunder conductor segments were applied to the substrate by print-dry-firing operations. A thick-film glaze was next applied as an insulating layer over part of the thick-film conductors, again by print-dry-firing operations. Finally, the crossover structure was completed using photolithographically defined thin-film conductor lines.

The first major application of the crossover technology was in support of the 1A processor, an improved processor for the 1ESS and 4ESS switches. The combination of thin-film technology and beam-lead SICs gave this processor a distinct performance advantage (see another volume in this series subtitled *Switching Technology (1925-1975)*, p. 291 and Chapter 2 of this volume). Several hundred hybrid IC codes were designed at the Bell Laboratories Indian Hill facility in Naperville, Illinois and manufactured at the Western Electric Hawthorne Works in Chicago, Illinois. Most circuits were designed in a standard format that became known as the 1A technology on a general-purpose computer with the help of a specially developed computer-aided design system. Before any actual circuit realization, many logic, encoding, and clerical errors were thereby avoided. Applications programs acting on this data base did the circuit partitioning, layout, and routing. Without these machine aids, large and complex switching systems could not have been realized.

An example of an FA-type circuit pack using the 1A technology is shown in Fig. 9-9. With 52 standardized locations (to facilitate testing) for beam-lead SICs on the 3.25- by 4.00-inch ceramic substrate, the ground distribution requirements are met by a low-resistance backplane metalization connected to the SIC sites through via holes. The power and signal distributions are made through hundreds of inches of front-side thin-film conductors with a minimum line width of 5 mils, except for the 2-mil-wide SIC bond fingers. A maximum of 21,112 conductor crossovers is provided by up to 841 crossover arrays, also at standardized locations. A typical circuit has over 40 ICs and over 3000 crossovers to realize about 300 gates. The SICs, power filter capacitors, and the connector are attached by thermocompression bonding. After an application of a protective coating of RTV silicone rubber, the completed circuit is mounted on an aluminum support plate.

A few very complex full-size FA codes incorporated resistors to accommodate analog functions in addition to the random logic. For functions not available in beam-lead SIC technology, such as delay lines, a second hybrid was mounted on the support plate and connected to the main

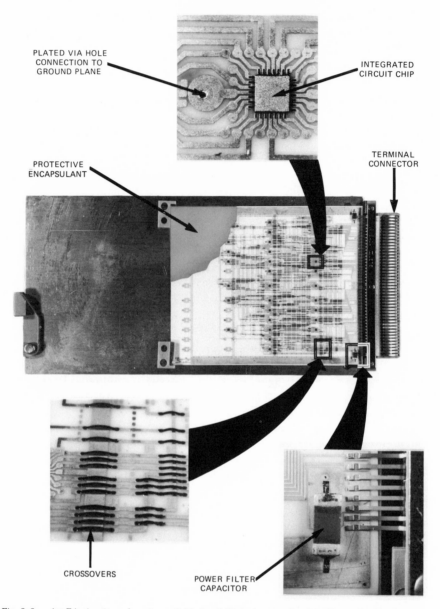

PLATED VIA HOLE
CONNECTION TO
GROUND PLANE

INTEGRATED
CIRCUIT CHIP

PROTECTIVE
ENCAPSULANT

TERMINAL
CONNECTOR

CROSSOVERS

POWER FILTER
CAPACITOR

Fig. 9-9.   An FA circuit pack, using a 3.25- by 4.00-inch ceramic with 52 standardized sites for ICs and a fixed pattern for crossovers. The back of the ceramic contains a metalized ground plane connected to the front through via holes, one of which appears in the magnified insert showing the IC site. In addition, power filter capacitors are mounted on the ceramic. The circuit shown uses batch-bonded crossovers. The ceramic is mounted on an aluminum base plate that also carries the terminal connector. The leads of the connector are bonded directly to the ceramic substrate.

circuit. The need for smaller circuits was met by several families of modular hybrid integrated circuits with leads directly bonded to the ceramic, some of which were configured in the dual inline format, but otherwise made in the same technology. Through use of a limited number of standard sizes, the preparation, testing, and handling of these circuits were greatly facilitated.

By 1975, over 400 codes of FA and small modular circuits were in manufacture by Western Electric, with an annual production volume of over one-half million circuits. The manufacturing information for the generation of masks and testing is obtained from the computerized data base generated by Bell Laboratories at the time the circuits are designed. Constant improvements in various manufacturing technologies resulted not only in ever-decreasing manufacturing cost but also in improved quality and reliability. (The production of these circuits reached a peak level of over one million per year towards the end of the 1970s. In 1984, the production still exceeded one-half million circuits per year.)

## V. TRANSFER OF FILM TECHNOLOGY INTO PRODUCTION

In order to smooth the way for the production of thin-film circuits in the Bell System, a new type of organizational unit was set up in Bell Laboratories: a process capability laboratory (PCL). Its function as an integral part of the components organization was to assist in coordinating the efforts of the device development, systems development, and manufacturing organizations, and to help the engineers of Western Electric put new designs and processes into production. PCLs began operation in 1967 in several Western Electric locations close to production lines and near the systems development organizations responsible for locally manufactured designs.[48] One such unit was set up at the Indianapolis Works in Indianapolis, Indiana for customer equipment, one at Merrimack Valley for transmission, and one at Hawthorne for switching. Each PCL was directed by a resident department head who reported to the components organization located at Allentown. The PCL concept worked well, and as of the mid-1970s, two of the three original PCLs were still operating.

Within ten years, from 1965 to 1974, the dollar volume of thin-film ICs manufactured by Western Electric increased from $105,000 to close to $80 million. Major contributions to the successful introduction of the thin-film technology into manufacture were made by the Western Electric Engineering Research Center and Western Electric in Allentown, as well as by the other thin-film circuit-producing Western Electric locations at Hawthorne, Indianapolis, Merrimack Valley, and the North Carolina Works in Winston-Salem, North Carolina. These contributions resulted in the introduction of mass-production machinery and other extensive process automation.

One important example is the design of in-line sputtering machines by the Engineering Research Center. Machines of this type began operation at Allentown in 1964. There are two designs, categorized as open-end and closed-end machines. The open-end machine consisted of a multichamber, differentially pumped arrangement in which uncoated glass or ceramic substrates entered at one end and came out at the other with the specified coating of tantalum.[49] In the closed-end design, the substrates entered and left the vacuum chambers through loading and unloading locks. Although the open-end design was more elegant, the closed-end machine was found to be more economical in manufacture.[50]

The production needs for conductor films were met initially by two types of evaporators, one using filament and the other using electron-beam-heated evaporation sources. In 1968, a large electron beam evaporator was installed at Western Electric in Allentown. This machine could be operated continuously through the use of loading and unloading locks. Five different metals could be deposited without breaking vacuum, and all its operations, except the evaporation rate control, were automatically sequenced. When magnetron sputtering machines became available commercially, sputtering in combination with electroplating became the preferred method for the deposition of conductor films. Other Western Electric developments included automated trimming of many resistors initially by anodization and later by laser machining; laser drilling of via holes into ceramic substrates; and laser scoring of substrates as part of the separation process.

## VI. CONCLUSIONS

Starting with resistors and capacitors, thin-film hybrid technology was developed to realize simple as well as very complex hybrid circuits, a spectrum that can meet many systems requirements successfully. The combination of versatility, small size, excellent reliability, and cost effectiveness has stimulated the use of hybrids in many applications, such as digital and analog circuits for transmission, switching, PBX, and station apparatus. Hybrids also play an important role in new applications in the microwave and lightwave areas. System designers have taken advantage of the many features that hybrid ICs provide, including precision resistors and capacitors, high interconnection densities, controlled impedance, high stability, and excellent reliability. The most recent trends in thin-film technology are toward increased circuit density, which will be valuable for new, more sophisticated telecommunications applications in the future. While the increasing integration of functions on SICs may reduce the importance of some of these advantages, the place of thin-film hybrids will remain assured for many years.

## REFERENCES

1. "Metal Sputtering: A Promising New Technique for Printed Circuitry," *Bell Lab. Rec.* **36** (November 1958), p. 426.

2. R. W. Berry and D. J. Sloan, "Tantalum Printed Capacitors," *Proc. IRE* **47** (June 1959), pp. 1070-1075.

3. H. Basseches, "The Oxidation of Sputtered Tantalum Films and Its Relationship to the Stability of the Electrical Resistance of These Films," *IRE Trans. Compon. Parts* **CP-8** (June 1961), pp. 51-56.

4. H. Basseches, "The Oxidation of Sputtered Tantalum Films," *J. Electrochem. Soc.* **109** (June 1962), pp. 475-479.

5. H. Basseches, P. L. McGeough, and D. A. McLean, U.S. Patent No. 3,148,129; filed October 12, 1959; issued September 8, 1964.

6. D. A. McLean, "Microminiaturization with Refractory Metals," *1959 IRE WESCON Conv. Rec.—Pt. 6*, San Francisco, California (August 18-21, 1959), pp. 87-91.

7. M. H. Bowie and R. D. Wiese, "Integrated Circuits For No. 101 ESS," *Bell Lab. Rec.* **44** (October/November 1966), pp. 334-339.

8. D. Gerstenberg and C. J. Calbick, "Effects of Nitrogen, Methane, and Oxygen on Structure and Electrical Properties of Thin Tantalum Films," *J. Appl. Phys.* **35** (February 1964), pp. 402-407.

9. D. Gerstenberg and E. H. Mayer, "Properties of Tantalum Sputtered Films," *Proc. Electron Compon. Conf.* **57**, Washington, D.C. (May 8-10, 1962), pp. 57-61.

10. G. I. Parisi, "Control of Temperature Coefficient of Resistance by Reactive Sputtering of Tantalum with Nitrogen and Oxygen Simultaneously," *Proc. 19th Electron. Compon. Conf.*, Washington, D.C. (April 30-May 2, 1969), pp. 367-371.

11. C. A. Steidel and D. Gerstenberg, "Component Properties of Co-Sputtered Ta-Al Alloy Films," *Proc. Electron. Compon. Conf.*, Washington, D.C. (April 30-May 2, 1969), pp. 372-377.

12. C. A. Steidel, "Electrical and Structural Properties of Co-Sputtered Tantalum Aluminum Films," *J. Vacuum Sci. Technol.* **6** (July/August 1969), pp. 694-698.

13. F. Vratny, B. H. Vromen, and A. J. Harendza-Harinxma, "Anodic Tantalum Oxide Dielectrics Prepared From Body-Centered-Cubic Tantalum and Beta-Tantalum Films," *Electrochem. Technol.* **5** (May/June 1967), pp. 283-287.

14. M. H. Read and C. Altman, "A New Structure in Tantalum Thin Films," *Appl. Phys. Lett.* **7** (August 1, 1965), pp. 51-52.

15. L. G. Feinstein and R. D. Huttemann, "Factors Controlling the Structure of Sputtered Ta Films," *Thin Solid Films* **16** (May 1973), pp. 129-145.

16. B. H. Vromen and J. Klerer, "Properties and Performance of Tantalum Oxide Thin Film Capacitors," *Proc. Electron. Compon. Conf.*, Washington, D.C. (May 5-7, 1965), pp. 194-204.

17. D. Gerstenberg and J. Klerer, "Anodic Tantalum Oxide Capacitors Prepared From Reactively Sputtered Tantalum," *Proc. Electron. Compon. Conf.*, Washington, D.C. (May 3-5, 1967), pp. 77-83.

18. J. Klerer, W. H. Orr, and D. Farrell, "The Effect of Moisture Upon Tantalum Oxide Thin Film Capacitors," *Proc. Electron. Compon. Conf.*, Washington, D.C. (May 5, 1966), pp. 348-360.

19. D. Gerstenberg, "Properties of Anodic Films Formed on Reactively Sputtered Tantalum," *J. Electrochem. Soc.* **113** (June 1966), pp. 542-547.

20. D. A. McLean and F. E. Rosztoczy, "Use of Manganese Oxide Counterelectrodes in Thin Film Capacitor. The TMM Capacitor," *Electrochem. Technol.* **4** (November/December 1966), pp. 523-525.

21. H. N. Keller, C. T. Kemmerer, and C. L. Naegele, "Tantalum Oxide-Silicon Oxide Duplex Dielectric Thin-Film Capacitors," *IEEE Trans. Parts Mater. Packag.* **PMP-3** (September 1967), pp. 97-104.

22. R. L. Whalin and E. D. Tidd, "Equipment Development of a Miniaturized Data Set," *Proc. 1963 Equip. Develop. Symp.*, Holmdel, New Jersey (November 13-14, 1963), pp. B6-1—B6-16.

23. W. H. Orr, D. O. Melroy, R. J. Moore, F. P. Pelletier, W. J. Pendergast, and W. H. Yocom, "Integrated Tantalum Film RC Circuits," *Proc. 20th Electron. Compon. Conf.*, Washington, D.C. (May 13-15, 1970), pp. 602-612.

24. J. E. Clark, "Wobble Table for Thermocompression Bonding Beam Lead Silicon Integrated Circuits," paper presented at *WESCON*, Los Angeles, California (August 19, 1968), paper 21.

25. J. S. Fisher and P. M. Hall, "Termination Materials for Thin Film Resistors," *Proc. IEEE* **59** (October 1971), pp. 1418-1424.

26. D. Hensler, N. A. Soos, E. A. Haas, R. W. Frankson, and D. A. Rott, "Optical Technique for Ceramic Surface Diagnostics," *Amer. Ceramic Soc. Bull.* **50**, Abstract (April 1971), p. 461.

27. D. J. Shanefield and R. E. Mistler, "The Manufacture of Fine-Grained Alumina Substrates for Thin Films," *W. Elec. Eng.* **15** (April 1971), pp. 26-31.

28. R. C. Sundahl and E. J. Sedora, "Effects of Alumina Substrate Surface Defects on Thin-Film Interconnect Patterns," *Proc. IEEE* **59** (October 1971), pp. 1462-1467.

29. M. L. White, "Encapsulation of Integrated Circuits," *Proc. IEEE* **57** (September 1969), pp. 1610-1615.

30. H. Basseches, "Interconnection Systems for Solid-State Components in Hybrid Integrated Circuits," *Proc. IEEE* **59** (October 1971), pp. 1468-1473.

31. D. Jaffe, "Encapsulation of Integrated Circuits Containing Beam Leaded Devices with a Silicone RTV Dispersion," *IEEE Trans. Parts Hybrid Packag.* **PHP-12** (September 1976), pp. 182-187.

32. R. W. Berry, P. Miller, and R. M. Rickert, "A Tone-Generating Integrated Circuit," *Bell Lab. Rec.* **44** (October/November 1966), pp. 318-323.

33. L. H. Steiff, "Applying Integrated Circuits to Transmission Systems," *Bell Lab. Rec.* **48** (November 1970), pp. 286-291.

34. J. J. Friend and W. Worobey, "STAR: A Universal Active Filter," *Bell Lab. Rec.* **57** (September 1979), pp. 232-236.

35. M. H. Rottersman, M. J. Bill, and D. Gerstenberg, "Tantalum Film Capacitors with Improved AC Properties," *IEEE Trans. Compon. Hybrids Manufacturing Technol.* **CHMT-1** (June 1978), pp. 137-142.

36. O. J. Duff, G. J. Koerckel, R. A. DeLuca, E. H. Mayer, and W. Worobey, "A High Stability RC Circuit Using High Nitrogen Doped Tantalum," *Proc. 28th Electron. Compon. Conf.*, Anaheim, California (April 24-26, 1978), pp. 229-233.

37. W. Worobey and J. Rutkiewicz, "Tantalum Thin-Film RC Circuit Technology for a Universal Active Filter," *IEEE Trans. Parts Hybrid Packag.* **PHP-12** (December 1976), pp. 276-282.

38. M. P. Lepselter, "Air-Insulated Beam-Lead Crossovers for Integrated Circuits," *Bell Syst. Tech. J.* **47** (February 1968), pp. 269-271.

39. H. Basseches and A. Pfahnl, "Crossovers for Interconnections on Substrates," *Proc. 19th Electron. Compon. Conf.*, Washington, D.C. (April 30-May 2, 1969), pp. 78-82.

40. A. Pfahnl and W. Worobey, U.S. Patent No. 3,890,177; filed March 8, 1973; issued June 17, 1975.

41. D. P. Brady and A. Pfahnl, "Reliability of Conductors and Crossovers for Film Integrated Circuits," *Proc. 11th Reliability Phys. Symp.*, Las Vegas, Nevada (April 3, 1973), pp. 89-90.

42. R. L. Patton, "Automatic Diagnosis of Logic Faults in Ceramic Circuit Packs," *W. Elec. Eng.* **20** (January 1976), pp. 16-19.

43. H. N. Keller and A. Pfahnl, U.S. Patent No. 3,783,056; filed June 20, 1972; issued January 1, 1974.

44. A. Pfahnl and J. M. Schuller, U.S. Patent No. 4,118,595; filed June 6, 1977; issued October 3, 1978.

45. J. H. Thomas, III, J. M. Morabito, and N. G. Lesh, "Ti-Cu-Ni-Au (TCNA) Compatibility with Resistor and Bilevel Crossover Circuit Processing," *J. Vacuum Sci. Technol.* **13** (January/ February 1976), pp. 152-155.

46. J. M. Morabito, J. H. Thomas, III, and N. G. Lesh, "Material Characterization of Ti-Cu-Ni-Au (TCNA)—A New Low Cost Thin Film Conductor System," *IEEE Trans. Parts Hybrid Packag.* **PHP-11** (December 1975), pp. 253-262.

47. D. H. Klockow, "Design of Economical Hybrid Integrated Circuits Utilizing Combined Thin and Thick Film Technologies," *Proc. Int. Microelectron. Symp.*, Orlando, Florida (October 27-29, 1975), pp. 139-141.

48. R. W. Berry and J. J. Degan, Jr., "PCL—A New Thin Film Organization," *Bell Lab. Rec.* **47** (April 1969), pp. 108-112.

49. S. S. Charschan, R. W. Glenn, and H. Westgaard, "A Continuous Vacuum Processing Machine," *W. Elec. Eng.* **7** (April 1963), pp. 9-17.

50. A. M. Hanfmann and F. J. Viola, "A Comparative Study of Open-End and Closed-End Continuous Sputtering Systems," *Proc. 4th Int. Vacuum Congress—Pt. 2*, Manchester, England (April 17-20, 1968), pp. 549-553.

# Credits

Figure 1-1 from W. Shockley, *Electrons and Holes in Semiconductors* (1950), p. 30. Copyright 1950 by D. Van Nostrand Co., Princeton, New Jersey.

Figure 2-3(a) from L. A. D'Asaro, *IRE WESCON Convention Record, Part 3—Electron Devices* (1959), p. 42. Copyright 1959 by the Institute of Radio Engineers, Inc.

Figure 3-33 from J. W. Gewartowski and H. A. Watson, *Principles of Electron Tubes* (1965), p. 556. Copyright 1965 by D. Van Nostrand Co., Princeton, New Jersey.

Figure 4-3 from A. A. Bergh and J. P. Dean, *Light-Emitting Diodes* (1976), p. 36. Copyright 1976 by Clarendon Press, Oxford, England. Reprinted with permission.

Figure 4-4 from A. A. Bergh and J. P. Dean, *Light-Emitting Diodes* (1976), p. 531. Copyright 1976 by Clarendon Press, Oxford, England. Reprinted with permission.

Figure 4-5 from J. E. Ripper, J. C. Dyment, L. A. D'Asaro, and T. L. Paoli, *Applied Physics Letters* **18** (February 15, 1971), p. 155. Copyright 1971 by the American Institute of Physics. Reprinted with permission.

Figure 4-7 from R. T. Denton in *Laser Handbook*, Volume 1, ed. F. T. Arecchi and E. O. Shulz-Dubois (1972), p. 706. Copyright 1972 by North-Holland Publishing Co., Amsterdam. Reprinted with permission.

Figure 4-8 from R. T. Denton in *Laser Handbook*, Volume 1, ed. F. T. Arecchi and E. O. Shulz-Dubois (1972), p. 720. Copyright 1972 by North-Holland Publishing Co., Amsterdam. Reprinted with permission.

Figure 4-9 from H. Melchior in *Laser Handbook*, Volume 1, ed. F. T. Arecchi and E. O. Shulz-Dubois (1972), p. 771. Copyright 1972 by North-Holland Publishing Co., Amsterdam. Reprinted with permission.

Figure 4-11 from A. A. Bergh and J. P. Dean, *Light-Emitting Diodes* (1976), p. 563. Copyright 1976 by Clarendon Press, Oxford, England. Reprinted with permission.

Figure 4-14 from J. R. Maldonado and D. B. Fraser, *Proceedings of the IEEE* **61** (July 1973), p. 975. Copyright 1973 by the Institute of Electrical and Electronics Engineers, Inc. Reprinted with permission.

Figure 4-15 from D. Maydan, *Proceedings of the IEEE* **61** (July 1973), p. 1011. Copyright 1973 by the Institute of Electrical and Electronics Engineers, Inc. Reprinted with permission.

Figure 4-16 from *Physics Today* **29** (May 1976), cover. Copyright 1976 by the American Institute of Physics. Reprinted with permission.

Figure 5-11 from L. W. Stammerjohn, *IEEE Transactions on Communications Electronics* **83** (November 1964), p. 817. Copyright 1964 by the Institute of Electrical and Electronics Engineers, Inc. Reprinted with permission.

Figure 6-5 from R. A. Heising, ed., *Quartz Crystals for Electrical Circuits* (1952), p. 223. Copyright 1952 by D. Van Nostrand Co., Inc.

Figure 6-8 from F. M. Smits and E. K. Sittig, *Ultrasonics* **7** (July 1969), p. 168. Copyright 1969 by Butterworth Scientific Ltd., Surrey, England. Reprinted with permission.

Figure 6-10 adapted from R. A. Sykes, *Proceedings of the IRE* **36** (January 1948), p. 6. Copyright 1948 by the Institute of Radio Engineers, Inc.

Figure 6-16 from P. Kartaschoff and J. A. Barnes, *Proceedings of the IEEE* **60** (May 1972), p. 498. Copyright 1972 by the Institute of Electrical and Electronics Engineers, Inc. Reprinted with permission.

Figure 6-18 from E. A. Gerber and R. A. Sykes, *Proceedings of the IEEE* **54** (February 1966), p. 107. Copyright 1966 by the Institute of Electrical and Electronics Engineers, Inc. Reprinted with permission.

# Index

**DATE DUE**

| JUN 16 1998 | | | | | |
|---|---|---|---|---|---|
| OCT. 05 2000 | | | | | |
| | | | | | |
| | | | | | |
| | | | | | |
| | | | | | |
| | | | | | |
| | | | | | |
| | | | | | |

DEMCO 38-301